ANIMAL SCIENCE, ISSUES AND PROFESSIONS

SYMBIOSIS

EVOLUTION, BIOLOGY AND ECOLOGICAL EFFECTS

ANIMAL SCIENCE, ISSUES AND PROFESSIONS

Additional books in this series can be found on Nova's website
under the Series tab.

Additional e-books in this series can be found on Nova's website
under the e-book tab.

ANIMAL SCIENCE, ISSUES AND PROFESSIONS

SYMBIOSIS

EVOLUTION, BIOLOGY AND ECOLOGICAL EFFECTS

ALEJANDRO F. CAMISÃO
AND
CELIO C. PEDROSO
EDITORS

New York

LIBRARY OF CONGRESS CATALOGING-IN-PUBLICATION DATA

Symbiosis : evolution, biology, and ecological effects / editors, Alejandro F. Camisco and Celio C. Pedroso.
    p. cm.
Includes index.
ISBN 978-1-62257-211-3 (hardcover)
1. Symbiosis. I. Camisco, Alejandro F. II. Pedroso, Celio C.
QH548.S94 2012
577.8'5--dc23
                                    2012021149

Published by Nova Science Publishers, Inc. † *New York*

CONTENTS

PREFACE

Symbiosis refers to the biological interaction between two organisms or species, living in close association. In this book, the authors present current research in the study of the evolution, biology and ecological effects of symbiosis. Topics discussed include the evolution of arbuscular mycorrhiza; legume properties and symbiosis; sea anemones and hermit crab symbiosis in temperate seas; the agronomic and ecological importance between legumes and rhizobia; the behavioral, physiological and ecological effects of organisms in symbiotic associations; and a neuronal model with symbiotic interactions.

Mycorrhizas are universally distributed fungus-plant (root), mutualistic symbioses known to endow plants with an enhanced capacity to support environmental stresses. Arbuscular mycorrhizal (AM) associations are the most widespread mycorrhizal type formed by 70-80% of land plant species. The AM fungi, integrated in the phylum Glomeromycota, are ubiquitous microscopic soil-borne fungi which establish symbiosis with the plant by colonizing the root cortex without eliciting any significant defence reaction from the host plant. Evidences from fossil records, paleobotanical data and molecular approaches indicate that the origin and divergence of AM fungi date back more than 500 million years. Actually, AM associations evolved as a symbiosis highly specialized to facilitate plant adaptation to the primitive harsh terrestrial environment, being commonly accepted that AM fungi played a crucial role in land colonisation by plants. AM fungi colonize the root cortex and develop an external mycelium which acts as an extension of the root system and increases the uptake of mineral nutrients. In addition AM fungi enhance plant performance through an increased resistance/tolerance to diverse types of stress either biotic (pathogen attack), or abiotic (drought, salinity, contamination by heavy metals or organic pollutants, etc.). The cellular and molecular dynamics regulating AM formation and functioning are currently being elucidated. In Chapter 1, particular emphasis will be placed on discussing topics related to the exchange of signal molecules between symbionts during AM development, the plant symbiotic genes essential for the intracellular accommodation of AM fungi, the functional processes involved in the coordinated nutrient exchange between symbionts, and on the physiological and molecular basis for stress alleviation. The ecological impact of AM symbioses is based on their recognized effects at affecting rates and patterns of nutrient cycling in agro-ecosystems, thereby helping plant development in nutrient deficient soils, their role at enhancing plant performance through an increased resistance to diverse types of stress either biotic or abiotic, and at improving soil quality through the formation of stable aggregates. Strategic and applied research has demonstrated that AM fungi can be managed to help sustainable environmentally-friendly agro-biotechnological practices.

Chapter 2 aspires to summarize the major advancements in the experimental study and understanding of the internal regulation of the intensity of rhizobial symbiosis. This kind of symbiosis takes place in the roots of predominantly legume plants in the form of specialized organs, root nodules, that are colonized by the soil nodule bacteria (rhizobia). Although the bacterial microsymbionts belong to several taxonomic groups, they share the ability of initiating root nodule development and fixation of atmospheric dinitrogen, which is subsequently utilized by the host plant. The latter feature predetermines the economic importance of rhizobial symbiosis and of the regulation of its intensity. While the symbiotic nodule induction is mediated by highly specific bacterial lipochitooligosaccharides, called Nod factors, further regulation of plant response is strictly controlled by plant endogenous mechanisms. Surprisingly, the major regulatory circuit includes systemic signaling from the root to the shoot and backwards. The nature of the involved molecules has been unknown until recently, except for the shoot-acting protein HAR1, described in 2003, and its homologs. Starting from 2008, a number of additional shoot-acting or root-acting regulatory mutants have been isolated. Their ongoing characterization has identified a series of new genes and products involved in the systemic signal generation, transduction and perception. Genomic approach allowed to identify low-molecular weight peptides of the CLE family as symbiotic signal messengers. The new experimental data enable a preliminary reconstruction of the systemic circuit and prediction of the nature of yet unknown components. A special interest represents the mechanism by which the inhibitory signal of ambient nitrate is integrated into the symbiotic circuit.

As explained in Chapter 3, improved nitrogen (N) management is needed to optimize economic returns to farmers and minimize environmental concerns associated with N use. Symbiotically N2 fixed is of particular significance in sustainable agriculture as it allows reducing the use of chemical N in the production of field crops. The legume-bacteria symbiosis is the most important N2 fixing system. This relationship is between bacteria collectively called Rhizobium and annual or perennial leguminous plants. In general, the rhizobia infect the plant via the root hair and invade the tissues. The cells are infected intracellular, divided, and enlarged to produce the characteristic nodules (determinate or indeterminate). The importance of this relationship relies in the ability of the rhizobia to fix (reduce) the atmospheric N2 to the host-plant, which in turn provides soluble carbohydrates to the microorganisms. Nitrogen fixation depends on legume performance, environmental conditions, soil nutrients availability, contaminants, bacteria abundance and diversity in soil, and bacteria specificity and infectivity for the legume. Under favourable conditions, legumes fix more than 50% of plant N.

Symbiosis, according to its initial meaning, refers to the biological interaction between two organisms living in close association. However, this definition is rather controversial, with the term being often used generically, since the outcome can vary across a continuum from negative to positive interactions. Symbiosis is a widespread phenomenon in temperate marine communities, and the association between sea anemones and hermit crabs belongs to the most common cases, being a familiar example of mutualism. In these latter specific cases of interactions gastropod shells are involved as prerequisite, since they provide both refuge for hermit crabs and substratum for the settlement of sea anemones; thus, shell resource availability is crucial for the establishment of this particular type of symbiosis. Within this context Chapter 4 aims to integrate the results of various studies to provide a general review about the symbiotic interactions of sea anemones and hermit crabs in temperate seas, addressing the following issues: (1) clarify the relevant terminology, which is differently interpreted by various authors; (2) provide a general description of the sea anemone - hermit crab association, as most studies examine separately the species involved and not the symbiosis as a whole; (3) assess the diversity and distribution of sea anemone - hermit crab associations in temperate seas, also incorporating gastropod shells and their availability, which although crucial, has been only little investigated; (4) address the behavioural patterns of both symbionts for the establishment of the symbiosis, including as well the behavioural plasticity of hermit crab related to shell resource utilization, and (5) report relevant information about co-evolution of the participant species, referring to the existing hypotheses on the evolution of the symbiosis, underlining its importance.

Legumes are able to fix nitrogen because of the bacterial symbionts (rhizobia) that inhabit nodules on their roots. The amount of ammonia produced by rhizobial fixation of nitrogen rivals that of the world's entire fertilizer industry. Consequently, this symbiotic relationship between legumes and rhizobia is of great agronomic and ecological importance.

Typical environmental stresses faced by the legume and their symbiotic partner may include, water stress, salinity and temperature and influence the survival in the soil.

In the Rhizobia-legume symbiosis, the host plant also influence rhizobial survival. In Arachis hypogaea rhizobia symbiosis is known that different abiotic stresses affect the viability, trehalose and membrane components content of rhizobia. Also, the attachment ability of peanut rhizobia is affected under abiotic stresses.

Chapter 5 addresses the idea that the rhizobia and the plants must be able to adapt to survive to the environmental conditions. Our hypothesis is that rhizobia survival in the soil environmental because they are able to modify fatty acid and phospholipid components of their membranes, as well as other molecules with important roles in stress tolerance.

As discussed in Chapter 6, the significance of coevolutionary adaptations by associated organisms at the cellular and molecular level has been the primary focus of much research in symbiology (e.g., endosymbiotic hypothesis for development of eukaryotes). Studies on the behavioral and physiological ecology of organisms involved in symbiotic associations have also demonstrated extraordinary examples of adaptation. These associations represent tremendous potential in demonstrating alternatives to competition as major evolutionary selective forces. Interactions vary tremendously within the commensalism, mutualism and parasitism subdivisions of symbiosis, and divergent examples will be discussed. For example, many organisms are limited by their anatomy to remove parasites and necrotic tissues, etc., and must rely on allogrooming by others. While terrestrial animals are usually better equipped for autogrooming, the "self-cleaning" problem is especially significant in marine environments where cohorts of animals switch from potential predators to symbiotic cleaning "clients." Organisms that possess significant, innate adaptations for protection are potentially attractive commensalistic or mutualistic hosts if a potential symbiont can inhibit or withstand the defensive mechanisms. Parasitism has also demonstrated bizarre behavioral outcomes, including examples where a larval form can "force" an intermediate host to alter its behavior to facilitate completion of the parasite's life history. Finally, the characteristics or phenoptypes of many organisms are actually based on composite genotypes of the host and any significant symbionts.

More than 200 species of non-legume dicotyledonous plants, mostly trees and schrubs, belonging to eight different families and 24 genera can enter actinorhizal symbioses with the nitrogen-fixing actinomycete Frankia. Actinorhizal nodules arise from divisions occurring in the pericyle and display a lateral root structure with a central vascular system and infected cells in the expanded cortex. Recently, the development of genomics both in Frankia and in actinorhizal plants, together with the possibility to obtain transgenic actinorhizal plants following Agrobacterium gene transfer, offered valuable tools to achieve significant progress in the understanding of actinorhizal symbiosis. Molecular data indicates that, like in legumes, actinorhizal symbiosis is a highly controlled process, involving plant signals perceived by the actinomycete and specific Frankia factors perceived by plant roots, that activates a symbiotic signalling pathway. In Chapter 7, current knowledge on the actinorhizal symbiosis will be discussed in the context of plant root endosymbioses evolution.

As explained in Chapter 8, improved nitrogen (N) management is needed to optimize economic returns to farmers and minimize environmental concerns associated with the N use. Symbiotically N2 fixed is of particular significance in sustainable agriculture as it allows reducing the use of chemical N in the production of field crops. Intercropping or crop rotation including legumes promises a more sustainable plant production in many agricultural systems through the N transfer and N release from legume residue. Sharing of N sources between the intercropped N2 fixing and non-fixing plants contribute to a better overall N use. In a crop rotation, the final contribution of fixed N2 to the soil depends upon the crop N balance, environmental conditions and agricultural practices.

In Chapter 9 the authors investigate various ecosystems containing always a symbiotic pair. In the first model, the latter is subject to the action of a predator. In the second system, the mutualistic association at the trophic level is located above a common prey population. In the third model a three level food chain is investigated, with the symbiotic populations constituting the intermediate trophic level. These are situations that can well arise in nature.

Finally, the authors make considerations on a symbiotic system in which a population X predates on another population Z; a third population Y lives in symbiosis with X; finally Z predates on Y. Such an association is rather difficult to find in nature, but an interpretation of it could come from the world of finance, in which biological "populations" are substituted by corporations. At times they interact among each other in search for synergies, but at the same time it may occur that part of these associations compete with each other in other fields, giving rise to rather intricated situations as the one described above.

In general, the behavior of large and complex aggregates of elementary components can not be understood nor extrapolated from the properties of a few components. The brain is a good example of this type of networked systems where some patterns of behavior are observed independently of the topology and of the number of coupled units. Following this insight, in Chapter 10 the authors have studied the dynamics of different aggregates of logistic maps according to a particular *symbiotic* coupling scheme that imitates the neuronal excitation coupling. All these aggregates show some common dynamical properties, concretely a bistable behavior that is reported here with a certain detail. Thus, the qualitative relationship with neural systems is suggested through a naive model of many of such networked logistic maps whose behavior mimics the wakingsleeping bistability displayed by brain systems. Due to its relevance, some regions of multistability are determined and sketched for all these logistic models.

In: Symbiosis: Evolution, Biology and Ecological Effects   ISBN: 978-1-62257-211-3
Editors: A. F. Camisão and C. C. Pedroso   © 2013 Nova Science Publishers, Inc.

Chapter 1

EVOLUTION, BIOLOGY AND ECOLOGICAL EFFECTS OF ARBUSCULAR MYCORRHIZA

José-Miguel Barea and Concepción Azcón-Aguilar
Departamento de Microbiología del Suelo y Sistemas Simbióticos
Estación Experimental del Zaidín, CSIC, Granada, Spain

Abstract

Mycorrhizas are universally distributed fungus-plant (root), mutualistic symbioses known to endow plants with an enhanced capacity to support environmental stresses. Arbuscular mycorrhizal (AM) associations are the most widespread mycorrhizal type formed by 70-80% of land plant species. The AM fungi, integrated in the phylum Glomeromycota, are ubiquitous microscopic soil-borne fungi which establish symbiosis with the plant by colonizing the root cortex without eliciting any significant defence reaction from the host plant. Evidences from fossil records, paleobotanical data and molecular approaches indicate that the origin and divergence of AM fungi date back more than 500 million years. Actually, AM associations evolved as a symbiosis highly specialized to facilitate plant adaptation to the primitive harsh terrestrial environment, being commonly accepted that AM fungi played a crucial role in land colonisation by plants. AM fungi colonize the root cortex and develop an external mycelium which acts as an extension of the root system and increases the uptake of mineral nutrients. In addition AM fungi enhance plant performance through an increased resistance/tolerance to diverse types of stress either biotic (pathogen attack), or abiotic (drought, salinity, contamination by heavy metals or organic pollutants, etc.). The cellular and molecular dynamics regulating AM formation and functioning are currently being elucidated. In this chapter, particular emphasis will be placed on discussing topics related to the exchange of signal molecules between symbionts during AM development, the plant symbiotic genes essential for the intracellular accommodation of AM fungi, the functional processes involved in the coordinated nutrient exchange between symbionts, and on the physiological and molecular basis for stress alleviation. The ecological impact of AM symbioses is based on their recognized effects at affecting rates and patterns of nutrient cycling in agro-ecosystems, thereby helping plant development in nutrient deficient soils, their role at enhancing plant performance through an increased resistance to diverse types of stress either biotic or abiotic, and at improving soil quality through the formation of stable aggregates. Strategic and applied research has demonstrated that AM fungi can be managed to help sustainable environmentally-friendly agro-biotechnological practices.

1. Introduction

Mycorrhizas are symbiotic, generally mutualistic, associations established between certain soil fungi and most vascular plants where both partners exchange nutrients and energy (Brundrett, 2002). Basically, the host plant receives mineral nutrients via the fungal mycelium (mycotrophism), while the heterotrophic fungus obtains carbon compounds from the host's photosynthates. It is generally accepted that mycorrhizal symbioses, which can be found in most soils, biomes and terrestrial ecosystems worldwide (Brundrett, 2009), are fundamental to improve plant fitness and soil quality through key ecological processes (Smith and Read, 2008; Azcón-Aguilar et al., 2009). Mycorrhizal fungi colonize the root and develop an extraradical mycelium which overgrows the rhizosphere soil, the volume of soil in the vicinity and surrounding plant roots. This hyphal net is a structure specialized for the acquisition of mineral nutrients from the soil, particularly those whose ionic forms have poor mobility or are present in low concentrations in the soil solution, as it is the case with phosphate and ammonium (Barea et al., 2005a).

There are two main types of mycorrhizas, ecto- and endomycorrhizas, which have considerable differences in structure and in the physiological relationships between symbionts (Smith and Read, 2008). In ectomycorrhizas, the fungus develops a sheath or mantle around the feeder roots. The mycelium penetrates the root and develops between the epidermal and cortical cells forming the so-called 'Hartig net' that constitutes the site of nutrient exchange between partners. About 3% of higher plants, mainly forest trees in the Fagaceae, Betulaceae, Pinaceae, *Eucalyptus,* and some woody legumes, form ectomycorrhizas. The fungi involved are mostly Basidiomycetes and Ascomycetes. In endomycorrhizas, the fungi colonize the root cortex with typical intracellular penetrations. Some endomycorrhizal types are restricted to species in the Ericaceae ("ericoid" mycorrhiza) or Orchidaceae ("orchid" mycorrhiza), whilst the commonest mycorhiza type, the arbuscular mycorrhizas, is widely distributed throughout the plant kingdom. An intermediate mycorrhizal type, the ectendomycorrhiza, is formed by plants in families other than the Ericaceae, but in the Ericales, and in the Monotropaceae and Cistaceae. In the arbuscular mycorrhizal (AM) associations the fungi form both a sheath and intracellular penetrations (Smith and Read, 2008; Brundrett, 2009). In the AM symbioses, the fungi involved are microscopic, belonging to the phylum Glomeromycota (Schüßler et al., 2001; Schüßler and Walker, 2011). This review will focus only on AM symbioses and fungi. However, the ecological importance of other mycorrhizal types, although restricted to specific plant families, must be recognized.

Most of the major plant families (70-80% of extant plant species), including non-vascular ones, form AM associations (Brundrett, 2009). AM fungi are ubiquitous soil-borne microscopic fungi, members of the phylum Glomeromycota, whose origin and divergence dates back more than 500 million years, thus being considered the oldest extant asexual lineage of eukaryotes (Redecker et al., 2000; Schüßler and Walker, 2011). The host plant for Glomeromycotan fungi includes angiosperms, gymnosperms and sporophytes of pteridophytes, all of which have roots, as well as the gametophytes of some hepatics and pteridophytes which do not (Smith and Read, 2008).

The widespread and ubiquitous AM symbioses are characterized by the tree-like structures termed "arbuscules" that the fungus develops within the root cortical cells, and where most of the nutrient exchange between the fungus and the plant is thought to occur.

AM fungi, by acting as a bridge connecting the root to the surrounding soil microhabitats, contribute to nutrients, particularly P but also N, Cu and Zn, acquisition and supply to plants, affecting consequently rates and patterns of nutrient cycling in both agricultural and natural ecosystems (Jeffries and Barea, 2012). In addition, the AM symbiosis improves plant health through increased protection against harsh environmental and cultural conditions including biotic (e.g. pathogen attack) or abiotic (e.g. drought, salinity and contamination by heavy metals or organic pollutants) stresses, and enhances soil structure through the formation of the aggregates necessary for good soil tilth (Jeffries et al., 2003). As the current scenario of global change is generating a great array of stress situations impacting negatively on plant-soil systems, it is expected that the establishment of AM associations be considered as an adaptive strategy for increasing the resilience of degraded agro- ecosystems (Barea et al., 2011; Jeffries and Barea, 2012). Either degradation of natural ecosystems or excessive agrochemical application can reduce AM fungal inoculum potential. However, strategic and applied research has demonstrated that AM fungi can be managed to help sustainable environmentally-friendly agro-biotechnological practices. Accordingly, this review summarizes and discusses some key aspects of the AM symbiosis including: (i) analysis of the origin and evolution of both the AM fungi and the symbiosis; (ii) biology of AM fungi and the processes involved in the formation and functioning of the AM symbiosis; and (iii) ecological effects of the AM symbiosis and its appropriate management to improve plant growth, health and productivity, as well as soil quality in either agro- and ecosystems. The main conclusions and future trends for this research area are then outlined.

2. Evolution of AM Fungi and Symbiosis

In this section the most significant aspects of the available information on the origin and evolution of Glomeromycota fungi and their symbioses with land plants (Embryophyta), are summarized and discussed. For that, information from fossil records and from molecular approaches is analyzed to provide a comprehensive insight into: (i) the origin of glomeromycotan fungi and land plants; (ii) the co-evolution of Glomeromycota fungi, land plants and AM associations; and (iii) the ecological meaning of the co-evolution of plant and their mycorrhizal symbioses.

2.1. Origin of Glomeromycota Fungi and Land Plants

Land colonization by plants is accepted to have started during the Early Paleozoic with plants at a bryophyte status, apparently having a liverwort grade of organization (Steemans et al., 2009). The fact that AM associations are found in extant species from most groups of bryophytes suggested an early origin of the AM symbiosis with primitive bryophytes (Schüßler and Walker, 2011). Fossilized plant structures discovered in the Early Devonian Rhynie chert beds, Scotland, include well preserved AM fungal fossils associated to bryophyte-like plants (Remy et al., 1994). These findings, fundamental to our understanding of early terrestrial life (Kidston and Lang, 1917; Taylor et al., 2004), support that AM fungi have an ancient fossil record history. Soon after the discovery of these fossil records the possible biological significance of the Rhynie chert fungi was recognized, and their role in nutrient cycling and their biotrophic interactions with the first plants suggested (Kidston and Lang, 1917). Early literature describes the Rhynie chert flora as associated to microbial components in form of fungal aseptate hyphae and thick-walled resting spores. These fungal structures were firstly detected inside rhizome tissues, particularly in the vascular plant *Nothia aphylla* (Krings et al., 2007). Furthermore, fungal structures, similar to the arbuscules of extant AM, were found colonizing plant cells of the early vascular land plant *Aglaophyton major* (Remy et al., 1994; Taylor et al., 2004). This information is considered the first direct evidence of a 400-million years (MY)-old plant-AM fungal symbiosis. The presence of Glomeromycota fungi associated with plants in the Rhynie chert (Remy et al., 1994) was later confirmed in this ecosystem by the discovery of *Scutellospora-* and *Acaulospora*-like AM spores (Dotzler et al., 2009).

Later on, presence of fungal spores clearly resembling those from modern AM fungi was evidenced in a ~460 MY old Ordovician dolomite rock of Wisconsin (Redecker et al., 2000), pushed back the evolutionary origin of glomeromycotan fungi to a period when land flora was likely to have consisted of plants similar to the modern-day groups of mosses, liverworts and hornworts (Bonfante and Genre, 2008). This new fossil evidence, considered the oldest known fossils of AM fungi, offered a calibration point for stem lineage of Glomeromycota in phylogenetic tress (Berbee and Taylor, 2007). Taken into account other previously found fossil fungi Redecker et al. (2000) estimated that the Ascomycetes and Basidiomycetes diverged from Glomeromycota about 600 to 620 MY ago.

Similarly, further studies also pushed back the origin of land plant. For example, fossil records of diverse cryptospore assemblage were found in Argentina, eastern Gondwana (Rubinstein et al., 2010). This Dapingian, Early Middle Ordovician fossil material, has embryophyte traces and share characters with those of extant liverworts, suggesting that land plants could be ~470 MY old. Since early vascular plants were found to exist ~420 MY ago (Cai et al., 1996), a minimum age of 420 MY for the liverwort–vascular plant divergence (Middle Silurian) must be assumed (Schüßler and Walker, 2011).

Based on the analysis from Small Sub-Unit (SSU) rDNA of several AM fungi Simon et al. (1993) reported the first molecular estimates for the age of the Glomeromycota, giving a radiation date of 460 MY ago. They also assumed land plant origin occurred about 420 MY ago. Further studies based on phylogenetic analysis pushed back these numbers for both AM fungi and land plants (Berbee and Taylor, 2007). However, a recent molecular clock study (Smith et al., 2010) suggested an origin of land plants around ~477 MY.

In summary, fossil records and molecular clock estimates considered together allow to date the origin of the AM fungal lineage as to be at least 50, possibly more than 200 MY earlier than that of land plants (Schüßler and Walker, 2011). A strong evidence, suggesting that plant–AM fungal symbiosis was one of the key processes that contributed to the origin of land flora, was the demonstration that common sym genes, functionally conserved during land plant evolution within all major plant lineages, were present in the common ancestor of land plants (Wang et al., 2010).

In a recent study, Bidartondo et al. (2012) present evidence that several species representing the earliest groups of land plants are symbiotic with fungi of the Mucoromycotina, and that these fungi evolved earlier than, and give way to glomeromycotan fungi. However, there are not the fossil evidences necessary to date accurately these diversification events and a test of this hypothesis using molecular clock data cannot be carried out.

2.2. Glomeromycota Fungi, AM Associations and Land Plant Co-evolution

Evidences discussed before, based on information from fossil records, paleobotanical studies and phylogenetic analyses of DNA sequences, altogether indicate that terrestrial AM fungi evolved before vascular land plants had emerged. These estimates supports the hypothesis of Pirozynski and Malloch (1975) suggesting that a mycotrophic lifestyle might be fundamental for early land plant to evolve and invade the harsh terrestrial environment. Therefore, it seems acceptable that the AM association could have evolved as a mutualistic symbiosis, facilitating the adaptation of plants to the terrestrial environment (Schüßler, 2002). The findings of Bidartondo et al. (2012) suggest the possibility that terrestrialization was facilitated by these fungi rather than by members of the Glomeromycota. However, whether or not the origin of Mucoromycotina predated that of Glomeromycota, the observation that vascular plant fossils of the early Devonian (400 MY) show arbuscule-like structures support a fundamental role of the fungi of the Glomeromycota allowing rootless early plants to invade poorly developed primaeval soils.

Acceptance that the origin of land AM fungi, at least for some of their lineages, have taken place 50–200 MY earlier than that of land plants implies that Glomeromycota fungi could have established other types of associations, whether saprophytic, parasitic, or mutualistic, before land plants existed. A feasible hypothesis is to assume a symbiotic lifestyle of glomeromycotan fungi associated with a photoautotrophic organism (Schüßler and Walker, 2011). To account for that the case with a glomeromycotan fungus, *Geosiphon pyriformis,* known to establish a symbiotic association with *Nostoc*, a photoautotrophic prokaryotic cyanobacteria (Schüßler, 2002), could be a model for such an ancestral partnership. Actually, *Geosiphon* is phylogenetically integrated, together with AM fungi, in the monophyletic phylum Glomeromycota (Schüßler et al., 2001) and is being investigated whether or not it form AM association with vascular plants or bryophytes (Schüßler and Walker, 2011).

In this context, several evolutionary scenarios for land plant-AM evolution/co-evolution have been proposed and discussed by Schüßler and Walker (2011). These scenarios include that:

1. Ancestors of either AM fungi or early land plants evolved independently and later interacted, perhaps firstly as saprobes or parasites, to finally become mutually beneficial symbionts.
2. Glomeromycotan precursors formed symbiosis with green-algae in an aquatic or semi-aquatic environment and later colonized land together (terrestrialization) as a symbiotic entity.
3. A kind of *Geosiphon/Nostoc* endosymbiosis could have evolved before plant evolved. Later the mechanisms involved in endosymbiosis establishment were acquired by either algae or plants.

In the three cases the primitive plant-AM fungi symbiosis later co-evolved to develop the AM associations as we know now (Brundrett, 2002). All members of the Glomeromycota phylum require a photosynthetic partner to complete their life cycle. This, together with current information on new identifications of AM fungal partners in lower plants, suggests that the ancestral fungus was already an obligate biotroph (Bonfante and Genre, 2008).

All in all, a symbiotic mechanism had to be developed by AM fungi, probably by recruiting already existing ones, perhaps derived from associations similar to the *Geosiphon–Nostoc* symbiosis (Schüßler, 2002). The Rhynie chert fossils contain diverse fungal endophytes, including potential parasites, that probably provoked the formation of structural 'defenses' by the plant, supporting the ancient nature of symbiosis and defense program in land plants (Parniske, 2008). Similarly, this symbiosis program could be recruited by other plant symbiotic entities evolving later, as is the case with the rhizobia-legume symbiosis which evolved much later than the AM symbiosis (Provorov and Vorobyov, 2009). Thereby, the cellular and molecular events occurring during the development of legume–rhizobia symbiosis may have derived from those already established in the AM symbiosis by means of a set of pre-adaptations during co-evolution with AM fungi (Parniske, 2008; Chen et al., 2009; Horváth et al., 2011). Moreover, developmental genetics and evolution timing analysis of microbe-plant symbioses, including both mutualistic, either N_2-fixing or mycorrhizal, and pathogenic associations, have revealed common cellular and molecular developmental programs for all of these compatible microbe-plant associations, essential for the intracellular accommodation of symbionts (Markmann et al., 2008; Parniske, 2008).

A case of particular interest concerns the co-evolution of three types of organisms: Glomeromycota fungi, their intimately associated endobacteria, and plants (Bonfante and Anca, 2009). The presence of endobacteria inside AM fungal cytoplasm has long been documented by electron microscopy, which distinguished two endobacterial morphotypes. The first, restricted to the *Gigasporaceae*, is an uncultured taxon, *Candidatus* Glomeribacter gigasporarum, related to *Burkholderia* (Lumini et al., 2007). The other bacterial type has been detected inside AM fungal spores and hyphae colonizing plant roots sampled in the field called 'bacterium-like organism' (BLO) (MacDonald et al., 1982; Schüßler et al., 1994). Recently, Naumann et al. (2010) analyzed 28 cultured AM fungi, from diverse evolutionary lineages and four continents, showing that most of the AM fungal species investigated possess BLOs. Analyzing the 16S rDNA they found that BLO sequences from divergent lineages all clustered in a well supported monophyletic clade that was not closely related to any described bacterial group, but with the *Mollicutes*. The intracellular location of BLOs was revealed by confocal microscopy and FISH, and confirmed by pyrosequencing. These bacteria diverged from their sister group more than 400 MY ago, colonizing their fungal hosts before main AM fungal lineages separated. The BLO–AM fungal symbiosis can, therefore, be dated back at least to the time when AM fungi formed the ancestral symbiosis with emergent land plants.

2.3. The Ecological Meaning of the Co-evolution of Plant and Their AM Symbiosis

Fundamental evidence from diverse review studies, some of them detailed and discussed above (Kidston and Lang 1917; Simon et al., 1993; Malloch et al., 1980; Brundrett, 2002; Taylor et al., 2004; Beerling and Berner, 2005; James et al., 2006; Krings et al., 2007; Bonfante and Genre, 2008; Honrubia, 2009; Rubinstein et al. 2010; Schüßler and Walker, 2011) supports that the ancient AM fungi have co-evolved with plants for, at least, 400 MY and that during co-evolution helped the colonization of dry lands by the first "lower" plants, and facilitated the invasion of most terrestrial ecosystems by "higher" plants. Contrasting with their long co-evolution history, and probably because of their unculturable character, and because of they do not cause obvious harm to the plant, the consideration of the responsible fungus, nowadays classified as Glomeromycota, as plant root colonizer was not recognized until the beginning of the twentieth century (Gallaud, 1905), and their abilities to form AM symbiosis were not evidenced until more recently (Mosse, 1953).

In spite of the early plant sporophytes did not have roots further evolutionary plant forms developed root-like organs (Brundrett, 2002), thus, sooner or later they needed to thrive in soil poorly developed and deficient in available mineral nutrients. Therefore, assuming that fossil AM structures functioned similarly to extant AM forms, it is likely that an association with Glomeromycota fungi could be critical to the success of the autotrophic plants for invading terrestrial ecosystems (Smith and Read, 2008). It must be considered that AM symbiosis was a key mechanism of nutrient acquisition by early land plants since their origin co-evolved with AM fungi. Secondary mycorrhizal types such as the ecto-, ericoid-, or orchid-mycorrhizas, established by some evolutionary advanced plants, have developed secondary adaptation pattern to other modes of nutrient uptake (Schüßler and Walker, 2011).

The origin and evolution of Glomeromycota fungi and land plants (Embryophyta) are actually linked events universally recognized not only for their critical evolutionary implications but also for their crucial ecological meaning. The advent of the first land plants, as aided by their associated AM fungi, is considered to be one of the most important evolutionary steps in Earth history which allowed all eukaryotic terrestrial life to evolve and to invade nearly all Earth environments. These evolutionary events resulted in acceleration of weathering processes and in the formation of modern terrestrial environments, including soil structure stabilization and the development of microbial communities. Certainly, embryophyte radiation and its co-evolution with AM fungi changed climate and biogeochemical processes on a global scale.

3. Biology of Glomeromycotan Fungi and AM Symbiosis

3.1. Biology and Phylogeny of AM Fungi

Because of their very peculiar evolutionary history, underground lifestyle and genetic make-up AM fungi are endowed with unusual biological traits which make them a fascinating subject of study. AM fungi develop a typically aseptate and coenocytic mycelial network containing hundreds of nuclei sharing the same cytoplasm, and produce very large multinucleate spores having abundant storage lipids and resistant thick walls containing chitin (Smith and Read, 2008). Among other important characteristics AM fungi are asexual, unculturable and obligatorily biotrophic microbes (Parniske, 2008; Bonfante and Genre, 2008; Schüßler and Walker, 2011). The character of obligate symbionts, unabling them to complete their life cycle without colonizing a host plant, has hampered the study of the biology and the biotechnological applications of AM fungi (Bago and Cano, 2005).

The complete nuclear genome of the model AM fungus *G. intraradices* is being sequenced but, due to the difficulties encountered (Martin et al., 2008) the results are not available yet. In the absence of the complete genome sequence, novel insights into the molecular basis of symbiosis-associated traits in *Glomus intraradices* have been recently reported (Tisserant et al., 2012). According to this recent genome-wide analysis of the transcriptome from the AM fungus *Glomus intraradices* (Tisserant et al., 2012), the lack of a known sexual cycle is not a result of major deletions of genes essential for sexual reproduction and meiosis. In addition, as deduced from the same study, the obligate biotrophy in *G. intraradices* is not associated with a striking reduction of metabolic pathways as observed in other obligate biotrophs. In any case, it is necessary to wait for the complete *G. intraradices* genome sequence to clarify these important biological questions.

The genetics and genomics of AM fungi have been reviewed elsewhere (Ferrol et al., 2004; Gianinazzi-Person et al., 2004 and 2009; Pawlowska, 2005; Parniske, 2008; Sanders and Croll, 2010; Schüßler and Walker, 2011). However, there is still scarce information about important characteristics of AM fungi like their genome organization and DNA ploidy levels. Different polymorphic DNA-sequence variants are present within a single cell or distributed among different nuclei in the same spore (Rosendahl, 2008; Sanders and Croll, 2010; Schüßler and Walker, 2011). Recent results based on the transcriptome from *G. intraradices* have shown that polymorphism is widespread in the nuclear genome, and that more than one variant of each polymorphic gene is transcriptionally active, what implies that this sequence variation may be of functional importance (Tisserant et al., 2012). By contrast, the mitochondrial genome sequence appears homogeneous (Lee and Young, 2009).

Another basic question, still a matter of debate, is whether AM fungi are hetero- or homokaryotic. As discussed by Schüßler and Walker (2011) to have an answer for this question is essential to explain the ecological success of the apparent asexual AM fungi. Whilst some reports hypothesize a heterokaryotic nature of AM fungi others suggest homokaryosis. This controversy is related to the intra-specific genetic variability within an individual polykaryotic spore (Schüßler and Walker, 2011). Anastomosis between hyphae of the same AM fungal species has been described building up extensive hyphal networks where nuclei may be exchanged giving way to a potential DNA recombination (Giovannetti et al., 2004).

The SSU rDNA phylogeny of AM fungi clearly revealed that they have a monophyletic origin and were moved to the new phylum Glomeromycota (Schüßler et al., 2001). Analysis of other genes has also corroborated the Glomeromycota as monophyletic, whilst a multi-gene analysis confirmed them as a very old group of terrestrial fungi (James et al., 2006). A comprehensive view of AM fungi phylogenetic relationships, based on rDNA phylogenies, have recently been established (Schüßler and Walker, 2011). They recognized four Orders: *Glomerales* (with the families *Glomeraceae* and *Claroideoglomeraceae*); *Diversisporales* (with the families *Diversisporaceae, Acaulosporaceae, Entrosphosporaceae, Gigasporaceae* and *Pacisporaceae*); *Archeosporales* (with the families *Geosiphonaceae, Ambisporaceae* and *Archeosporaceae*); and *Paraglomerales* (with the family *Paraglomeraceae*).

From a taxonomic point of view and because of their extraordinary genomic features there are difficulties for defining clear species concepts of individuals and also boundaries within populations. However, recent incorporation of molecular approaches into species descriptions, particularly molecular phylogenetic and evolutionary analyses, is incorporating new knowledge on AM fungi speciation (Rosendahl, 2008; Sanders and Croll, 2010; Schüßler and Walker, 2011). These novel advances refer to topics such as the significance of different polymorphic DNA-sequence variants that are present within a single cell or distributed between genomes or nuclei, the genetic structure of populations, and the definition of an individual, genet or clone (Rosendahl, 2008; Sanders and Croll, 2010; Schüßler and Walker,

2011). The 'molecular species concept' was proposed for AM fungi based on the analysis of their rDNA regions (Schüßler and Walker, 2011). As stated by these authors, "…the number of described species in the Glomeromycota is low (~230; see www.amfphylogeny.com) and only a small proportion of the actual species richness is represented in the molecular database. At the species level, SSU rDNA exceeds its limits of phylogenetic resolution". The use of phylogenetic reference data for systematics and phylotaxonomy of AM fungi has been recently addressed (Krüger et al., 2012).

Diversity studies based on molecular approaches (see section 5 in this Chapter) allowed us to ascertain that individual fungal strains exhibit little host specificity, whilst a single plant can be colonized by many different AM fungal species within the same root. A certain degree of host preference (functional compatibility) was evidenced to occur and this has been shown to play an important role in regulating the diversity, stability and productivity of natural ecosystems (Barea et al., 2011). The diversity of fungal and plant communities are positively correlated with each other (Maherali and Klironomos, 2007), but little is known about the basis of host preferences. In the context of AM fungal diversity research, Helgason and Fitter (2009) suggested that the Darwin´s model of natural selection and the evolutionary ecology can be applied to AM fungi, particularly in that concerning the interaction between fungus and the soil that may play the major role in determining AM fungal diversity. They argued that, because of the emphasis in AM research was placed on the interaction between plant and fungus, considerably less attention has been devoted to evaluate the impact of the soil environment, where the bulk of the fungus is developing (and not inside the host). Therefore, Helgason and Fitter (2009) suggest that the selection pressures on the soil-based mycelia are the main responsible for a significant part of the diversity of AM fungi.

Undoubtedly, the more relevant biological characteristic of AM fungi is their capacity to form AM associations with members of all phyla of land plants, whatever their taxonomic position, life form or geographical distribution (Smith and Read, 2008). The events occurring from AM propagule activation until the intracellular accommodation of the fungal symbiont that culminates with the establishment of functional arbuscules, will be analysed in sub-section 4.2 (AM development), paying particular emphasis to the cellular, genetic and molecular interactions between symbionts. Functioning of AM symbioses with regard to the coordinated nutrient exchange and other effects ameliorating plant tolerance to drought, pathogen attacks or contaminating agents will be analysed in sub-section 4.3 (AM function).

3.2. Biology of AM Establishment and Development

The information generated during the last years on the cellular and molecular interactions between the plant and fungal symbionts before and during AM establishment have recently been reviewed (Bonfante and Genre, 2008; Parniske, 2008; Smith and Read, 2008; Gianinazzi-Pearson et al., 2009; Genre and Bonfante, 2010; Giovannetti et al., 2010). Accordingly, the main well-established conclusions from these review articles are considered and critically summarized here. This will constitute the basis to provide a synthesis of the new insights into the main developmental processes involved in establishing and maintaining the AM symbiosis.

The main features of AM development are firstly schematized, at a glance, to give a general view of the processes involved. The soil-borne AM fungal spores are able to germinate in the absence of any host root and produce a germ tube that elongates to form a restricted mycelial network expanding in the soil. Once the fungus approaches to a host root a chemical dialogue is initiated and fungal hyphae are stimulated to growth and branch by root exudates molecules. Upon contact with root epidermal cells, characteristic fungal structures called appressoria (also known as hyphopodia) develop. The symbiotic phase starts with the development of a penetrating hypha, which follows an intracellular route across epidermal and outer cortical cells. Once the inner cortex is reached, colonization can either proceed via intercellular hyphae that penetrate individual cells to form terminal arbuscules (finely branched tree-like structures) or via intracellular hyphal coils that subsequently differentiate into intercalary arbuscules. These two types of colonization patterns correspond to the called 'Arum/Paris' models. Outside the root the AM fungi form extensive mycelial networks where spores are developed (Bonfante and Genre, 2008; Smith and Read, 2008).

When a hypha from an asymbiotic, soil-based, AM-mycelium approaches to a host root an exchange of signalling molecules between both symbionts is produced (Parniske, 2008; Gianinazzi-Pearson et al., 2009; Genre and Bonfante, 2010). This molecular dialogue activates specific signalling pathways affecting fungal development and plant gene expression. At least seven plant genes are involved in reprogramming processes from a direct cell-to-cell contact on the root surface to the intracellular accommodation of the fungal symbiont. These genes are required for both the AM symbiosis and the root-nodule symbiosis with rhizobia and have been identified in legumes therefore they are called common symbiosis (*SYM*) genes. They encode proteins that are involved in a signal transduction network that mediates the development of intracellular accommodation structures for fungal and bacterial symbionts by the host cell (Parniske, 2008 and Genre and Bonfante, 2010).

In summary, the consolidation of a functional AM symbiosis is the result of a series of developmental steps some of them are host-independent, such as the relief of spore dormancy and the development of germlings, while others take place through a series of plant-controlled checkpoints which direct fungal penetration across the root ending with the intracellular accommodation of the fungal symbiont. The root cortical cells are the niche for AM fungi while root meristems, differentiating tissues, endodermis or vascular tissues are never colonized. The developmental switches in AM establishment involve a sequence of morphogenetic events consisting on: spore germination and asymbiotic hyphal growth, differential hyphal branching in the presence of host roots, appressorium (hyphopodium) formation, root colonization, arbuscule development, extraradical symbiotic mycelial growth and spore production (Giovannetti et al., 2010). These morphogenetic steps, which complete the fungal life cycle are analysed in the following sub-sections.

Spore Germination and Development of Asymbiotic Mycelium

Colonization of roots by AM fungi can arise from three main types of propagules in soil: spores, mycorrhizal root fragments and hyphae. For many years it was assumed that spores were the most important AM propagules, but this seems not to be true. Soil-borne spores, having different states of dormancy or quiescence, are able to germinate in response to different environmental conditions, but spore germination is not a host-regulated step. Germination of some species may be poor, slow and variable, but spores are able to persist for many years as a reservoir of propagules. In many habitats, persistent soil-based hyphae, together with those emerging from previously colonized root fragments, are the main responsible means for an early colonization of root systems, even when significant spore populations are also present (see Smith and Read, 2008; Giovannetti et al., 2010 for details and references).

The growth of hyphae from germinating spores, which occurs at the expenses of spore reserves, is very restricted and a saprotrophic growth has not been demonstrated (Giovannetti et al., 2010). Fluorescence microscopy and video-enhanced microscopy allowed the detection of nuclei moving along hyphae originating from germinated spores (Bago et al., 1998b).

Both germination of AM fungal spores and the subsequent asymbiotic growth are characterized by an increased metabolic activity, involving critical biochemical changes from a metabolically quiescent-like status to an active metabolism (Giovannetti et al., 2010). Since triacylglycerols (TAG) constitute a large proportion of AM fungal spores (Gaspar et al., 1994), their degradation seems central to the process of spores germination and germling growth. Ultrastructural data on the movement and disappearance of lipid globules in hyphae originating from germinating spores (Bago et al., 2002) support the hypothesis that storage lipids are used to provide precursors for anabolic activities during spore germination. In addition to these capabilities, other biochemical, cytological and genetic activities have been demonstrated to be operative during germination and asymbiotic growth of AM fungi (see Giovannetti et al., 2010 for details and references). In spite of that, spores of AM fungi are not able to carry out an independent hyphal development and, if there is no colonization of a susceptible host root, mycelial growth arrest within 8–20 days, long before the spore reserves are depleted. This has been explained as some nutritional deficiencies or loss of metabolic pathways during the asymbiotic and presymbiotic phases related to putative genome erosion (Ercolin and Reinhardt, 2011). However, the available information supports that AM fungal spores possess the genetic information necessary to allow them to grow and suggests that the fail for independent growth is not associated with a striking reduction of metabolic complexity (Tisserant et al., 2012). In any case, and whatever the molecular basis underling the obligate biotrophy of AM fungi, the strategy to stop mycelial growth in the absence of a host plant may compensate for the lack of host-regulated spore germination, and contribute to the survival of these obligate symbionts in the soil environment.

Differential Hyphal Branching in the Presence of Host Roots: Signalling and Recognition

Diffusible molecules released by AM fungi and plants in the rhizosphere are perceived by the reciprocal partner establishing a molecular dialogue that leads to their mutual recognition (presymbiotic phase) (Parniske, 2008; Gianinazzi-Pearson et al., 2009; Genre and Bonfante, 2010) and, through an "anticipation program", prepares the symbionts for an AM successful establishment (Paszkowski, 2006).

When an exploratory hypha, asymbiotically developed, approaches to a host root it results morphogenetically and functionally modified by the presence of plant-derived signals. Two classes of plant molecules, strigolactones and flavonoids, have been proposed as potential signal molecules involved in enhancing hyphal growth and/or branching during the early developmental stages of AM formation (Parniske, 2008; Gianinazzi-Pearson et al., 2009; Genre and Bonfante, 2010). Strigolacton perception by the fungus have been shown to induce the presymbiotic stage, including a continued hyphal growth, increased physiological activity and profuse branching of hyphae (Akiyama et al., 2005; Bouwmeester et al., 2007;

López-Ráez et al., 2011). The effects of root exuded flavonoids on hyphal growth are more controversial and not always reproducible (Vierheilig et al., 1998; Bécard et al., 2004).

On the other side, when the AM fungus starts to branch in the vicinity of the root, plants perceive diffusible fungal signals that induce symbiosis-specific responses in the host root, even in the absence of any physical contact (Parniske, 2008; Genre and Bonfante, 2010). By analogy to the Nod factors of nitrogen fixing rhizobia, the fungal diffusible symbiotic signals were called "Myc factors". The Myc factors were found to be diffusible molecules that induce both transcriptional activation of symbiosis-related genes (Kosuta et al., 2003) and calcium oscillations (calcium spiking) in root epidermal cells (Kosuta et al., 2008). The chemical structure of these Myc factors was identified as lipochitooligosaccharides with structural similarities with rhizobial Nod factors (Maillet et al., 2011). The discovery of the structure of diffusible Myc signals provide a better understanding of the evolution of signalling mechanisms involved in plant root endosymbioses.

Root Colonization Initiation (Hyphopodium and PPA Formation)

When a mature hypha contact a host root, an effective cell-to-cell adhesion to the root epidermis give rise to a special type of appressoria, called hyphopodia, thereby initiating the symbiotic phase of for AM establishment (Parniske, 2008; Genre and Bonfante, 2010). The differentiation of hyphae into hyphopodia indicates recognition of the host root surface by the fungal hyphae, being this the first step of a series of morphological re-arrangements in both partners (Gianninazzi-Pearson et al., 2009). Reception of the fungi in the root involves changes in the plant cells. The first visible response of the epidermal cell to hyphopodium formation is the migration of the nucleus from a peripheral position to situate underneath the contact point, remaining associated to the hyphopodium (Genre and Bonfante, 2010). Others cellular re-structurations taken place after hyphopodium formation involve alterations in cytoskeletal activity and membrane proliferation (Smith and Read, 2008). Subsequently, the nucleus initiates a second migration across the cell and a specific cytoplasmic tunnel is developed. This novel cellular structure has been called the pre-penetration apparatus (PPA), a longitudinal array of microtubules and microfilaments lined with endoplasmic reticulum cisternae that facilitates passage of AM fungi through root epidermal cells, predicting the trajectory of intracellular hyphal development (Genre et al., 2005). When this 'trans-cellular tunnel' is completed, a fungal hypha growing from the hyphopodium enters the PPA to penetrate the host cell. In Arum-type mycorrhizas, and upon reaching the cortex, the fungus leaves the plant cell and colonizes the intercellular spaces, later on branches and grows laterally along the root axis. Branches from these intercellular hyphae penetrate the cortex cells to form the arbuscules. At all stages of AM development the intracellular hyphae remain in the apoplast surrounded by the host plasma membrane, thus the symbiotic interfaces are bounded by both fungal and plant plasma membranes (Smith and Read, 2008).

The PPA response seems to be AM-specific, as supported by the absence of comparable structures in root epidermal cells upon the contact with fungal pathogens (Genre et al. 2009). Such an AM-specificity pattern concerns inter-symbiont compatibility issues and related this to host defence responses patterns (Gianinazzi-Pearson et al., 2009). Several plant protein-coding gene families associated with defence strategies against pathogens are induced in response to hyphopodia and before root penetration (Garcia-Garrido and Ocampo 2002). However, AM fungi, having these similarities with phytopatogenic microbes, fail to trigger major plant defences. In fact, AM interactions only induce weak and transient plant defence responses. There is great interest in ascertaining the mechanisms accounting for the reasons why AM fungi escape from the plant defence mechanisms usually expressed against pathogenic microbial invaders (Azcón-Aguilar and Barea, 1996). The related information, mostly based on using plant mutant genotypes resistant to AM fungi or non-host plants, has recently been reviewed (Smith and Read, 2008; Gianinazzi-Pearson et al., 2009). Some evidence suggests that hyphopodia formation, while activating symbiosis-related plant genes, seems to inactivate defence gene expression (García-Garrido and Ocampo, 2002). However, rather than inducing major inactivation of defence responses, AM formation appears to mobilize plant defence reactions for a short period but these are later suppressed. Nevertheless, these transient and small changes may be sufficient to confer a kind of immunity to the plants enough to reduce damage from a subsequent pathogen attack. The mechanisms involved in priming plant defence against pathogens by AM fungi (Pozo et al., 2009 and 2010) are discussed in section 4.3.

Hosting AM Fungi Inside a Plant Cell (Arbuscule Development)

Root colonization involves the formation of intercellular and/or intracellular hyphae, and particularly, intracellular coils and arbuscules. In Arum type mycorrhizas, hyphae growing through intercellular spaces induce the development of PPA-like structures in inner cortical cells and subsequently, the fungus enters these cells to form hyphal coils or to branch repeatedly to form the tree-shaped arbuscules. Each fungal branch within a plant cell is surrounded by a plant-derived periarbuscular membrane (PAM) which is continuous with the plant plasma membrane, and separates physically the fungus from the cell cytoplasm (Genre et al., 2008; Parniske, 2008). The apoplastic interface between the fungal plasma membrane and the PAM is called the periarbuscular space (PAS) which contains cell-wall materials of plant origin, as well as different enzymatic systems (Takeda et al., 2009), playing a fundamental role in the exchange of symbiotic signals and in the bidirectional transfer of nutrient between symbionts. Particularly, arbuscules are recognized as key structures for a co-ordinated functioning of the symbiosis and the fact that arbuscule formation represents a considerable increase in surface area of contact between symbionts soon suggested that they were involved in nutrient transfer in AM symbiosis, as later detailed (see 4.3.).

Arbuscule accommodation changes very much host cell architecture (Reinhart, 2007). Apart from plasma membrane proliferation and cell-wall deposition around the fungus, the nucleus of arbuscule-containing cells increases in size and migrates from the periphery close to the cell wall to a central position. Additionally, the vacuole is fragmented and there is a proliferation of plastids and mitochondria (Loshe et al., 2005).

Arbuscules have a short life span (something between 7-15 days), and consequently, a single host cell is thought to be competent for several rounds of successive fungal invasions (Parniske, 2008). In a comprehensive study Javot et al. (2007) found that arbuscules undergo a phase of growth until a certain maximum size is reached. Later on arbuscule degradation or senescence is initiated and arbuscular hyphae become separated from the remaining cytoplasm by septation. Arbuscules subsequently collapse over time and finally disappear.

Growth of the Extraradical Mycelium in Soil and Production of Spores

Once the AM fungus is well established in the host root an extensive development of the so-called external AM mycelium usually takes place in the root-associated soil environment. This extraradical mycelium (ERM) constitutes a fundamental source of propagules for further colonization of root systems either for the same plant or for others growing nearby. As Smith and Read (2008) discussed, it is commonly accepted that the development of the ERM in the rhizosphere of mycorrhizal plants mainly occurs by extension and branching of already existing hyphae outside the roots. Actually, there is little or no direct microscopical evidence supporting that intra-radical hyphae are growing out from the root cortex by penetrating root cell walls. Growth of the ERM does not begin until the root has been colonized. An important eco-physiological issue, matter of discussion in the past, is to ascertain which stage of intraradical AM colonization should be reached for a vigorous development of the ERM. It is generally accepted that formation of arbuscules is an absolute requirement for ERM growth (Bécard and Piché, 1989).

The extensive fungal growth in the soil surrounding the roots can account for the widely recognized role of AM fungi at searching for available sources of plant nutrients. The main hyphae growing outside the root give rise to the characteristic branching systems represented by the so-called BAS, for branched adsorbing structures (Bago et al., 1998a). The finely branched hyphal fans are clearly well adapted to explore soil pores, where their diameters may change in response to pore diameter. ERM development requires a considerable transfer of organic C from the roots to the soil for both uptake and translocation of nutrients as well as for fungal growth and respiration. At later stages in the AM symbiotic development, spores and auxiliary cells are formed on the ERM, This means that translocation of relatively large amounts of carbohydrate and, more especially lipid, into spores, are needed (Smith and Read, 2008)

The mycelial networks spreading from plants meet frequently and often fuse and anastomose facilitating nuclear migration through fusion bridges (Giovannetti et al., 2004). This suggests that, in the absence of sexual recombination, a genetic exchange could occur in AM fungi by means of anastomoses and nuclear exchange.

3.3. AM Functioning

The eco-physiological and molecular components of AM functioning have been the subject of diverse experimental and review studies during the last decade. The main conclusions from recent articles specialized in the different thematic areas involved are considered and critically summarized here. This will constitute a basis to provide a synthesis of the new insights into the main processes involved in the functioning of AM symbiosis. The selected topics of interest include:

Coordinated Nutrient Exchange (P, N and C)

As any heterotrophic microbe, and particularly because of their character of obligate biotrophs, AM fungi rely for their growth and activity on carbon provided by their host plant. In exchange, the fungus contributes supplying the plants mineral nutrient nutrients (P, N, Cu and Zn) that the AM mycelium capture and transport from the soil solution to the root (Smith and Read, 2008). The analysis of the biochemical and molecular mechanisms involved in nutrient transport processes in AM, and in the bidirectional nutrient exchange between symbionts at the symbiotic interfaces is matter of recent and current interest (see Barea et al., 2008; Ferrol and Pérez-Tienda, 2009; Franken, 2010; Harrison et al., 2010; Pérez-Tienda et al., 2011 for details and references). In summary, the progresses that have recently been made in the understanding of phosphate transport processes in AM symbiosis can be summarized as follows. Transport proteins putatively involved in Pi uptake by AM fungi have been identified. Pi taken up by the fungus is first incorporated into the cytosolic P pool while the excess cytosolic Pi is transported into the fungal vacuoles and condensed into polyphosphates (polyP). Once Pi is transferred to the arbuscules, it is released from the fungus and transported across the periarbuscular membrane into the cortical cell. The mechanism involved in the release of Pi from the arbuscules remains still unknown. Because Pi efflux across the fungal plasma membrane probably follows a concentration gradient, it could be facilitated by an as yet unidentified anion channel, carrier or pump. In the past few years, the plant transporters implicated in the uptake of the Pi exported across the membrane of the arbuscule have been identified in several plant species. Gene expression studies of the plant Pi transporters have revealed that development of the symbiosis induces the *novo* expression of the so-called mycorrhiza-specific Pi transporters, transporters which are exclusively expressed in mycorrhizal roots, and up-regulation of Pi transporters which have a basal expression in non-mycorrhizal roots, the mycorrhiza up-regulated transporters.

During the last years some advances have also been made in understanding the mechanisms of N transport. The model of N transfer proposes that inorganic N taken up by extraradical hyphae, either as NH_4^+ or NO_3^-, is assimilated in the fungal cytoplasm into arginine, transferred via the tubular vacuoles to the intraradical hyphae to be released as urea and either transported to the plant directly or after cleavage to as NH_4^+. Genes encoding ammonium transporters have been identified so far (López-Pedrosa et al., 2006; Pérez-Tienda et al., 2011).

Both the mechanism and site of carbon transfer remain unclear. It is likely that sucrose delivered into the apoplast at the arbuscular interface is hydrolyzed via a cell wall invertase and that the resulting hexoses, principally glucose, are then taken up by the plasma membrane of the fungus (Bago et al., 2002; Ferrol and Pérez-Tienda, 2009).

The schematic illustrations of: (i) the potential pathways of nutrient acquisition in AM roots (Smith and Read, 2008); (ii) the metabolic fluxes and long-distance transport in AM (Parniske, 2008); and (iii) the currently accepted model for the mechanisms involved in nutrient exchange between the plant and the fungus in the AM symbiosis (Ferrol and Pérez-Tienda, 2009) are recommended to get a general and integrative view of the general processes involved in a coordinated nutrient exchange in AM symbiosis.

Priming Plant Defense against Pathogens

Root colonization by AM fungi can improve plant resistance/tolerance to biotic stresses. Although this bioprotection has been widely described in different plant systems, some of the underlying mechanisms remain largely unknown. Alongside to mechanisms such as improved plant nutrition and competition, experimental evidence supports the involvement of plant defence mechanisms in the AM-mediated plant protection observed. As stated before during

AM establishment, modulation of plant defence responses occurs upon recognition of the AM fungi in order to achieve a functional symbiosis, consequently, a mild, but effective activation of the plant immune responses may occur, not only in local but also in systemic tissues. This activation leads to a primed state of the plant that allows a more efficient activation of defence mechanisms in response to attacks by potential enemies (Pozo, et al., 2009 and 2010; Jung et al., 2012).

Actually, AM symbioses have an important impact on plant interactions with pathogens and insects. The association leads generally to a reduction of damage caused by soil-borne pathogens, but effects on shoot targeting organisms depend greatly on the attacker lifestyle. Mycorrhiza induced resistance (MIR) in above-ground tissues seems effective against necrotrophic pathogens and generalist chewing insects, but not against biotrophs. Instead of constitutive activation of defences, MIR is associated with priming for an efficient activation of defence mechanisms upon attack. The spectrum of MIR efficiency correlates with a potentiation of jasmonate-dependent plant defences (Pozo and Azcón-Aguilar, 2007). This low-cost type of induced resistance may be among the reasons to explain why root associations with AM fungi have been conserved during evolution and are widespread among plant species worldwide.

Plant Response to Osmotic Stresses

The enhancement by AM symbiosis of host plant tolerance to water deficit is achieved by alteration of several physiological or ecological processes. One of the most common explanations for the improved water status and physiology in mycorrhizal plants is the strong increased absorbing surface caused by soil growing hyphae combined with the fungal capability to take up water from soils with low water potential (Ruiz-Lozano, 2003). The direct and indirect hyphal contribution to the total plant water uptake has been estimated to be up to 20%. In this process, the regulation of host plant aquaporins and root hydraulic properties is also involved. In fact, the induction or inhibition of particular aquaporins by AM symbiosis results in a better regulation of plant water status and contribute to the global plant resistance to the stressful conditions as evidenced by their better growth and water status under conditions of water deficit (Aroca et al., 2007 and 2011; Bárzana et al., 2012). The AM symbiosis may also improve the plant osmotic adjustment by accumulation of different compounds such as proline, sugars, free amino acids, etc, although this effect may differ according to the plant tissue considered (Sheng et al., 2011). Mycorrhizal plants exhibit an enhanced gas exchange and, usually, an improved the water use efficiency (WUE) properties, which also contributes to maintain plant growth and productivity under drought stress conditions (Augé et al., 2008; Ruiz-Lozano and Aroca, 2010). This effect has been closely related to the regulation of plant hormonal balance, being the absicic acid the plant hormone usually modulated by the AM symbiosis under drought conditions (Aroca et al., 2008; Ruíz-Lozano et al., 2009; Martín-Rodriguez et al., 2011). Finally, it is becoming clear that the AM symbiosis protects the host plant against the detrimental effects of *reactive oxygen species* (ROS) generated by drought. Thus, improved antioxidant enzymatic activities and/or accumulation of non-enzymatic antioxidant compounds result in reduced oxidative damage to AM plants (Porcel et al., 2003 and 2012; Ruíz-Sánchez et al., 2010).

Mechanisms Underlying Heavy Metals Tolerance

AM fungi are able to tolerate a wide range of metal concentrations in soils. The mechanisms evolved by arbuscular mycorrhizal (AM) fungi to survive in metal-contaminated environments have recently been review by (Ferrol et al., 2009; González-Guerrero et al., 2009). A number of passive and active molecular processes are undertaken by these fungi to maintain metal homeostasis. The main passive mechanism is the binding of metals to the fungal walls, responsible for a significant percentage of the metal retained. Meanwhile, in the cytosol, a number of chelators (metallothioneins, glutathione) bind the metals very efficiently. Heavy metal transporters collaborate with the intracellular chelators to actively reduce the levels of metal by pumping metal out of the cytosol. Additionally, the fungus struggles to reduce the free radicals produced by heavy metals. Recent progresses in the identification and characterization of the elements involved in maintaining metal homeostasis in AM fungi, as well as how the heavy metal control systems of the plant are affected by the development of the symbiosis, have been discussed (Ferrol et al., 2009; González-Guerrero et al., 2009). The information on the mechanisms of metal homeostasis generated in such as review articles can be summarized as follows. Although the negatively charged fungal wall binds metal ions, some of them are incorporated into the cytosol through specific metal transporters. In the cytosol, metals are sequestered by intracellular chelators, such as metallothioneins or glutathione (GSH). At higher concentrations, they induce the expression of efflux systems as well as they inhibit metallothioneins and probably the uptake systems. ROS are produced through metal-catalyzed Fenton reactions, or by the depletion of the GSH pool, which result in gene activation. The efflux systems will transport metals to the exterior or introduce them into the vacuoles, where they associate to polyphosphate. At the level of the fungal colony, AM fungi have also evolved compartmentalization strategies based on the accumulation of Cu into specific fungal structures, such as extraradical spores and intraradical vesicles.

AM Impact on Its Soil Biotic Environment

AM-colonization changes the chemical composition of root exudates, while the AM soil mycelium introduces physical modifications into the environment surrounding the roots thereby affecting microbial structure and diversity (Barea et al., 2005b). In addition, there are specific modifications in the soil habitat surrounding the AM mycelium itself, the mycorrhizosphere (Toljander et al., 2007; Finlay, 2008; Azcón and Barea, 2010; Jansa and Gryndler, 2010) where microbial populations can be stimulated for an active development.

As the AM extraradical mycelium occupies a central position in the rhizosphere of plants, many types of interactions involving this microbial symbiosis and significant microbial groups have been reported, as recently reviewed (Barea et al., 2012). These microbial components of rhizosphere population include: (i) N_2-fixing bacteria, either plant symbiotic or free-living; (ii) phosphate solubilising microorganisms; (iii) bacteria and fungi involved in the phytoremediation of heavy metal contaminated soils; (iv) microbial antagonists of root pathogens thus involved in biological control; and (v) bacteria producing substances able to aggregate soil particles thus cooperating to an improvement of physical soil properties through an improved structure stabilization. The results of the interaction between AM symbiosis and these microbial groups have a great agro-ecological impact on plant growth and health, and soil quality they are analysed in Section 5 accordingly.

A case of particular interest concerns the relationships between AM fungal structures and certain intimately associated endobacteria (Bonfante and Anca, 2009). Such biological interactions involving AM fungi were discussed in section 3 from an evolutionary-developmental perspective.

4. Ecological Effects of AM Symbiosis

Because of the well-recognized role of AM symbiosis on nutrient acquisition, water uptake and at enhancing plant tolerance against environmental stresses, this fungal-root association plays a key role in helping plants not only to survive but also to be productive under adversity (Jeffries et al., 2003). Such AM activities can account for the impact of AM fungi at facilitating colonization of primitive terrestrial habitat by land plants and continue being operative in modern terrestrial ecosystems where they exert crucial ecological impacts.

Adverse conditions of differing origin, particularly exacerbated in the current scenario of global change, generate a great array of stress situations which interferes the stability of both natural and agricultural systems. Plants must be able to cope with these stresses thus need from adaptive strategies able to increase their resilience to overcome negative impacts on agro- ecosystems, fundamental in maintaining their sustainability. AM formation can be considered one of these adaptive strategies based on AM activities able to confer the plant an increased tolerance to environmental stresses (Barea et al., 2011; Jeffries and Barea, 2012). As the biodiversity of AM fungal populations can influence plant community dynamics in both natural and agricultural environments the methods to analyze their diversity are first summarized. Then some examples are given to illustrate how a rational manipulation of the AM symbiosis can help restore sustainability to disturbed environments.

Concerning the analysis of the diversity of AM fungal communities, earlier studies were largely based on the morphology, wall characters and ontogeny of their large multinucleated spores. However, molecular tools are now available for a challenging dissection of AM fungal population dynamics (Robinson-Boyer et al., 2009). For molecular identification, the PCR-amplified rDNA fragments of the spores and/or the mycelia from AM fungi are usually subjected to cloning, fingerprinting and sequencing (Hempel et al., 2007; Öpik et al., 2008; Toljander et al., 2008; Rosendahl et al., 2009; Sonjak et al., 2009; Sánchez-Castro et al., 2012a o b). Alternative molecular tools now exist to quantitatively analyse the effect of environment, management or inoculation of soils on more diverse AM fungal communities. For example, Q-PCR can be used for simultaneous specific and quantitative investigations of particular taxa of AM fungi in roots and soils colonised by several taxa (Gamper et al., 2008; König et al., 2010). In addition, new techniques of high throughput sequencing (e.g. pyrosequencing) are now being used for AM fungi (Lumini et al., 2010). Despite the advancement in molecular techniques, the identification approaches employed for AM fungi based on morphological characteristics are still valid and used, being considered complementary to the molecular methods (Morton, 2009). A lack of relationship between genetic and functional diversity has been shown (Munkvold et al., 2004; Ehinger et al., 2009).

Managing AM fungi and their interactions with ecosystem beneficial microorganisms (mycorrhizosphere tailoring), is recognized as a feasible biotechnological tool to improve plant growth and health, and soil quality as a sustainable practice in agriculture or restoration of natural ecosystems. An increasing demand for low-input agriculture has resulted in greater interest in the manipulation and use of beneficial soil microorganisms. The strategic management of AM fungi and other beneficial soil microbes can reduce the use of chemicals and energy in agriculture leading to a more economical and sustainable production, while minimizing environmental degradation (Jeffries and Barea, 2012). These biological interventions are becoming more attractive as the use of chemicals for fumigation and disease control is progressively discouraged and fertilizers have become more and more expensive (Atkinson, 2009). Agro-biotechnological approaches include the use of microbial inoculants (Azcón and Barea, 2010). While the technology for the production of inexpensive rhizobial and free-living beneficial bacteria is commercially available, constraints on the production of inocula and the development of inoculation techniques have limited the use of AM fungal inoculants. The difficulty in culturing obligate symbionts such as AM fungi in the absence of their host plant is a major obstacle (Baar, 2008). Despite these problems, several companies worldwide are producing AM inoculum products which are now commercially available (Gianinazzi and Vosátka, 2004; Vosátka et al., 2008; IJdo et al., 2011).

Many co-inoculation experiments using selected AM fungi and rhizosphere microorganisms and their ecological impact have been reported (Barea et al., 2012). The main conclusions from these experiments are summarized here with emphasis on the ecological impacts of interactions related to: (1) biofertilization in nutrient deficient soils; (2) abiotic stress alleviation; (3) biological control of root pathogens; (4) restoration of degraded ecosystems; and (5) recovering endangered flora.

4.1. Biofertilization in Nutrient Deficient Soils

Soil microorganisms that increase the amount of nutrients available to plants (biofertilizers) include rhizobial bacteria and phosphate solubilising bacteria and fungi. The widespread presence of the AM symbiosis in legumes and its role in improving nodulation and N_2 fixation by legume-rhizobia associations are both universally recognized processes, as based on the supply of P by the AM fungi to satisfy the high P-demand of symbiotic N_2 fixation (Azcón and Barea, 2010). Methodologies based on the use of ^{15}N-enriched inorganic fertilizer allow to ascertain and quantify the amount of N which is actually fixed by legume-rhizobia consortia in a particular situation and to measure the contribution of the AM symbiosis to the process. A lower $^{15}N/^{14}N$ ratio in the shoots of rhizobia-inoculated AM plants with respect to those achieved by the same rhizobial strain in non-mycorrhizal plants was found. This indicated an enhancement of the N_2 fixation rates (an increase in ^{14}N from the atmosphere), as induced by the AM activity (Barea et al., 2005a).

The interactions between AM fungi and phosphate-solubilizing-microorganisms (PSM) are relevant to P cycling and plant nutrition, particularly in P-deficient soils. Because the Pi made available by PSM, acting on sparingly-soluble P sources, has limited diffusion in soil solution, the already available Pi may not reach the root surface. However, AM fungi could tap the phosphate ions solubilised by the PSM and translocate them into plant roots (Barea et al., 2005a; Azcón and Barea, 2010). The microbial interaction of AM fungi and PSM has been tested in experiments using ^{32}P-tracer methodologies (Barea et al., 2007). Upon adding a small amount of ^{32}P to label the exchangeable soil P pool, the isotopic composition, or "specific activity" (SA = $^{32}P/^{31}P$ quotient), was determined in plant tissues. It was found that dual inoculation reduced the SA of the host plant, indicating that these plants acquired P from sources which were not directly available to non-inoculated or singly-inoculated plants. Microbial inoculation improved biomass production and P accumulation in plants, demonstrating the interactive effects of PSM and AM-fungi on P capture, cycling and supply in a tailored mycorrhizosphere (Barea et al., 2007).

4.2. Abiotic Stress Alleviation

Abiotic stresses include drought and salinity, and contamination with pollutants such as heavy metals, radionuclides or polycyclic aromatic hydrocarbons.

Drought and salinity, together with extreme temperatures, all share a common osmotic component since they cause a dehydration of plant tissues. Osmotic stresses are known to have a major adverse effect on survival, normal development and productivity of crop plants. The eco-physiological and molecular bases for osmotic stress alleviation were described in sub-section 4.3. In diverse experiments have been demonstrated that the AM symbiosis often results in altered rates of water movement into, through and out of the host plants, with beneficial consequences for tissue hydration and plant physiology. Consequently, the AM symbiosis can protect crop plants against the detrimental effects of water deficit and that the AM contribution to plant drought tolerance results from a combination of physical, nutritional and cellular effects (Ruíz-Lozano and Aroca, 2010).

AM fungi improve phytoremediation of soils contaminated with heavy metals, radionuclides or polycyclic aromatic hydrocarbons (Leyval et al., 2002). Most phytoremediation assays involving mycorrhizosphere interactions concern heavy metals (HMs) and different strategies of phytoremediation have been investigated (Turnau et al., 2006; Ruíz-Lozano and Azcón, 2011). These studies mostly concentrated on Zn, Cu, Cd, Pb or Ni. Interactions between rhizobacteria and AM-fungi have been investigated in diverse experiments to ascertain whether they co-operate to benefit phytoremediation (Azcón et al., 2010). The main achievements resulting from these experiments using HM-multiple contaminated soils and *Trifolium* as test plants were: (i) a number of bacteria and the AM fungi were isolated from a HM-contaminated soil, and identified by 16S rDNA or 18S rDNA respectively; (ii) the target bacteria were able to accumulate large amounts of metals; (iii) co-inoculation with a HM-adapted autochthonous bacteria and AM fungi increased biomass, N and P content as compared to non-inoculated plants, and also enhanced the establishment of symbiotic structures (nodule number and AM colonization), which were negatively affected as the level of HM in soil increased; (iv) dual inoculation lowered HM concentrations in *Trifolium* plants, inferring a phytostabilization-based activity, however, as the total HM content in plant shoots was higher in dually-inoculated plants, due to the effect on biomass accumulation, a possible phytoextraction activity was suggested; and (v) inoculated HM-adapted bacteria increased dehydrogenase, phosphatase and □-gluconase activities, and auxin production, in the mycorrhizosphere, indicating an enhancement of microbial activities related to plant development.

4.3. Biological Control of Root Pathogens (and of Aggressive Agricultural Weeds)

AM establishment has been shown to reduce damage caused by soil-borne plant pathogens with an enhancement of plant resistance/tolerance in mycorrhizal plants (see sub-section 4.3 for molecular mechanisms). Since specific free-living microorganisms antagonistic to plant pathogens are being used as biological control agents, a major aim in rhizosphere biotechnology is to exploit the prophylactic ability of AM fungi in association with these antagonists (Barea et al., 2005b). The mycorrhizosphere has been hypothesized to constitute an environment conductive to microorganisms antagonistic to soil-borne pathogen proliferation. Indeed, various antagonistic bacteria have been identified associated to AM extraradical structures or in the mycorrhizosphere of several AM species, thus favouring biological control of pathogens (Lioussanne, 2010). Most co-inoculation studies involving AM fungi and biocontrol agents have dealt with bioprotection against soil-borne pathogens such as *Fusarium* or *Rhizoctonia* and other fungi (Saldajeno et al., 2008).

Whith regard to the control of aggressive agricultural weeds Rinaudo et al (2010) hypothesized that AM fungi may suppress weed growth a mycorrhizal attribute which has hardly been considered. Accordingly, these authors investigated the impact of AM fungi on weed growth. The presence of AM fungi reduced total weed biomass while sunflower benefitted from AM symbiosis via enhanced phosphorus nutrition. The results indicate that the stimulation of AM fungi in agro-ecosystems may suppress some aggressive weeds and suggest a possible applicability of the AM symbiosis in weed control, an agricultural practice in the context of sustainability issues. López-Ráez et al. (2012) discussed the potential use of some rhizosphere signal molecules as new biological control strategies against weeds. Particularly, they focus on the role of strigolactones, a new class of plant hormones emerging as important signal molecules for some rhizosphere processes (see section 4.2). Strigolactones are exuded into the soil, where they are known to act both as host detection signals for AM formation and as germination stimulants for root parasitic plant seeds (Bouwmeester et al., 2007). López-Ráez et al. (2011) found that strigolactone production is significantly reduced upon AM symbiosis establishment. Thus they suggested the potentiality of the AM symbiosis for controlling root parasitic weeds since these plants have a lower supply of the hormone needed to stimulate germination of their seed.

4.4. Restoration of Degraded Ecosystems

One of the most serious world problems affecting marginal agricultural lands is desertification. It is a complex and dynamic process which is claiming several hundred million hectares annually. Human activities can cause or accelerate desertification and the loss of most plant species and their corresponding symbionts (Requena et al., 2001). Revegetation of desertified ecosystems is problematical but management practices have been developed and recently reviewed (Barea et al., 2011).

As a result of desertification processes, disturbance of natural plant communities is often accompanied, or preceded by, loss of physical, chemical and biological soil properties, such as soil structure, plant nutrient availability, organic matter content, microbial activity, etc. (Jeffries and Barea, 2012), traits which are fundamental for soil quality and soil structure

(Miller and Jastrow, 2000; Buscot, 2005) and thus limit re-establishment of the natural plant cover. In particular, desertification causes disturbance of plant-microbe symbioses and thus the recovery of populations of AM fungi and rhizobial bacteria is essential to the integral restoration of a degraded area (Barea et al., 2011; Jeffries and Barea, 2012).

Accordingly, management of AM fungi, together with rhizosphere bacteria, was proposed for the integral restoration of degraded ecosystems. A model experiment in this context is that carried out in a desertified semi-arid ecosystem with *Anthyllis cytisoides*, a drought-tolerant legume, as the test plant (Requena et al., 2001). *Anthyllis* seedlings inoculated with indigenous rhizobia and AM fungi were transplanted to field plots for a five-year-trial. The tailored mycorrhizosphere enhanced seedling survival and growth, P-acquisition, N-fixation, and N-transfer from N-fixing to associated non-fixing species in the natural succession. The improvement in the physical-chemical properties in the soil around the *Anthyllis* plants was shown by the increased levels of N, organic matter and number of hydro-stable soil aggregates. Glomalin-related glycol-proteins, produced by the external hyphae of AM fungi, seen to be involved in the initiation and stabilization of water-stable soil aggregates, due to its glue-like hydrophobic nature (Miller and Jastrow, 2000; Rillig and Mummey, 2006; Bedini et al., 2009).

4.5. Recovering Endangered Flora

Diverse ecosystems in general and high mountains in particular often bear a reach plant species diversity but they include large numbers of threatened species (Azcón-Aguilar et al., 2012). Endangered plant species are threatened by human exploitations, alterations in their natural habitats or environmental changes. Especially global warming may result in the loss of species, particularly the high mountain species. It is well known that the *in situ* and *ex situ* conservation of endangered plant species has ecophysiological constraints which difficult regeneration in their natural habitats. One of these constraints can be mycorrhiza formation. In fact, AM colonization can affect vegetative and sexual reproduction of plants by impacting on the number of inflorescences, fruit and seed production, and offspring vigour (Bothe et al., 2010). Most plants depend on mycorrhizas to thrive, particularly in fragile and stressed environments, as those in certain areas of the high Mediterranean mountains of the Sierra Nevada National Park (Granada, Spain), that we summarize here as a model example. Sierra Nevada constitutes an exceptional refuge for the flora and one of the enclaves with higher biodiversity levels of the European continent. It presents about 2100 plant species and 80 exclusive endemisms, some of them threatened with extinction. Due to the importance of mycorrhizal symbioses for plant establishment and development in stress conditions, and since information concerning the mycorrhizal status of endangered plants is of major importance for their potential re-establishment (Fuchs and Haselwandter, 2004; Zubek et al., 2009; Bothe et al., 2010), a research project aiming at ascertaining the impact of mycorrhizal associations in facilitating the conservation of species from the threatened flora of Sierra Nevada. A series of endangered plant species have successfully been produced under nursery conditions as aided by a tailored mycorrhizal inoculation (specific mixes of suitable, autochthonous AM fungi). These AM-seedlings are endowed therefore with attributed to facilitate the successful re-establishment in their natural habitats.

Conclusion

Evidences from fossil records, paleobotanical data and recent molecular phylogenetic data indicate that the origin and divergence of AM fungi (Glomeromycota) date back more than 500 million years, and that they soon associated to early plant lineages. Such AM associations evolved as a symbiosis highly specialized in nutrient acquisition and plant protection thereby facilitating plant adaptation to the primitive harsh terrestrial environment, being commonly accepted that AM fungi played a crucial role in land colonisation by plants.

Genomic and transcriptomic analyses are introducing new insights for the characterization of the unculturable AM fungi. Molecular tools are now available for a challenging dissection of AM fungal population dynamics. The developmental events in AM establishment have been dissected with special emphasis in describing new advances on the knowledge of key switches such as: signalling and recognition between symbionts, appressorium (hyphopodium) and PPA formation for initiation of root colonization by the fungi, intracellular accommodation and arbuscule development. The identification of genes that are required for AM development and function is subject of current interest. Recent advances in AM functioning are shedding light on the underlying mechanisms of the coordinated nutrient exchange between symbionts, priming of plant defences against pathogens, the increased tolerance of mycorrhizal plants to abiotic stresses and the impact of AM symbiosis on its soil biotic environment.

Adverse conditions, particularly exacerbated in the current scenario of global change, can generate a great array of stress situations which interfere with the stability of both natural and agricultural systems. At this respect AM formation can be considered an adaptive strategy able to confer plants an increased tolerance to environmental stresses. Thus, mycorrhizal technology should be a component in sustainable strategies in the future, since application of AM fungi can reduce fertilizer and energy inputs yet promote healthy plant growth, particularly in a world of depleting non-renewable resources.

Acknowledgments

We are grateful to our colleagues from EEZ-CSIC Drs. R. Azcón, N. Ferrol, J.M. Ruíz-Lozano, M.J. Pozo, R. Aroca and J.A. López-Ráez for supplying us valuable material and information. We also thank support from the Spanish National Research Program (R and D+i)-European Union (Feder) CGL2009-08825/BOS, and from the Andalucian (Spain) Government, PAIDI (R and D+i) Program: P07-CVI-02952.

References

Akiyama, K., Matsuzaki, K. and Hayashi H. (2005). Plant sesquiterpenes induce hyphal branching in arbuscular mycorrhizal fungi. *Nature* 435, 824-827.

Aroca, R., Porcel, R. and Ruíz-Lozano, J. M. (2007). How does arbuscular mycorrhizal symbiosis regulate root hydraulic properties and plasma membrane aquaporins in phaseolus vulgaris under drought, cold or salinity stresses? *New Phytologist*, 173, 808-816.

Aroca, R., Porcel, R. and Ruíz-Lozano, J. M. (2011). Plant drought tolerance enhacement by arbuscular mycorrhizal symbiosis. In S. M. Fulton (Ed.) *Mycorrhizal fungi: Soil, agriculture and environmental implications* (pp. 229-240). New York, Nova Science Publishers, Inc.

Aroca, R., Vernieri, P. and Ruíz-Lozano, J. M. (2008). Mycorrhizal and non-mycorrhizal lactuca sativa plants exhibit contrasting responses to exogenous aba during drought stress and recovery. *Journal of Experimental Botany*, 59, 2029-2041.

Atkinson, D. (2009). Soil microbial resources and agricultural policies. In C. Azcón-Aguilar, J. M. Barea, S. Gianinazzi and V. Gianinazzi-Pearson (Eds.), *Mycorrhizas functional processes and ecological impact* (pp. 33-45). Berlin, Heidelberg, Springer-Verlag.

Augé, R. M., Toler, H. D., Sams, C. E. and Nasim, G. (2008). Hydraulic conductance and water potential gradients in squash leaves showing mycorrhiza-induced increases in stomatal conductance. *Mycorrhiza*, 18, 115-121.

Azcón, R. and Barea, J. M. (2010). Mycorrhizosphere interactions for legume improvement. In M. S. Khan, A. Zaidi and J. Musarrat (Eds.), *Microbes for legume improvement* (pp. 237-271). Vienna, New York, Springer-Verlag.

Azcón, R., Perálvarez, M. D., Roldán, A. and Barea, J. M. (2010). Arbuscular mycorrhizal fungi, bacillus cereus, and candida parapsilosis from a multicontaminated soil alleviate metal toxicity in plants. *Microbial Ecology*, 59, 668-677.

Azcón-Aguilar, C. and Barea, J. M. (1996). Arbuscular mycorrhizas and biological control of soil-borne plant pathogens – an overview of the mechanisms involved. *Mycorrhiza*, 6, 457-464.

Azcón-Aguilar, C., Barea, J. M., Gianinazzi, S. and Gianinazzi-Pearson, V. (2009). *Mycorrhizas functional processes and ecological impact*. Berlin, Heidelberg, Springer-Verlag.

Azcón-Aguilar, C., Palenzuela, J., Ferrol, N., Oehl, F. and Barea, J. M. (2012). Mycorrhizal status and arbuscular mycorrizal fungal diversity of endangered plant species in the sierra nevada national park. In R. Duponnois (Ed.) *The mycorrhizal symbiosis in mediterranean environment: Importance in ecosystem stability and in soil rehabilitation strategies* (pp. In press). Editions Nova Science Publishers.

Baar, J. (2008). From production to application of arbuscular mycorrhizal fungi in agricultural systems: Requirements and needs. In A. Varma (Ed.) *Mycorrhiza: State of the art, genetics and molecular biology, eco-function, biotechnology, eco-physiology, structure and systematics. 3rd ed* (pp. 361-373). Berlin, Heidelberg, Germany, Springer-Verlag.

Bago, B., Azcón-Aguilar, C., Goulet, A. and Piché, Y. (1998a). Branched absorbing structures (BAS): A feature of the extraradical mycelium of symbiotic arbuscular mycorrhizal fungi. *New Phytologist*, 139, 375-388.

Bago, B., Zipfel, W., Williams, R.M., Chamberland, H., Lafontaine, J.G., Webb, W.W. and Piche Y (1998). Invivo studies on the nuclear behavior of the arbuscular mycorrhizal fungus *Gigaspora rosea* grown under axenic conditions. *Protoplasma* 203, 1–15.

Bago, B. and Cano, C. (2005). Breaking myths on arbuscular mycorrhizas in vitro biology. In S. Declerck, F. G. Strullu and J. A. Fortin (Eds.), *In vitro culture of mycorrhizas* vol. 4. *Soil biology* (pp. 111-138). Berlin, Heidelberg, Springer-Verlag.

Bago, B., Pfeffer, P. E., Zipfel, W., Lammers, P. and Shachar-Hill, Y. (2002). Tracking metabolism and imaging transport in arbuscular mycorrhizal fungi. Metabolism and transport in AM fungi. *Plant and Soil*, 244, 189-197.

Barea, J. M., Azcón, R. and Azcón-Aguilar, C. (2005a). Interactions between mycorrhizal fungi and bacteria to improve plant nutrient cycling and soil structure. In F. Buscot and A. Varma (Eds.), *Microorganisms in soils: Roles in genesis and functions* (pp. 195-212). Berlin, Heidelbert, Springer-Verlag.

Barea, J. M., Ferrol, N., Azcón-Aguilar, C. and Azcón, R. (2008). Mycorrhizal symbioses. In P. J. White and J. P. Hammond (Eds.), *The ecophysiology of plant-phosphorus interactions. Series: Plant ecophysiology, vol. 7* (pp. 143-163). Dordrecht, Springer.

Barea, J. M., Palenzuela, J., Cornejo, P., Sánchez-Castro, I., Navarro-Fernández, C., Lopéz-García, A., Estrada, B., *et al.* (2011). Ecological and functional roles of mycorrhizas in semi-arid ecosystems of southeast spain. *Journal of Arid Environments*, 75, 1292-1301.

Barea, J. M., Pozo, M. J., Azcón, R. and Azcón-Aguilar, C. (2005b). Microbial co-operation in the rhizosphere. *Journal of Experimental Botany*, 56, 1761-1778.

Barea, J. M., Pozo, M. J., Azcón, R. and Azcón-Aguilar, C. (2012). Microbial interactions in the rhizosphere. In F. de Bruijn (Ed.) *Molecular microbial ecology of the rhizosphere* (pp. in press). USA, Wiley-Blackwell.

Barea, J. M., Toro, M. and Azcón, R. (2007). The use of ^{32}p isotopic dilution techniques to evaluate the interactive effects of phosphate-solubilizing bacteria and mycorrhizal fungi at increasing plant p availability. In E. Velázquez and C. Rodríguez-Barrueco (Eds.), *First international meeting on microbial phosphate solubilization. Series: Developments in plant and soil sciences* (pp. 223-227). Dordrecht, The Netherlands, Springer.

Bárzana, G., Aroca, R., Paz, J. A., Chaumont, F., Martínez-Ballesta, M. C., Carvajal, M. and Ruíz-Lozano, J. M. (2012). The arbuscular mycorrhizal symbiosis increases relative apoplastic water flow in roots of the host plant under both well-watered and drought stress conditions. *Annals of Botany*, doi:10.1093/aob/mcs1007. (In press).

Bécard, G., Kosuta, S., Tamasloukht, M., Séjalon-Delmas, N. and Roux, C. (2004). Partner communication in the arbuscular mycorrhizal interaction. *Canadian Journal of Botany-Revue Canadienne De Botanique*, 82, 1186-1197.

Bécard, G. and Piché, Y. (1989). Fungal growth-stimulation by co$_2$ and root exudates in vesicular-arbuscular mycorrhizal symbiosis. *Applied and Environmental Microbiology*, 55, 2320-2325.

Bedini, S., Pellegrino, E., Avio, L., Pellegrini, S., Bazzoffi, P., Argese, E. and Giovannetti, M. (2009). Changes in soil aggregation and glomalin-related soil protein content as affected by the arbuscular mycorrhizal fungal species glomus mosseae and glomus intraradices. *Soil Biology and Biochemistry*, 41, 1491-1496.

Beerling, D. J. and Berner, R. A. (2005). Feedbacks and the coevolution of plants and atmospheric co$_2$. *Proceedings of the National Academy of Sciences of the United States of America*, 102, 1302-1305.

Berbee, M. L. and Taylor, J. W. (2007). Rhynie chert: A window into a lost world of complex plant-fungus interactions. *New Phytologist*, 174, 475-479.

Bidartondo, M. I., Read, D. J., Trappe, J. M., Merckx, V., Ligrone, R. and Duckett, J. G. (2012). The dawn of symbiosis between plants and fungi. *Biology Letters*, (In press).

Bonfante, P. and Anca, I.-A. (2009). Plants, mycorrhizal fungi, and bacteria: A network of interactions. *Annual Review of Microbiology*, 63, 363-383.

Bonfante, P. and Genre, A. (2008). Plants and arbuscular mycorrhizal fungi: An evolutionary-developmental perspective. *Trends in Plant Science*, 13, 492-498.

Bothe, H., Turnau, K. and Regvar, M. (2010). The potential role of arbuscular mycorrhizal fungi in protecting endangered plants and habitats. *Mycorrhiza*, 20, 445-457.

Bouwmeester, H. J., Roux, C., López-Ráez, J. A. and Becard, G. (2007). Rhizosphere communication of plants, parasitic plants and am fungi. *Trends in Plant Science*, 12, 224-230.

Brundrett, M. C. (2002). Coevolution of roots and mycorrhizas of land plants. *New Phytologist*, 154, 275-304.

Brundrett, M. C. (2009). Mycorrhizal associations and other means of nutrition of vascular plants: Understanding the global diversity of host plants by resolving conflicting information and developing reliable means of diagnosis. *Plant and Soil*, 320, 37-77.

Buscot, F. (2005). What are soils? In F. Buscot and S. Varma (Eds.), *Microorganisms in soils: Roles in genesis and functions* (pp. 3-18). Springer-Verlag, Heidelbert, Germany.

Cai, C. Y., Shu, O. Y., Wang, Y., Fang, Z. J., Rong, J. Y., Geng, L. Y. and Li, X. X. (1996). An early silurian vascular plant. *Nature*, 379, 592-592.

Chen, C., Fan, C., Gao, M. and Zhu, H. (2009). Antiquity and function of castor and pollux, the twin ion channel-encoding genes key to the evolution of root symbioses in plants. *Plant Physiology*, 149, 306-317.

Dotzler, N., Walker, C., Krings, M., Hass, H., Kerp, H., Taylor, T. N. and Agerer, R. (2009). Acaulosporoid glomeromycotan spores with a germination shield from the 400-million-year-old rhynie chert. *Mycological Progress*, 8, 9-18.

Ehinger, M., Koch, A. M. and Sanders, I. R. (2009). Changes in arbuscular mycorrhizal fungal phenotypes and genotypes in response to plant species identity and phosphorus concentration. *New Phytologist*, 184, 412-423.

Ercolin, F. and Reinhardt, D. (2011). Successful joint ventures of plants: Arbuscular mycorrhiza and beyond. *Trends in Plant Science*, 16, 356-362.

Ferrol, N., Azcón-Aguilar, C., Bago, B., Franken, P., Gollotte, A., González-Guerrero, M., Harrier, L. A., *et al.* (2004). Genomics of arbuscular mycorrhizal fungi. In D. K. Arora and G. G. Khachatourinas (Eds.), *Applied mycology and biotechnology, vol 4 fungal genomics* (pp. 379-403). Amsterdam, Elsevier Science.

Ferrol, N., González-Guerrero, M., Valderas, A., Benabdellah, K. and Azcón-Aguilar, C. (2009). Survival strategies of arbuscular mycorrhizal fungi in Cu-polluted environments. *Phytochemistry Reviews*, 8, 551-559.

Ferrol, N. and Pérez-Tienda, J. (2009). Coordinated nutrient exchange in arbuscular mycorrhiza. In C. Azcón-Aguilar, J. M. Barea, S. Gianinazzi and V. Gianinazzi-Pearson (Eds.), *Mycorrhizas functional processes and ecological impact* (pp. 73-87). Berlin, Heidelberg, Springer-Verlag.

Finlay, R. D. (2008). Ecological aspects of mycorrhizal symbiosis: With special emphasis on the functional diversity of interactions involving the extraradical mycelium. *Journal of Experimental Botany*, 59, 1115-1126.

Franken, P. (2010). Molecular-physiological aspects of the am symbiosis post penetration. In H. Koltai and Y. Kapulnik (Eds.), *Arbuscular mycorrhizas: Physiology and function* (pp. 93-116). Netherlands, Springer.

Fuchs, B. and Haselwandter, K. (2004). Red list plants: Colonization by arbuscular mycorrhizal fungi and dark septate endophytes. *Mycorrhiza*, 14, 277-281.

Gallaud, I. (1905). Etudes sur les mycorrhizes endotrophes. *Revue Générale de Botanique*, 17, 5–48, 66–83, 123–135, 223–239, 313–325, 425–433, 479–500.

Gamper, H. A., Young, J. P. W., Jones, D. L. and Hodge, A. (2008). Real-time pcr and microscopy: Are the two methods measuring the same unit of arbuscular mycorrhizal fungal abundance? *Fungal Genetics and Biology*, 45, 581-596.

García-Garrido, J. M. and Ocampo, J. A. (2002). Regulation of the plant defence response in arbuscular mycorrhizal symbiosis. *Journal of Experimental Botany*, 53, 1377-1386.

Gaspar, M. L., Pollero, R. J. and Cabello, M. N. (1994). Triacylglycerol consumption during spore germination of vesicular-arbuscular mycorrhizal fungi. *Journal of the American Oil Chemists Society*, 71, 449-452.

Genre, A. and Bonfante, P. (2010). The making of symbiotic cells in arbuscular mycorrhizal roots. In H. Koltai and Y. Kapulnik (Eds.), *Arbuscular mycorrhizas: Physiology and function* (pp. 57-71). Netherlands, Springer.

Genre, A., Chabaud, M., Faccio, A., Barker, D. G. and Bonfante, P. (2008). Prepenetration apparatus assembly precedes and predicts the colonization patterns of arbuscular mycorrhizal fungi within the root cortex of both medicago truncatula and daucus carota. *Plant Cell*, 20, 1407-1420.

Genre, A., Chabaud, M., Timmers, T., Bonfante, P. and Barker, D. G. (2005). Arbuscular mycorrhizal fungi elicit a novel intracellular apparatus in *medicago truncatula* root epidermal cells before infection. *Plant Cell*, 17, 3489-3499.

Gianinazzi, S. and Vosátka, M. (2004). Inoculum of arbuscular mycorrhizal fungi for production systems: Science meets business. *Canadian Journal of Botany-Revue Canadienne De Botanique*, 82, 1264-1271.

Gianinazzi-Pearson, V., Azcón-Aguilar, C., Bécard, G., Bonfante, P., Ferrol, N., Franken, P., Gollote, A., *et al.* (2004). Structural and functional genomics of symbiotic arbuscular mycorrhizal fungi. In J. S. Tkacz and L. Lange (Eds.), *Advances in fungal biotechnology for industry, medicine and agriculture* (pp. 405-424). New York, Boston, Kluwer Academic/Plenum Publishers.

Gianinazzi-Pearson, V., Tollot, M. and Seddas, P. M. A. (2009). Dissection of genetic cell programmes driving early arbuscular mycorrhiza interactions. In C. Azcón-Aguilar, J. M. Barea, S. Gianinazzi and V. Gianinazzi-Pearson (Eds.), *Mycorrhizas functional processes and ecological impact* (pp. 33-45). Berlin, Heidelberg, Springer-Verlag.

Giovannetti, M., Avio, L. and Sbrana, C. (2010). Fungal spore germination and pre-symbiotic mycelial growth – physiological and genetic aspects. In H. Koltai and Y. Kapulnik (Eds.), *Arbuscular mycorrhizas: Physiology and function* (pp. 3-32). Netherlands, Springer.

Giovannetti, M., Sbrana, C., Avio, L. and Strani, P. (2004). Patterns of below-ground plant interconnections established by means of arbuscular mycorrhizal networks. *New Phytologist*, 164, 175-181.

González-Guerrero, M., Benabdellah, K., Ferrol, N. and Azcón-Aguilar, C. (2009). Mechanisms underlying heavy metal tolerance in arbuscular mycorrhizas. In C. Azcón-Aguilar, J. M. Barea, S. Gianinazzi and V. Gianinazzi-Pearson (Eds.), *Mycorrhizas functional processes and ecological impact* (pp. 107-122). Berlin, Heidelberg, Springer-Verlag.

Harrison, M. J., Pumplin, N., Breuillin, F. J., Noar, R. D. and Park, H. J. (2010). Phosphate transporters in arbuscular mycorrhizal symbiosis. In H. Koltai and Y. Kapulnik (Eds.), *Arbuscular mycorrhizas: Physiology and function* (pp. 117-135). Netherlands, Springer.

Helgason, T. and Fitter, A. H. (2009). Natural selection and the evolutionary ecology of the arbuscular mycorrhizal fungi (phylum glomeromycota). *Journal of Experimental Botany*, 60, 2465-2480.

Hempel, S., Renker, C. and Buscot, F. (2007). Differences in the species composition of arbuscular mycorrhizal fungi in spore, root and soil communities in a grassland ecosystem. *Environmental Microbiology*, 9, 1930-1938.

Honrubia, M. (2009). The mycorrhizae: A plant-fungus relation that has existed for more than 400 million years. *Anales Del Jardin Botanico De Madrid*, 66, 133-144.

Horváth, B., Yeun, L. H., Domonkos, A., Halasz, G., Gobbato, E., Ayaydin, F., Miro, K., *et al.* (2011). Medicago truncatula ipd3 is a member of the common symbiotic signaling pathway required for rhizobial and mycorrhizal symbioses. *Molecular Plant-Microbe Interactions*, 24, 1345-1358.

Ijdo, M., Cranenbrouck, S. and Declerck, S. (2011). Methods for large-scale production of am fungi: Past, present, and future. *Mycorrhiza*, 21, 1-16.

James, T. Y., Kauff, F., Schoch, C. L., Matheny, P. B., Hofstetter, V., Cox, C. J., Celio, G., *et al.* (2006). Reconstructing the early evolution of fungi using a six-gene phylogeny. *Nature*, 443, 818-822.

Jansa, J. and Gryndler, M. (2010). Biotic environments of the arbuscular mycorrhizal fungi in soil. In H. Koltai and Y. Kapulnik (Eds.), *Arbuscular mycorrhizas: Physiology and function* (pp. 209-236). Netherlands, Springer.

Javot, H., Penmetsa, R. V., Terzaghi, N., Cook, D. R. and Harrison, M. J. (2007). A medicago truncatula phosphate transporter indispensable for the arbuscular mycorrhizal symbiosis. *Proceedings of the National Academy of Sciences of the United States of America*, 104, 1720-1725.

Jeffries, P. and Barea, J. M. (2012). Arbuscular mycorrhiza – a key component of sustainable plant-soil ecosystems. In B. Hock (Ed.) *The mycota* (pp. In press). Berlin, Heidelberg, Springer-Verlag.

Jeffries, P., Gianinazzi, S., Perotto, S., Turnau, K. and Barea, J. M. (2003). The contribution of arbuscular mycorrhizal fungi in sustainable maintenance of plant health and soil fertility. *Biology and Fertility of Soils*, 37, 1-16.

Jung, S., Martínez-Medina, A., López-Ráez, J. A. and Pozo, M. J. (2012). Mycorrhiza-induced resistance and priming of plant defences. *Journal of Chemical Ecology*, In press.

Kidston, R. and Lang, W. H. (1917). On old red sandstone plants showing structure from the rhynie chert bed, aberdeenshire. Part i. *Rhynia gwynne-vaughanii*, kidston and lang. *Transactions of the Royal Society of Edinburgh*, 5, 761-784.

König, S., Wubet, T., Dormann, C. F., Hempel, S., Renker, C. and Buscot, F. (2010). Taqman real-time pcr assays to assess arbuscular mycorrhizal responses to field manipulation of grassland biodiversity: Effects of soil characteristics, plant species richness, and functional traits. *Applied and Environmental Microbiology*, 76, 3765-3775.

Kosuta, S., Chabaud, M., Lougnon, G., Gough, C., Denarie, J., Barker, D. G. and Bécard, G. (2003). A diffusible factor from arbuscular mycorrhizal fungi induces symbiosis-specific mtenod11 expression in roots of medicago truncatula. *Plant Physiology*, 131, 952-962.

Kosuta, S., Hazledine, S., Sun, J., Miwa, H., Morris, R. J., Downie, J. A. and Oldroyd, G. E. D. (2008). Differential and chaotic calcium signatures in the symbiosis signaling pathway of legumes. *Proceedings of the National Academy of Sciences of the United States of America*, 105, 9823-9828.

Krings, M., Taylor, T. N., Hass, H., Kerp, H., Dotzler, N. and Hermsen, E. J. (2007). Fungal endophytes in a 400-million-yr-old land plant: Infection pathways, spatial distribution, and host responses. *New Phytologist*, 174, 648-657.

Krüger, M., Krüger, C., Walker, C., Stockinger, H. and Schüßler, A. (2012). Phylogenetic reference data for systematics and phylotaxonomy of arbuscular mycorrhizal fungi from phylum to species level. *New Phytologist*, 193, 970-984.

Lee, J. and Young, J. P. W. (2009). The mitochondrial genome sequence of the arbuscular mycorrhizal fungus glomus intraradices isolate 494 and implications for the phylogenetic placement of glomus. *New Phytologist*, 183, 200-211.

Leyval, C., Joner, E. J., del Val, C. and Haselwandter, K. (2002). Potential of arbuscular mycorrhizal fungi for bioremediation. In S. Gianinazzi, H. Schüepp, J. M. Barea and K. Haselwandter (Eds.), *Mycorrhizal technology in agriculture* (pp. 175-186). Basel, Switzerland, Birkhäuser Verlag.

Lioussanne, L. (2010). The role of the arbuscular mycorrhiza-associated rhizobacteria in the biocontrol of soilborne phytopathogens. *Spanish Journal of Agricultural Research*, 8, S51-S61.

Lohse, S., Schliemann, W., Ammer, C., Kopka, J., Strack, D. and Fester, T. (2005). Organization and metabolism of plastids and mitochondria in arbuscular mycorrhizal roots of Medicago truncatula. *Plant Physiology*, 139, 329-340.

López-Pedrosa, A., González-Guerrero, M., Valderas, A., Azcón-Aguilar, C. and Ferrol, N. (2006). Gintamt1 encodes a functional high-affinity ammonium transporter that is expressed in the extraradical mycelium of glomus intraradices. *Fungal Genetics and Biology*, 43, 102-110.

López-Ráez, J. A., Bouwmeester, H. and Pozo, M. J. (2012). Communication in the rhizosphere, a target for pest management In E. Lichtfouse (Ed.) *Sustainable agriculture reviews* vol 8. *Agroecology and strategies for climate change* (pp. 109-133). Springer Netherlands.

López-Ráez, J. A., Pozo, M. J. and García-Garrido, J. M. (2011). Strigolactones: A cry for help in the rhizosphere. *Botany-Botanique*, 89, 513-522.

Lumini, E., Bianciotto, V., Jargeat, P., Novero, M., Salvioli, A., Faccio, A., Bécard, G., *et al.* (2007). Presymbiotic growth and sporal morphology are affected in the arbuscular mycorrhizal fungus gigaspora margarita cured of its endobacteria. *Cellular Microbiology*, 9, 1716-1729.

Lumini, E., Orgiazzi, A., Borriello, R., Bonfante, P. and Bianciotto, V. (2010). Disclosing arbuscular mycorrhizal fungal biodiversity in soil through a land-use gradient using a pyrosequencing approach. *Environmental Microbiology*, 12, 2165-2179.

Macdonald, R. M., Chandler, M. R. and Mosse, B. (1982). The occurrence of bacterium-like organelles in vesicular-arbuscular mycorrhizal fungi. *New Phytologist*, 90, 659-663.

Maherali, H. and Klironomos, J. N. (2007). Influence of phylogeny on fungal community assembly and ecosystem functioning. *Science*, 316, 1746-1748.

Maillet, F., Poinsot, V., Andre, O., Puech-Pages, V., Haouy, A., Gueunier, M., Cromer, L., *et al.* (2011). Fungal lipochitooligosaccharide symbiotic signals in arbuscular mycorrhiza. *Nature*, 469, 58-64.

Malloch, D. W., Pirozynski, K. A. and Raven, P. H. (1980). Ecological and evolutionary significance of mycorrhizal symbioses in vascular plants. *Proceedings of the National Academy of Sciences of the United States of America-Biological Sciences*, 77, 2113-2118.

Markmann, K., Giczey, G. and Parniske, M. (2008). Functional adaptation of a plant receptor-kinase paved the way for the evolution of intracellular root symbioses with bacteria. *Plos Biology*, 6, 497-506.

Martin, F., Gianinazzi-Pearson, V., Hijri, M., Lammers, P., Requena, N., Sanders, I. R., Shachar-Hill, Y., *et al.* (2008). The long hard road to a completed Glomus intraradices genome. *New Phytologist*, 180, 747-750.

Martín-Rodríguez, J. A., León-Morcillo, R., Vierheilig, H., Ocampo, J. A., Ludwig-Mueller, J. and García-Garrido, J. M. (2011). Ethylene-dependent/ethylene-independent aba regulation of tomato plants colonized by arbuscular mycorrhiza fungi. *New Phytologist*, 190, 193-205.

Miller, R. M. and Jastrow, J. D. (2000). Mycorrhizal fungi influence soil structure. In Y. Kapulnik and D. D. Douds (Eds.), *Arbuscular mycorrhizas: Physiology and function* (pp. 3-18). Dordrecht, The Netherlands, Kluwer Academic Publishers.

Morton, J. B. (2009). Reconciliation of conflicting phenotypic and rrna gene phylogenies of fungi in glomeromycota based on underlying patterns and processes. In C. Azcón-Aguilar, J. M. Barea, S. Gianinazzi and V. Gianinazzi-Pearson (Eds.), *Mycorrhizas functional processes and ecological impact* (pp. 137-154). Berlin, Heidelberg, Springer-Verlag.

Mosse, B. (1953). Fructifications associated with mycorrhizal strawberry roots. *Nature*, 171, 974-974.

Munkvold, L., Kjøller, R., Vestberg, M., Rosendahl, S. and Jakobsen, I. (2004). High functional diversity within species of arbuscular mycorrhizal fungi. *New Phytologist*, 164, 357-364.

Naumann, M., Schuessler, A. and Bonfante, P. (2010). The obligate endobacteria of arbuscular mycorrhizal fungi are ancient heritable components related to the mollicutes. *Isme Journal*, 4, 862-871.

Öpik, M., Saks, Ü., Kennedy, J. and Daniell, T. (2008). Global diversity patterns of arbuscular mycorrhizal fungi-community composition and links with functionality. In A. Varma (Ed.) *Mycorrhiza: State of the art, genetics and molecular biology, eco-function, biotechnology, eco-physiology, structure and systematics. 3rd ed* (pp. 89-111). Berlin, Heidelberg, Germany, Springer-Verlag.

Parniske, M. (2008). Arbuscular mycorrhiza: The mother of plant root endosymbioses. *Nature Reviews Microbiology*, 6, 763-775.

Paszkowski, U. (2006). A journey through signaling in arbuscular mycorrhizal symbioses 2006. *New Phytologist*, 172, 35-46.

Pawlowska, T. E. (2005). Genetic processes in arbuscular mycorrhizal fungi. *Fems Microbiology Letters*, 251, 185-192.

Pérez-Tienda, J., Testillano, P. S., Balestrini, R., Fiorilli, V., Azcón-Aguilar, C. and Ferrol, N. (2011). Gintamt2, a new member of the ammonium transporter family in the arbuscular mycorrhizal fungus glomus intraradices. *Fungal Genetics and Biology*, 48, 1044-1055.

Pirozynski, K. A. and Malloch, D. W. (1975). Origin of land plants: A matter of mycotropism. *Biosystems*, 6, 153-164.

Porcel, R., Aroca, R. and Manuel Ruiz-Lozano, J. (2012). Salinity stress alleviation using arbuscular mycorrhizal fungi. A review. *Agronomy for Sustainable Development*, 32, 181-200.

Porcel, R., Barea, J. M. and Ruíz-Lozano, J. M. (2003). Antioxidant activities in mycorrhizal soybean plants under drought stress and their possible relationship to the process of nodule senescence. *New Phytologist*, 157, 135-143.

Pozo, M. J. and Azcón-Aguilar, C. (2007). Unraveling mycorrhiza-induced resistance. *Current Opinion in Plant Biology*, 10, 393-398.

Pozo, M. J., Jung, S. C., López-Ráez, J. A. and Azcón-Aguilar, C. (2010). Impact of arbuscular mycorrhizal symbiosis on plant response to biotec stress: The role of plant defence mechanisms. In H. Koltai and Y. Kapulnik (Eds.), *Arbuscular mycorrhizas: Physiology and function* (pp. 193-207). Netherlands, Springer.

Pozo, M. J., Verhage, A., García-Andrade, J., García, J. M. and Azcón-Aguilar, C. (2009). Priming plant defence against pathogens by arbuscular mycorrhizal fungi. In C. Azcón-Aguilar, J. M. Barea, S. Gianinazzi and V. Gianinazzi-Pearson (Eds.), *Mycorrhizas functional processes and ecological impact* (pp. 123-135). Berlin, Heidelberg, Springer-Verlag.

Provorov, N. A. and Vorobyov, N. I. (2009). Interspecies altruism in plant-microbe symbioses: Use of group selection models to resolve the evolutionary paradoxes. In C. Azcón-Aguilar, J. M. Barea, S. Gianinazzi and V. Gianinazzi-Pearson (Eds.), *Mycorrhizas functional processes and ecological impact* (pp. 17-31). Berlin, Heidelberg, Springer-Verlag.

Redecker, D. (2002). Molecular identification and phylogeny of arbuscular mycorrhizal fungi. *Plant and Soil*, 244, 67-73.

Redecker, D., Kodner, R. and Graham, L. E. (2000). Glomalean fungi from the ordovician. *Science*, 289, 1920-1921.

Reinhardt, D. (2007). Programming good relations - development of the arbuscular mycorrhizal symbiosis. *Current Opinion in Plant Biology*, 10, 98-105.

Remy, W., Taylor, T. N., Hass, H. and Kerp, H. (1994). Four hundred-million-year-old vesicular arbuscular mycorrhizae. *Proceedings of the National Academy of Sciences of the United States of America*, 91, 11841-11843.

Requena, N., Pérez-Solis, E., Azcón-Aguilar, C., Jeffries, P. and Barea, J. M. (2001). Management of indigenous plant-microbe symbioses aids restoration of desertified ecosystems. *Applied and Environmental Microbiology*, 67, 495-498.

Rillig, M. C. and Mummey, D. L. (2006). Mycorrhizas and soil structure. *New Phytologist*, 171, 41-53.

Rinaudo, V., Barberi, P., Giovannetti, M. and van der Heijden, M. G. A. (2010). Mycorrhizal fungi suppress aggressive agricultural weeds. *Plant and Soil*, 333, 7-20.

Robinson-Boyer, L., Grzyb, I. and Jeffries, P. (2009). Shifting the balance from qualitative to quantitative analysis of arbuscular mycorrhizal communities in field soils. *Fungal Ecology*, 2, 1-9.

Rosendahl, S. (2008). Communities, populations and individuals of arbuscular mycorrhizal fungi. *New Phytologist*, 178, 253-266.

Rosendahl, S., McGee, P. and Morton, J. B. (2009). Lack of global population genetic differentiation in the arbuscular mycorrhizal *fungus glomus mosseae suggests* a recent range expansion which may have coincided with the spread of agriculture. *Molecular Ecology*, 18, 4316-4329.

Rubinstein, C. V., Gerrienne, P., de la Puente, G. S., Astini, R. A. and Steemans, P. (2010). Early middle ordovician evidence for land plants in argentina (eastern gondwana). *New Phytologist*, 188, 365-369.

Ruíz-Lozano, J. M. (2003). Arbuscular mycorrhizal symbiosis and alleviation of osmotic stress. New perspectives for molecular studies. *Mycorrhiza*, 13, 309-317.

Ruiz-Lozano, J. M., Alguacil, M. M., Barzana, G., Vernieri, P. and Aroca, R. (2009). Exogenous aba accentuates the differences in root hydraulic properties between mycorrhizal and non mycorrhizal maize plants through regulation of pip aquaporins. *Plant Molecular Biology*, 70, 565-579.

Ruíz-Lozano, J. M. and Aroca, R. (2010). Host response to osmotic stresses: Stomatal behaviour and water use efficiency of arbuscular mycorrhizal plants. In H. Koltai and Y. Kapulnik (Eds.), *Arbuscular mycorrhizas: Physiology and function, 2nd ed.* (pp. 239-256). Dordrecht, The Netherlands, Springer Science+Business Media B.V. .

Ruíz-Lozano, J. M. and Azcón, R. (2011). Brevibacillus, arbuscular mycorrhizae and remediation of metal toxicity in agricultural soils. In N. A. Logan and P. de Vos (Eds.), *Endospore-forming soil bacteria, soil biology* (pp. 235-258). Berlin, Heidelberg, Germany, Springer-Verlag.

Ruíz-Sánchez, M., Aroca, R., Munoz, Y., Polon, R. and Ruíz-Lozano, J. M. (2010). The arbuscular mycorrhizal symbiosis enhances the photosynthetic efficiency and the antioxidative response of rice plants subjected to drought stress. *Journal of Plant Physiology*, 167, 862-869.

Saldajeno, M., Chandanie, W., Kubota, M. and Hyakumachi, M. (2008). Effects of interactions of arbuscular mycorrhizal fungi and beneficial saprophytic mycoflora on plant growth and disease protection. In Z. Siddiqui, M. Akhtar and K. Futai (Eds.), *Mycorrhizae: Sustainable agriculture and forestry* (pp. 211-226). Springer.

Sánchez-Castro, I., Ferrol, N. and Barea, J. M. (2012a). Analyzing the community composition of arbuscular mycorrhizal fungi colonizing the roots of representative shrubland species in a mediterranean ecosystem. *Journal of Arid Environments*, 80, 1-9.

Sánchez-Castro, I., Ferrol, N., Cornejo, P. and Barea, J.-M. (2012b). Temporal dynamics of arbuscular mycorrhizal fungi colonizing roots of representative shrub species in a semi-arid mediterranean ecosystem. *Mycorrhiza*, (DOI 10.1007/s00572-011-0421-z).

Sanders, I. R. and Croll, D. (2010). Arbuscular mycorrhiza: The challenge to understand the genetics of the fungal partner. In A. Campbell, M. Lichten and G. Schupbach (Eds.), *Annual review of genetics, vol 44* (pp. 271-292).

Schüßler, A. (2002). Molecular phylogeny, taxonomy, and evolution of geosiphon pyriformis and arbuscular mycorrhizal fungi. *Plant and Soil*, 244, 75-83.

Schüßler, A., Mollenhauer, D., Schnepf, E. and Kluge, M. (1994). Geosiphon-pyriforme, an endosymbiotic association of fungus and cyanobacteria - the spore structure resembles that of arbuscular mycorrhizal (am) fungi. *Botanica Acta*, 107, 36-45.

Schüßler, A., Schwarzott, D. and Walker, C. (2001). A new fungal phylum, the *glomeromycota,* phylogeny and evolution. *Mycological Research*, 105, 1413-1421.

Schüßler, A. and Walker, C. (2011). Evolution of the 'plant-symbiotic' fungal phylum, glomeromycota. In S. Pöggeler and J. Wöstemeyer (Eds.), *Evolution of fungi and fungal-like organisms* (pp. 163-185). Berlin Heidelberg, Springer-Verlag.

Sheng, M., Tang, M., Zhang, F. and Huang, Y. (2011). Influence of arbuscular mycorrhiza on organic solutes in maize leaves under salt stress. *Mycorrhiza*, 21, 423-430.

Simon, L., Bousquet, J., Levesque, R. C. and Lalonde, M. (1993). Origin and diversification of endomycorrhizal fungi and coincidence with vascular land plants. *Nature*, 363, 67-69.

Smith, S. A., Beaulieu, J. M. and Donoghue, M. J. (2010). An uncorrelated relaxed-clock analysis suggests an earlier origin for flowering plants. *Proceedings of the National Academy of Sciences of the United States of America*, 107, 5897-5902.

Smith, S. E. and Read, D. J. (2008). *Mycorrhizal symbiosis* (3rd). New York, Elsevier, Academic Press.

Sonjak, S., Beguiristain, T., Leyval, C. and Regvar, M. (2009). Temporal temperature gradient gel electrophoresis (ttge) analysis of arbuscular mycorrhizal fungi associated with selected plants from saline and metal polluted environments. *Plant and Soil*, 314, 25-34.

Steemans, P., Le Herisse, A., Melvin, J., Miller, M. A., Paris, F., Verniers, J. and Wellman, C. H. (2009). Origin and radiation of the earliest vascular land plants. *Science*, 324, 353-353.

Takeda, N., Sato, S., Asamizu, E., Tabata, S. and Parniske, M. (2009). Apoplastic plant subtilases support arbuscular mycorrhiza development in lotus japonicus. *Plant Journal*, 58, 766-777.

Taylor, T. N., Klavins, S. D., Krings, M., Taylor, E. L., Kerp, H. and Hass, H. (2004). Fungi from the rhynie chert: A view from the dark side. *Transactions of the Royal Society of Edinburgh-Earth Sciences*, 94, 457-473.

Tisserant, E., Kohler, A., Dozolme-Seddas, P., Balestrini, R., Benabdellah, K., Colard, A., Croll, D., *et al.* (2012). The transcriptome of the arbuscular mycorrhizal fungus glomus intraradices (daom 197198) reveals functional tradeoffs in an obligate symbiont. *New Phytologist*, 193, 755-769.

Toljander, J. F., Lindahl, B. D., Paul, L. R., Elfstrand, M. and Finlay, R. D. (2007). Influence of arbuscular mycorrhizal mycelial exudates on soil bacterial growth and community structure. *FEMS Microbiol. Ecol.*, 61, 295-304.

Toljander, J. F., Santos-González, J. C., Tehler, A. and Finlay, R. D. (2008). Community analysis of arbuscular mycorrhizal fungi and bacteria in the maize mycorrhizosphere in a long-term fertilization trial. *Fems Microbiology Ecology*, 65, 323-338.

Turnau, K., Jurkiewicz, A., Língua, G., Barea, J. M. and Gianinazzi-Pearson, V. (2006). Role of arbuscular mycorrhiza and associated microorganisms in phytoremediation of heavy metal-polluted sites. In M. N. V. Prasad, K. S. Sajwan and R. Naidu (Eds.), Trace elements in the environment. *Biogeochemistry, biotechnology and bioremediation* (pp. 235-252). Boca Raton, Florida, CRC/Taylor and Francis.

Vierheilig, H., Bago, B., Albrecht, C., Poulin, M. J. and Piché, Y. (1998). Flavonoids and arbuscular-mycorrhizal fungi. In J. A. Manthey and B. S. Buslig (Eds.), *Flavonoids in the living system* (pp. 9-33). New York, Plenum Press.

Vosátka, M., Albrechtová, J. and Patten, R. (2008). The international marked development for mycorrhizal technology. In A. Varma (Ed.) *Mycorrhiza: State of the art, genetics and molecular biology, eco-function, biotechnology, eco-physiology, structure and systematics.* 3rd ed (pp. 419-438). Berlin, Heidelberg, Germany, Springer-Verlag.

Wang, B., Yeun, L. H., Xue, J.-Y., Liu, Y., Ane, J.-M. and Qiu, Y.-L. (2010). Presence of three mycorrhizal genes in the common ancestor of land plants suggests a key role of mycorrhizas in the colonization of land by plants. *New Phytologist*, 186, 514-525.

Zubek, S., Turnau, K., Tsimilli-Michael, M. and Strasser, R. J. (2009). Response of endangered plant species to inoculation with arbuscular mycorrhizal fungi and soil bacteria. *Mycorrhiza*, 19, 113-123.

In: Symbiosis: Evolution, Biology and Ecological Effects   ISBN: 978-1-62257-211-3
Editors: A. F. Camisão and C. C. Pedroso   © 2013 Nova Science Publishers, Inc.

Chapter 2

RECENT DEVELOPMENTS IN THE INTERNAL REGULATION OF SYMBIOTIC NODULE NUMBER BY LEGUME PLANTS

Karel Novák[*]

Institute of Animal Science, Dept. of Molecular Genetics, Czech Republic

Abstract

The present review aspires to summarize the major advancements in the experimental study and understanding of the internal regulation of the intensity of rhizobial symbiosis. This kind of symbiosis takes place in the roots of predominantly legume plants in the form of specialized organs, root nodules, that are colonized by the soil nodule bacteria (rhizobia). Although the bacterial microsymbionts belong to several taxonomic groups, they share the ability of initiating root nodule development and fixation of atmospheric dinitrogen, which is subsequently utilized by the host plant. The latter feature predetermines the economic importance of rhizobial symbiosis and of the regulation of its intensity. While the symbiotic nodule induction is mediated by highly specific bacterial lipochitooligosaccharides, called Nod factors, further regulation of plant response is strictly controlled by plant endogenous mechanisms. Surprisingly, the major regulatory circuit includes systemic signaling from the root to the shoot and backwards. The nature of the involved molecules has been unknown until recently, except for the shoot-acting protein HAR1, described in 2003, and its homologs. Starting from 2008, a number of additional shoot-acting or root-acting regulatory mutants have been isolated. Their ongoing characterization has identified a series of new genes and products involved in the systemic signal generation, transduction and perception. Genomic approach allowed to identify low-molecular weight peptides of the CLE family as symbiotic signal messengers. The new experimental data enable a preliminary reconstruction of the systemic circuit and prediction of the nature of yet unknown components. A special interest represents the mechanism by which the inhibitory signal of ambient nitrate is integrated into the symbiotic circuit.

Keywords: Legumes, nitrogen fixation, *Rhizobium*, *Agrobacterium rhizogenes*, supernodulation, symbiosis

---

[*] E-mail address: novak.karel@vuzv.cz

1. Introduction

The aim of the present review is to summarize the major progress in the experimental study and interpretation of the internal regulation of rhizobial symbiosis that we observed during the last years. This type of symbiosis, formed mostly between legume plants (*Fabaceae*) on one side and nodule bacteria (rhizobia) on the other side, is a convenient model to discover the general principles of plant-microbe interactions. Its position of a popular experimental system is supported by the easy management of rhizobial infection, the possibility to grow and study rhizobia in pure culture outside the host (in contrast to arbuscular mycorrhiza-forming fungi) and, not the last, fully sequenced model species of hosts and microsymbionts. Many principles and concepts developed in the field of rhizobial symbiosis have been transferred to the other areas of plant microbiology, such as the role of low-molecular weight mediators in activating gene expression in symbionts (Peters et al., 1986), symbiotic suppression of host defense reactions (Vasse et al., 1993) or systemic regulation of symbiosis intensity (Staehelin et al., 2011). Rhizobial symbiosis is also the main source of nitrogen for the world crop production since nodule bacteria are capable of fixing atmospheric N_2. It is obvious that the exploitation of the possibilities provided by symbiotic nitrogen fixation will create conditions for further increase in world crop yields, reduction of costs and energetic demands and a decreased impact on the environment (Bohlool et al., 1992). Therefore, an additional impulse supporting the studies aimed at the intensity of rhizobial symbiosis comes from the demands of plant breeders and the producers of bacterial inoculants on the knowledge guiding their programs.

2. Formation of Rhizobial Symbiosis

2.1. Bases of Symbiotic Nodule Initiation and Development

Rhizobial symbiosis takes place predominantly in the roots of legume (*Fabaceae*) hosts where bacterial action induces formation of specialized organs, root nodules, in most species. In the course of their development, the root nodules are colonized by soil bacteria with nodulation ability (rhizobia). Although colonization is typically intracellular in the advanced legume species (subfamily *Faboideae*), intercellular symbiosis is spread in the primitive subfamilies *Caesalpinoideae* and *Mimosoideae* (Sprent and James, 2007).

The primary trigger inducing the symbiotic nodules are nodulation (Nod) factors, which are the final products of bacterial nodulation (*nod*, *nol*, *noe*) genes. Chemically, they represent abundantly substituted lipo-chitooligosaccharides derived from N-acetyl glucosamine backbone (Spaink, 2000).

On the plant side, the key molecules participating in the process of symbiotic signal transduction have been identified using plant symbiotic mutants. Symbiotic nodule development can be interrupted by early plant mutations already at nodule initiation. The resulting phenotype is then classified as non-nodulating (Nod⁻). On the other hand, the late mutations prevent nodule growth and differentiation, resulting in the non-fixing (Fix⁻) phenotype (Engvild, 1987). In view of the minute amounts of interacting molecules, the induced mutations turned to be an essential tool for dissecting subtle recognition events. The Nod factors are first recognized by presumed Nod-factor receptors (Figure 1), LysM domain-containing receptor-like kinases (Bek et al., 2010), which are localized to the plasma-membrane. In the model legumes *Lotus japonicus* (Regel) Larsen and *M. truncatula* Gaertn., the Nod factor receptors are coded for by symbiotic genes *NFR5* and *NFR1* (Radutoiu et al., 2003) and *LYK* family genes with *NFP* (Limpens et al., 2003; Arrighi et al., 2006), respectively. The subsequent members of the signaling pathway (Figure 1) have also been shown to be essential for nodulation by mutational analysis.

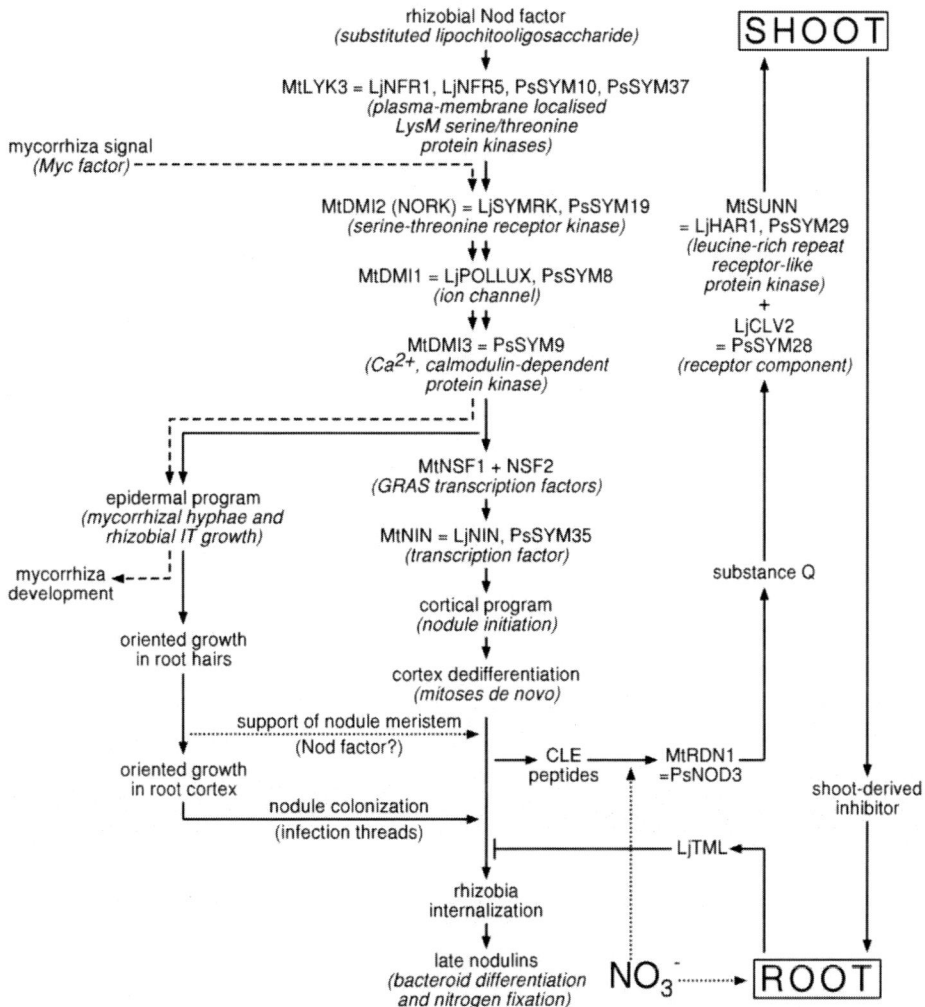

Figure 1. Systemic regulation of nodule number by symbiotic and nitrate signal related to the symbiotic signal transduction chain. The denotation of gene products follows the *Medicago truncatula* (Mt) nomenclature. The *Lotus japonicus* (Lj) and *Pisum sativum* (Ps) orthologous gene products are included. The scheme employs the data from the referenced works.

In the assumed functional order, they comprise a serine-threonine protein kinase SYMRK in *L. japonicus* with its ortholog NORK in *Medicago* (Stracke et al., 2002; Endre et al., 2002) and an ionic channel DMI1 in *M. truncatula* (Ané et al., 2004) with its counterpart POLLUX in *L. japonicus* (Ané et al., 2004; Imaizumi-Anraku et al., 2005). They are followed by Ca^{2+}, calmodulin-dependent protein kinase (CCPK) coded by *M. truncatula DMI3* and its orthologs in other legumes (Lévy et al., 2004; Gleason et al., 2006). Colocalization of the ion channel and CCPK in the perinuclear envelope (Riely et al., 2007) supports their complementary role in calcium signaling. The early response to rhizobia can be monitored as calcium level oscillations in root hair cells upon Nod factor treatment, the so-called "calcium spiking" (Downie and Walker, 1999). The transfer of the signal to the nuclear level leads to the activation of legume-specific transcription factors of GRAS family (Smit et al., 2005) and of the symbiotic transcription factor NIN (Schauser et al., 1999). The subsequent development of symbiosis can be divided into two parallel programs: nodule initiation as such, taking place in the root cortex, and the intracellular growth of the so-called bacterial infection threads (Guinel and Geil, 2002). Bacterial growth usually starts from the hooks of the root hairs that are deformed in response to the Nod factor treatment. The processes or root hair deformation ("root hair curling") and the mechanism of infection thread initiation and tip growth have recently become subject of focused mutational dissection (Murray et al., 2011). The first of the two pathways that leads to the induction of mitotic zones in the already differentiated root cortex has been shown to be associated with cytokinin relay and cytokinin perception (Murray et al., 2006; Tirichine et al., 2007). However, the patterning mechanism that leads to discrete meristematic zones still has to be elucidated.

At later stages of nodule development, the epidermal and cortical processes interact (Oldroyd and Downie, 2008). Although the initiation of new nodules usually occurs randomly throughout the root system, their further growth is conditional on the presence of bacteria underneath the active nodule meristem. The growth of the corresponding infection thread in the root cortex continues towards the initial nodule, which is subsequently colonized by the infection threads. In the nodules of indeterminate type, a well distinguished zone of cells interwoven with branching infection threads is formed behind the apical nodule meristem and functions as a reservoir of rhizobial infection for the whole time of nodule functioning. Individual bacteria are released from the tips of infection threads that enter the cells and can grow intercellularly. Nevertheless, the released bacteria remain separated from the host cell cytoplasm by a peribacteroid membrane as a derivative of host plasmalemma. In the early symbiotic zone, the intracellularly located bacteria multiply up to several hundreds per host cell. With advancing differentiation of nodule tissues and the plant-regulated drop in free oxygen concentration, nodule bacteria differentiate into bacteroids (Brewin, 1991). In addition to the size and morphological differences, the bacteroids develop enzymatic apparatus for atmospheric nitrogen fixation with nitrogenase (EC 1.18.6.1) as a central enzyme. The microaerobic niche is created thanks to a high O_2-diffusion resistance of the nodule peripheral tissues (Van de Wiel et al., 1990). The necessary respiration rate supporting the energetically-demanding process of nitrogen fixation is supported by the bound oxygen transport, which is enabled by the leghemoglobin synthesis in the infected cells of symbiotic tissues. The fixed nitrogen is transported into the host cells in the form of glutamate (Prell et al., 2009) and subsequently utilized by the host plant. Since the symbiotic intracellular growth of rhizobia is separated only slightly from pathogenic invasion, it is not surprising that the scale of plant responses is very wide and includes those resembling defense reactions. It has been shown that the signal towards activation of plant secondary metabolism branches very early from the main symbiotic pathway (Kévei et al., 2007).

2.2. Rhizobial Diversity Unified in Action

The synthesis of Nod factors on the bacterial side of symbiosis is controlled by the nodulation (*nod*, *noe*, *nol*) genes (Spaink, 2000). They comprise common genes shared by all nodule-inducing bacteria (*nodA*, *-B*, *-C*, *-D*, *-I*, *-J*, *-K*, *-M*, and *-N*) and host-specific genes, defined as those required for nodulation in a particular host plant. Common nodulation genes control the synthesis of the core structure of Nod factors that represents several N-acetyl-D-glucosamine units connected by β-1,4-bonds. Therefore, common nodulation genes originating from different bacterial species can functionally complement each other. In contrast, host-specific genes control the variable length of N-acetyl glucosamine backbone and its substitution pattern. Other specific genes control the fatty acids attached to the N-acylated non-reducing terminal glycosidic residue. Partly, in the alfalfa-specific rhizobial group (*Ensifer meliloti*), they include *nodH*, *nodP*, and *nodQ* that control the sulfurylation of Nod factor (Spaink, 2000). Consistent with their role in symbiosis, nodulation genes are regulated by flavonoid (including isofavonoid) compounds produced in and released from the host tissues (Spaink, 2000). The set of the flavonoid compounds is an important factor limiting the host range at the first stage of symbiosis. Flavonoids are perceived by rhizobial receptors, products of *nodD* genes, with a defined specificity towards molecular species. On the other side, NodD protein/flavonoid complexes serve as transcriptional activators of nodulation operons thus potentiating the flavonoid action through positive feedback. In light of the key role of a set of nodulation genes in the symbiosis establishment and host specificity, it is very surprising that, according to the recent advance in the molecular taxonomy of nodule bacteria, the symbiotic ability is spread to a number of distant bacterial groups comprising at least 13 genera (Willems, 2006). In addition to the traditional nodule bacteria of the *Rhizobiaceae* family from the genera *Rhizobium*, *Ensifer* (*Sinorhizobium*), A*zorhizobium*, and *Mesorhizobium*, nodulation ability has been reported in such unrelated groups as β-proteobacteria including the genus *Burkholderia* (Sprent and James, 2007). The well-established genus *Bradyrhizobium*, a traditional member of *Rhizobiaceae*, now forms a separate family *Bradyrhizobiaceae* (Garrity et al., 2001; Willems, 2006). In spite of their classification into unrelated taxonomic groups, all microsymbionts share the ability to induce root nodule development and to fix atmospheric dinitrogen. It has been shown that non-rhizobial genera share the common *nod* genes responsible for the production of Nod factor lipopolysaccharide, albeit with certain structural differences (Chen et al., 2003). The distribution of *nod* gene variants illustrates their horizontal transfer in bacterial populations in soil and rhizosphere (Spaink, 2000; Willems, 2006). Another embarrassing exception from the traditional legumes vs. *Rhizobiaceae* concept is the symbiosis with nodule bacteria observed in the *Parasponia* genus from the *Ulmaceae* family (Op den Kamp et al., 2011). Although this case is isolated from other nodulating species, it is still placed in the *Fabids* clade like the family *Fabaceae*. As shown only recently, the *Parasponia andersonii* primary receptor for the rhizobial Nod factor developed independently from legume family. However, in both cases the molecules required for the more ancient mycorrhiza served as a start point for the evolution of rhizobial symbiosis. In *Parasponia*, the Nod receptor is less diverged and still indispensable for mycorrhiza establishment (Op den Kamp et al., 2011).

3. Major Regulatory Circuit of Nodulation

3.1. Evidence of a Large-Scale Regulation

In contrast to the symbiotic nodule establishment, which is induced by external bacterial stimuli, the degree of plant response has been shown to be regulated by plant endogenous factors. The major regulatory circuit is systemic and includes distantly acting messengers transported from the root to the shoot and vice versa. Like the basic process of symbiotic nodule induction and establishment, nodule number regulation has been dissected by specific host mutations. It should be noted that this line of experimental study reflected a low productivity of efforts to change nodule number by genetic variation on the side of microsymbiont. The bacterial manipulations enhanced nodule number only exceptionally and by a mechanism that presumably involved the plant internal circuits by disturbing the symbiotic nodule development and thus preventing the production of endogenous plant signals. In this way, the action of *nifA* mutations in *S. meliloti* (Paau et al., 1985) and *Bradyrhizobium japonicum* (Studer et al., 1987) should be interpreted. Accordingly, the efforts of enhancing nodule number by mutational inactivation of the host or bacterial nitrate reductase and interrupting the hypothetical inhibitory feedback were not successful (Nelson, 1987).

First host symbiotic mutations that affected the intensity of symbiosis, as characterized by nodule number per root or nodule density per root length unit, have been reported by Jacobsen and Feenstra (1984) for pea (*Pisum sativum* L.) and by Carroll et al. (1985) for soybean (*Glycine max* [L.] Merr.). The mutational interruption of the main regulatory circuit demonstrates itself as abundant nodulation where nodules are distributed evenly throughout the root system. This is a characteristic difference compared to the wild-type nodulation pattern, which demonstrates suppression of nodulation in the distant parts of the root system (Jacobsen and Feenstra, 1984; Carroll et al., 1985; Kokubun and Akao, 1994). This feature is reflected in the term supernodulation, suggested fort this type of mutants (Carroll et al., 1985). The supernodulators typically form 5 – 20 times more nodules than the wild-type plants when they are assayed without inhibitory factors such as a high concentration of ambient nitrate. The full degree of phenotypic change can be masked by allelic variation, since many leaky mutants have been observed in the *nts* locus of soybean (Delves et al., 1988) and *SYM29* of pea (Sagan and Duc, 1996).

3.2. Association with Nitrate Perception

In addition to the primary trait of changed nodule number and nodulation pattern, the supernodulation mutations also bring about the trait of nitrate-tolerant symbiosis (Nts).

Actually, the first supernodulating mutants were revealed in the course of screening for the desirable trait of nitrate-tolerant nodulation (Jacobsen and Feenstra, 1984; Carroll et al., 1985). In the wild-type plants of legumes, nitrate acts as a strong nodulation inhibitor.

This is biologically reasonable due to the fact that nitrate is an energetically cheaper N source than the fixation of dinitrogen (Streeter, 1988). Already in 1989, Day et al. (1989) considered the coincidence of the two phenotypic traits as the evidence that the symbiotic signaling shares part of the pathway with the signaling for ambient nitrate (Forde, 2002).

Consistently with the shared pathway hypothesis, supernodulation and Nts traits have never been separated by genetic recombination in mapping or breeding crosses (e.g., Novák et al., 1997).

3.3. Systemic Character of the Major Circuit

Soon after its discovery, the main regulatory circuit has been shown to be systemic, i.e., acting between root and shoot (Figure 1). This basic information was obtained by means of shoot/root grafting in soybean (Delves et al., 1987) and pea (Duc and Messager, 1989; Sagan and Duc, 1996) using the combinations of mutant and wild-type scions and rootstocks. The first investigated soybean supernodulation mutation nts382 was expressed only in the chimerical plants with a mutant shoot. The systemic nature of supernodulation has subsequently been demonstrated in the model legumes *M. truncatula* (Penmetsa et al., 2003) and *L. japonicus* (Wopereis et al., 2000) where the mutations in loci SUNN and HAR1 are also expressed only when present in the shoot. In the working model of the regulation of nodulation, the presence of an ascending messenger denoted as "substance Q" has been postulated from the experiments in soybean. The messenger is supposed to be generated by the initial nodules in the root (Kinkema et al., 2006). After being transported to the shoot, substance Q is perceived and transformed into the descending signal, the so called "shoot-derived inhibitor" of nodulation (SDI). In its turn, this substance acts in the root, where it signals the level of achieved nodulation combined with the information about the state of the shoot and finally suppresses nodulation. The described negative feedback is supposed to optimize total nodule number according to the shoot photosynthetic capacity and/or expected requirements in atmospheric N.

3.4. Two Types of Root-Acting Mutations

Given that this model is correct, it can be assumed that the negative feedback can be mutationally interrupted both in the shoot and in the root, both variants resulting in supernodulation. Accordingly, in addition to the shoot-acting mutations like *L. japonicus* har1 the root localization of mutant actions has been demonstrated for some supernodulation mutants of pea. They were all ascribed to the locus *NOD3* (Postma et al., 1988; Sidorova and Shumnyi, 2003; Li et al., 2009; Novák, 2010), subsequently found to be orthologous to the root-acting locus *RDN1* ("root-determined nodulation 1") of *M. truncatula* (Schnabel et al., 2011). In the model legumes, additional root-determined supernodulators have been described, namely *L. japonicus* rdh ("root-determined hypernodulation"; Ishikawa et al., 2008), tml ("too much love"; Magori et al., 2009) and PLENTY (Yoshida et al., 2010). The involvement of root-acting supernodulation genes in the main circuit is supported by the same phenotype as observed in the shoot-determined mutants, including the nitrate tolerance of nodulation (Nts).

The root localization of mutant action is ambiguous with respect to the function of the mutant gene product since this class of mutations can affect either ascending signal generation in the root or descending signal perception in the end of the circuit. The latter possibility has been suggested for the pea line nod3 (Oka-Kira and Kawaguchi, 2006) and the tml mutant of *L. japonicus* (Magori et al., 2009). The experimental tests employed the fact that the two variants of root-determined supernodulators should differ by the predicted presence of downstream mediator (SDI) in the shoot. This factor is expected to be present both in the wild-type plants and the supernodulators with the terminal block in signaling. On the other hand, the substance will be absent in the mutants blocked at the early stages of nodulation. In the line tml of *L. japonicus*, the terminal position of the mutational block was persuasively deduced from the local action of the mutation in the split-root system (Magori et al., 2009). On the other hand, the terminal localization of the block has not found experimental support in grafting experiments with pea *nod3* mutants (Li et al., 2009; Novák, 2010). The presence of the systemic signal was tracked in the experimental arrangement of the longitudinal stem grafts ("approach grafts") of pea (Li et al., 2009). The root of a reporter plant joined with a challenged (sensor) plant by a longitudal graft is inoculated with rhizobia with a delay after the challenged plant. Subsequently, the final nodule number developed in the reporter is used as a measure of the level of SDI, which is formed by the challenged plant and distributed into both root systems.

In the alternative assay, the shoot/root grafted plants of pea were used (Novák, 2010). This arrangement employs the spontaneous formation of adventitious roots in scions upon shoot grafting that has been observed in other legumes including *M. truncatula* (Cros-Arteil et al., 2006) and even in *A. thaliana* (Turnbull et al., 2002). The density of nodules that were formed in the adventitious roots growing spontaneously from the basal part of the scions indicated the level of SDI, as influenced by the rootstock and scion genotypes. Unfortunately, a large-scale application of this approach is limited by low frequency of spontaneous rhizogenesis. While the frequency of adventitious roots in scions of the *M. truncatula* grafts has reached 30% (Cros-Arteil et al., 2006), the frequency in pea was only 16% (Novák, 2010). The spontaneous rhizogenesis also shows the pronounced effect of growing conditions and arbitrary factors. For this reasons, any modification enhancing the rhizogenesis in the scions of chimerical plants would be helpful in further applications of this method.

The response to grafting by root initiation has been ascribed to the local increase in the concentration of auxin, which is transported basipetally, due to the transport resistance in the graft joining site (McComb and McComb, 1970). However, the exogenous auxin application in grafted plant plants was inefficient (Novák, unpublished data). On the other hand, the frequency of adventitious root formation in pea scions could be enhanced by agrobacterial transformation. The transformed roots developing from *A. rhizogenes*-induced tumors are considered to be equivalent to the normal root system with respect to the symbiosis formation, as supported by their use in the studies on functional complementation of symbiotic genes (Limpens et al., 2004; Floss et al., 2008) in the so-called composite plants. The toolbox for rhizogenic legume transformation includes agrobacterial strains with approved broad-range virulence in a set of legume species like strain R1000 (White et al., 1985; Cros-Arteil et al., 2006) and its derivative ArQua1 (Quandt et al., 1993).

Using the *A. rhizogenes*-transformed roots, the action of the mutation of the pea supernodulator RisfixC could be placed in the regulatory circuit (Novák, unpublished data) independently of the indications obtained from the spontaneously-formed adventitious roots (Novák, 2010). Since the pea line RisfixC is presumably affected in the *NOD3* locus, as indicated by genetic mapping (Novák, 2010; Novák et al., 2012), this result further supports the role of *NOD3* in the systemic signaling as obtained with the parallel graft approach (Li et al., 2009; Reid et al., 2011). In view of the *NOD3* orthology with *M. truncatula RDN1*, this assumption can be extended on the role of *RDN1* (Table 1). There is no doubt that any procedures enabling to assay the SDI level in planta will find its use in the validation of the anticipated molecular markers of systemic regulation of nodulation in future.

3.5. Mutational Dissection of Root Signal Generation

The approach graft technique combined with a set of nodulation mutants of pea also allowed to correlate the strength of the nodulation regulatory signal with a nodule developmental stage. Assuming that the SDI intensity is proportional to the intensity of the primary root signal (substance Q) generated in the root and migrating upwards to the shoot, the root signal intensity was evaluated in a series of longitudal grafts where early nodulation mutants of pea that served as a sensor plant. Since the chosen panel of pea nodulation mutants was blocked at different stages of nodule initiation and formation, as inferred from histological observations (Tsyganov et al., 2002), the results allowed to relate the developmental stage with the production of the primary nodule-derived signal. Its intensity was increasing starting from the stage of meristematic divisions in the root cortex, however, no sharp stepwise raise in the root activity associated with a particular stage in nodule development has been revealed (Li et al., 2009; Ferguson et al., 2010; Reid et al., 2011).

3.6. Lessons from the Pleiotropy of Supernodulation Mutations

Many supernodulating mutations, like host symbiotic mutations in general, have been reported to be associated with pleiotropic changes to non-symbiotic traits which involve plant metabolism, growth and architecture. The alternative that the non-symbiotic traits result from multiple mutations in other loci affected by random mutagenesis is excluded by the coincidence of similar changes in independently obtained lines and even in different legume species. The pleiotropic changes co-segregated with the symbiotic trait in the crossing schemes, when followed (Novák et al., 1997; Krusell et al., 2011), thus supporting a relation to the primary mutation.

Shoot growth depression, shortened internodes, underdeveloped root and fasciation of the stem are common in the lines demonstrating supernodulation/Nts (Postma et al., 1988; Duc and Messager, 1989; Sagan and Duc, 1996; Sidorova and Shumnyi, 2003). Partly, pea mutants in the root-acting *NOD3* locus comprising lines nod3 (Postma et al., 1988), P79 (Duc and Messager, 1989), K10a, K11a and K12a (Sidorova and Shumnyi, 2003) and presumably RisfixC (Novák et al., 1993a, b; Novák, 2010) exhibited stunted growth and short internodes in nodulated plants. Increased nitrogen content in the shoot has been detected in pea lines RisfixC and K10a (Novák et al., 1993b; Nazaryuk et al., 2006; Novák et al., 2011).

Table 1. List of supernodulating and hypernodulating lines of legumes

Species	Parental genotype	Mutant line	Locus	Allele	Site of action	Nodulation resistant to	Other pleiotropic traits	Gene product	Mutation	Reference
Gm	Bragg	nts1007	NARK=NTS1	nts1007	shoot	NO$_3^-$	compact shoot	LRR serine-threonine kinase	nonsense Q106* of GmNARK	Delves et al. (1987), Delves et al. (1988), Searle et al. (2003)
Gm	Bragg	nts246	NARK=NTS1	nts246	shoot	NO$_3^-$		LRR serine-threonine kinase		Delves et al. (1988)
Gm	Bragg	nts382	NARK=NTS1	nts382	shoot	NO$_3^-$		LRR serine-threonine kinase		Delves et al. (1988)
Gm	Bragg	nts1116	NARK=NTS1	nts1116	shoot	NO$_3^-$		LRR serine-threonine kinase		Delves et al. (1988)
Gm	Enrei	en6500	NARK = NTS1	en6500	shoot	NO$_3^-$		LRR serine-threonine kinase	nonsense K606* of GmNARK	Kokubun and Akao (1994), Arai et al. (2005)
Gm	Sinpaldal-kong 2	SS2-2	NARK=NTS1	ss2-2	shoot	NO$_3^-$		LRR serine-threonine kinase	A to T transition in NARK	Lestari et al. (2006)
Gm	Williams	NOD1-3	rj7	nod1-3		NO$_3^-$ partially				Lee et al. (1997), Fujikake et al. (2003)
Gm	Williams	NOD4	rj7	nod4		NO$_3^-$ partially				Lee et al. (1997)
Gm	Williams	NOD2-4	rj8	nod2-4		NO$_3^-$ partially				Lee et al. (1997)
Lj	Gifu B-129	ASTRAY	BZF (SYM77)	astray	root		gravity and light response	leucine zipper protein with a RING-finger motif	G to A transition in a splice donor site	Nishimura et al. (2002a, b)
Lj	Gifu	har1	SYM78=HAR1	har1-1	Shoot	NO$_3^-$	shortened roots, branching	LRR serine-threonine kinase	Trp676 stop	Wopereis et al. (2000), Nishimura et al. (2002a), Krusell et al. (2002)
Lj	Gifu	har1-2	SYM78=HAR1	har1-2	shoot	NO$_3^-$		Gln919 stop		Wopereis et al. (2000), Nishimura et al. (2002a), Krusell et al. (2002)
Lj	Gifu	har1-3	SYM78=HAR1	har1-3	shoot	NO$_3^-$		aa964-969 deletion		Wopereis et al. (2000), Nishimura et al. (2002a), Krusell et al. (2002)
Lj	Miyakojima MG20	klv (klavier)	KLV		shoot	NO$_3^-$ partially	dwarf plants, fasciation, venation, late flowering	homolog of har1 kinase		Oka-Kira et al. (2005), Miyazawa et al. (2010)
Lj	Miyakojima MG20	rdh ("root-determined hypernodulation")	RDH1	rdh1	root	NO$_3^-$	none			Ishikawa et al. (2008), Yokota et al. (2009)

Species	Parental genotype	Mutant line	Locus	Allele	Site of action	Nodulation resistant to	Other pleiotropic traits	Gene product	Mutation	Reference
Lj	Miyakojima MG20	tml ("too much love")	TML		root	partially NO$_3^-$				Magori et al. (2009)
Lj	Miyakojima MG20	plenty	PLENTY		root	partially NO$_3^-$		putatively MtRDN1 ortholog		Yoshida et al. (2010), Schnabel et al. (2011)
Mt		LSS	LSS		shoot	NO$_3^-$, ethylene	short roots, symbiosis-independent	cis-regulating element of SUNN		Frugoli et al. (2008), Schnabel et al. (2010)
Mt		RDN	RDN1		root			new family with unknown function, PsNOD3 ortholog		Frugoli et al. (2008), Schnabel et al. (2011), Novák et al. (2012)
Mt	Jemalong A17	efd-1	EFD	efd-1						Vernié et al. (2008)
Mt	Jemalong A17	Sickle	SKL	skl1-1	root	ethylene	long roots, nodule positioning	ortholog of At ethylene signaling protein EIN2	Gln894 stop	Penmetsa and Cook (1997), Penmetsa et al. (2008)
Mt	Jemalong A17	DD2	SKL	skl1-2	root	ethylene	long roots, nodule positioning	ortholog of At ethylene signaling protein EIN2.	Cys1137Tyr in EIN2 domain	Penmetsa and Cook (1997), Penmetsa et al. (2008)
Mt	Jemalong A17	EMS T1	SKL	skl1-2b	root	ethylene	long roots, nodule positioning	ortholog of At ethylene signaling protein EIN2		Penmetsa and Cook (1997), Penmetsa et al. (2008)
Mt	Jemalong A17	L309	SKL	skl1-3	root	ethylene	long roots, nodule positioning	ortholog of At ethylene signaling protein EIN2	Trp1143 stop in EIN2 domain	Penmetsa and Cook (1997), Penmetsa et al. (2008)
Mt	Jemalong A17	12884	SKL	skl1-4	root	ethylene	long roots, nodule positioning	ortholog of At ethylene signaling protein EIN2.	Trp313 stop in N-terminal domain	Penmetsa and Cook (1997), Penmetsa et al. (2008)
Mt	Jemalong A17	D40-7J-v3	SKL	skl1-5	root	ethylene	long roots, nodule positioning	ortholog of At ethylene signaling protein EIN2.	Gln768 stop	Penmetsa and Cook (1997), Penmetsa et al. (2008)
Mt	Jemalong A17	sunn-1	SUNN	sunn-1	shoot		shortened roots, symbiosis-independent	LRR serine-threonine kinase	Arg950Lys in kinase domain	Schnabel et al. (2005), Schnabel et al. (2010)
Lj	Miyakojima MG20	tml ("too much love")	TML		root	partially NO$_3^-$				Magori et al. (2009)
Lj	Miyakojima MG20	plenty	PLENTY		root	partially NO$_3^-$		putatively MtRDN1 ortholog		Yoshida et al. (2010), Schnabel et al. (2011)

Table 1. Continued

Species	Parental genotype	Mutant line	Locus	Allele	Site of action	Nodulation resistant to	Other pleiotropic traits	Gene product	Mutation	Reference
Mt		LSS	*LSS*		shoot	NO$_3^-$; ethylene	short roots, symbiosis-independent	cis-regulating element of *SUNN*		Frugoli et al. (2008), Schnabel et al. (2010)
Mt		RDN	*RDN1*		root			new family with unknown function, PsNOD3 ortholog		Frugoli et al. (2008), Schnabel et al. (2011), Novák et al. (2012)
Mt	Jemalong A17	efd-1	*EFD*	*efd-1*						Vernié et al. (2008)
Mt	Jemalong A17	Sickle	*SKL*	*skl1-1*	root	ethylene	long roots, nodule positioning	ortholog of At ethylene signaling protein EIN2	Gln894 stop	Penmetsa and Cook (1997), Penmetsa et al. (2008)
Mt	Jemalong A17	DD2	*SKL*	*skl1-2*	root	ethylene	long roots, nodule positioning	ortholog of At ethylene signaling protein EIN2.	Cys1137Tyr in EIN2 domain	Penmetsa and Cook (1997), Penmetsa et al. (2008)
Mt	Jemalong A17	EMS T1	*SKL*	*skl1-2b*	root	ethylene	long roots, nodule positioning	ortholog of At ethylene signaling protein EIN2		Penmetsa and Cook (1997), Penmetsa et al. (2008)
Mt	Jemalong A17	L309	*SKL*	*skl1-3*	root	ethylene	long roots, nodule positioning	ortholog of At ethylene signaling protein EIN2	Trp1143 stop in EIN2 domain	Penmetsa and Cook (1997), Penmetsa et al. (2008)
Mt	Jemalong A17	12884	*SKL*	*skl1-4*	root	ethylene	long roots, nodule positioning	ortholog of At ethylene signaling protein EIN2.	Trp313 stop in N-terminal domain	Penmetsa and Cook (1997), Penmetsa et al. (2008)
Mt	Jemalong A17	D40-7J-v3	*SKL*	*skl1-5*	root	ethylene	long roots, nodule positioning	ortholog of At ethylene signaling protein EIN2.	Gln768 stop	Penmetsa and Cook (1997), Penmetsa et al. (2008)
Mt	Jemalong A17	sunn-1	*SUNN*	*sunn-1*	shoot		shortened roots, symbiosis-independent	LRR serine-threonine kinase	Arg950Lys in kinase domain	Schnabel et al. (2005), Schnabel et al. (2010)
Ps	Rondo	K12a	*NOD3*		root		fasciation, compact	*MtRDN1* ortholog		Jacobsen and Feenstra (1984), Sidorova and Shumnyi (2003)

Species	Parental genotype	Mutant line	Locus	Allele	Site of action	Nodulation resistant to	Other pleiotropic traits	Gene product	Mutation	Reference
Ps	Rondo	nod3	NOD3		root		compact shoot	MtRDN1 ortholog		Jacobsen and Feenstra (1984), Postma et al. (1988), Schnabel et al. (2011)
Ps	Ramonskii 77	K301	NOD4		shoot		fasciation, compact			Sidorova and Shumnyi (1998), Sidorova and Shumnyi (2003)
Ps	Ramonskii 77	Torsdag	NOD5	NOD5	shoot					Sidorova and Shumnyi (1998), Sidorova and Shumnyi (2003)
Ps	Rondo	K21a	NOD6		shoot		fasciation, compact			Sidorova and Shumnyi (1998), Sidorova and Shumnyi (2003)
Ps	Rondo	K22a	NOD6		shoot		fasciation, compact			Sidorova and Shumnyi (1998), Sidorova and Shumnyi (2003)
Ps	Frisson	P64	SYM28		shoot		stem fasciation			Sagan and Duc (1996), Sinjushin et al. (2008), Krusell et al. (2011)
Ps	Frisson	P77	SYM28		shoot		fasciation			Sagan and Duc (1996)
Ps	Frisson	P87	SYM29		shoot			LRR serine-threonine kinase	Gly831Arg	Sagan and Duc (1996), Krusell et al. (2002)
Ps	Frisson	P88	SYM29		shoot			LRR serine-threonine kinase	Leu290Phe	Sagan and Duc (1996), Krusell et al. (2002)
Ps	Frisson	P89	SYM29		shoot			LRR serine-threonine kinase	Gln910 stop	Sagan and Duc (1996), Krusell et al. (2002)
Ps	Frisson	P90	SYM29		shoot			LRR serine-threonine kinase	Gly695Arg	Sagan and Duc (1996), Krusell et al. (2002)
Ps	Frisson	P91	SYM29		shoot			LRR serine-threonine kinase	Gly698Glu	Sagan and Duc (1996), Krusell et al. (2002)
Ps	Frisson	P93	SYM29		shoot			LRR serine-threonine kinase	Leu290Phe	Sagan and Duc (1996), Krusell et al. (2002)
Ps	Frisson	P94	SYM29		shoot			LRR serine-threonine kinase	Gln910 stop	Sagan and Duc (1996), Krusell et al. (2002)

List of Pssym29 alleles is reduced.

Abbreviations: At, *Arabidopsis thaliana*; Gm, *Glycine max*; Lj, *Lotus japonicus*; Mt, *Medicago truncatula*; Ps, *Pisum sativum*; EMS, ethyl methanesulfonate; NEU, N-nitroso-N-ethylurea; IVD, intervarietal difference; HN, hypernodulation; SN, supernodulation. LRR, leucine-rich repeat. Site of action as determined by grafting.

It seems that these properties are independent of the symbiosis development and represent a true pleiotropy of the mutated locus (Novák et al., 2011). The independence of symbiosis development suggests early branching of the pathway leading to the pleiotropic traits, consequently an early localization of the mutational block in systemic signaling. Consistently with this model, no pleiotropic changes were detected in the rdh1 mutant of *L. japonicus*, which is supposedly blocked at the late stage of SDI perception in the root (Yokota et al., 2009).

The inherent pleiotropy of most of the supernodulating mutants is relevant to their potential use in breeding programs for nitrogen fixation improvement. Nevertheless, the tuning of nodule number and plant architecture by minor genes can be accompanied by the correction of pleiotropic traits as well. The alleviation of the pleiotropic growth depression in distant crosses of a pea supernodulator RisfixC was ascribed to the effect of modifier genes occurring in the background of recipient lines (Novák et al., 2009).

4. Identification of Molecules Involved in Long-Distance Signaling

4.1. Positional Cloning

Until recently, the nature of the molecules involved in the systemic regulation of nodulation has been unknown, except for the shoot-acting protein HAR1, described in 2002, and its homologs. The description of HAR1 coincided with the beginning of the serial identification and characterization of legume symbiotic mutants. Although the strategy of mapping and positional cloning of soybean symbiotic (partly supernodulation) mutants was formed as early as in 1991 (Landau-Ellis et al., 1991), the final results appeared in 2002 and subsequent years, starting with cloning of *NORK* of *Medicago* and *SYMRK* of *L. japonicus* (Stracke et al., 2002; Endre et al., 2002). The mutational identification of *NIN* ("nodule inception") gene coding for an early transcription factor of *L. japonicus* three years earlier (Schauser et al., 1999) was enabled by the method of insertional tagging with a Ts element derivative. To date, the symbiotic signaling pathway leading from the Nod factor perception to the establishment of nitrogen-fixing nodules has been characterized (Figure 1) using almost saturating mutagenesis of model host plants, screening for asymbiotic phenotype and subsequent cloning of causal point mutations (Stracke et al., 2002; Endre et al., 2002).

The general scheme of symbiotic mutant identification involves delimitation of the region containing the candidate gene and its subsequent sequencing. The fully sequenced genomes of model species like *M. truncatula*, *L. japonicus* and *G. max* represent an advantage. The candidate gene showing difference in the nucleotide sequence between the wild-type and mutant plants is subsequently isolated and tested for functional complementation in *Agrobacterium*-transformed plants. Alternatively, multiple mutations in the same gene obtained independently and conditioning the same phenotype can be considered as evidence for candidate gene function (Edwards et al., 2007; Krusell et al., 2011). The proof of the function by direct complementation does not necessarily involve whole plant transformation and regeneration. Usually the transformed roots formed upon *A. rhizogenes* infection of stems or roots can be used instead of the true root system, as far as the symbiosis formation is considered. The *Agrobacterium*-transformed roots are used as models for the functional complementation of symbiotic genes in the so-called composite plants (Limpens et al., 2004;

Floss et al., 2008). The first molecular characterization of a supernodulation gene demonstrated described *L. japonicus HAR1* (Krusell et al., 2002). Subsequently, its homologs in pea *SYM29* (Krusell et al., 2002), *M. truncatula SUNN* (Schnabel et al., 2005) and soybean *NARK* (Searle et al., 2003) have been identified. By this finding, the key role of the legume homologs of the meristem growth-supporting protein CLV1 of *Arabidopsis thaliana* (L.) Heynh. in the signal relay has been shown. The action of all legume *HAR1* orthologs is confined to the shoot, as shown by grafting studies (Delves et al., 1987; Sagan and Duc, 1996; Wopereis et al., 2000). Therefore, the *HAR1* orthologs were assigned a role in the transformation of the root-derived signal into the SDI molecule.

Surprisingly, the *LSS* ("like sunn supernodulator") locus of *M. truncatula* close to *SUNN* has been shown to be a cis-acting element reducing *SUNN* expression in the mutated form. Its action was associated with the degree of methylation, suggesting the involvement of epigenetic changes in the regulation of nodulation intensity (Schnabel et al., 2010). This mechanism of action might explain the variable expression of supernodulation trait bordering on phenotypic reversion in crosses with related and distant lines of pea (Sidorova and Shumnyi, 1998; Novák et al., 2009). The recently cloned *RDN1* (*ROOT DETERMINED NODULATION*) locus in *M. truncatula* with its ortholog *NOD3* of pea (Schnabel et al., 2011) might shed light on the initial stages of nodulation signal generation in the nodulating root provided that the assumption on its early location in the major signaling circuit (Li et al., 2009; Novák, 2010) is fully confirmed.

4.2. Candidate Gene Approach

The current entangling of the middle stages of systemic signal spread and processing is mostly based on the candidate gene approach. This is encouraged by the success of this strategy in the identification of the astray hypernodulation mutation (Nishimura et al., 2002a) in the gene *Bzf* of *L. japonicus* using solely phenotypic resemblance of pleiotropic features with the *A. thaliana* mutants in *HY5*. The genes encode basic leucine zipper proteins with highly homologous C-terminal halves (Nishimura et al., 2002b). Similarly, the identification of the shoot-acting KLV (KLAVIER; Oka-Kira et al., 2005) gene product was facilitated by assuming and testing the homology with HAR1-like proteins (Miyazawa et al., 2010).

Using the anticipated similarity between the CLV1 - CLV2 - CLV3 system controlling shoot meristems in *Arabidopsis* and HAR1-associated components in legumes, the search for corresponding legume components has been suggested by Downie and Parniske (2002) immediately upon the HAR1 discovery. Similarly, the discovery of the role of CLE peptides in symbiotic signaling was based on the CLE peptides origin from the CLV3 peptide processing, already known in *A. thaliana* (Oelkers et al., 2008; Okamoto et al., 2009).

Surprisingly, although the *Arabidopsis thaliana* homolog CLV1 also controls stem cell proliferation, its role consists in short-distance signaling in shoot apices, in contrast to the whole plant-level communication exhibited by the legume symbiotic kinases of HAR1 and KLV type (Krusell et al., 2002; Miyazawa et al., 2010). Nevertheless, assumed functional homology between the *A. thaliana* CLV2 and a pea symbiotic gene product has led to the identification of the *SYM28* gene and its *L. japonicus* ortholog (Krusell et al., 2011), although the search was supported by mapping data in parallel (Sinjushin et al., 2008; Krusell et al.,

2011). Pea SYM28 was originally described simultaneously with *SYM29* as another shoot-acting supernodulation locus (Sagan and Duc, 1996).

Consistent with the function of the *A. thaliana* homolog CLAVATA2, it probably interacts with *SYM29* completing the receptor structure. I must be noted that the predicted function of SYM28 in meristem support was also based on the stem fasciation as a pleiotropic trait. However, this feature is absent in *L. japonicus* orthologous mutants (Krusell et al., 2011).

The candidate gene approach also helped to identify the symbiotic versions of the CLAVATA - WUSCHEL regulatory feedback loop. While the *A. thaliana* WUSCHEL gene expression supports the meristematic state of the shoot meristem cells, the concurrent activation of CLAVATA complex with negative effect on WUSCHEL limits the meristematic region extent. The feedback circuit has been shown to be operational in embryogenic *M. truncatula* cultures (Chen et al., 2009), opening the possibility for its studies in the nodulation of the same model plant. The pleiotropic fasciation in pea supernodulating mutant sym28 corresponding to the excessive growth of meristematic region also pointed at the possibility of a WUSCHEL homolog deregulation, as subsequently proved (Krusell et al., 2011). This line of experimental research has demonstrated the involvement of WUSCHEL-RELATED HOMEOBOX5 gene expression in the regulation of nodulation of *M. truncatula* (Osipova et al., 2012).

4.3. Low Molecular Weight Components

Obviously, the elucidation of the structure of the *SYM28* gene of pea (Krusell et al., 2011) should in turn contribute to the identification of the biochemical function of HAR1-like proteins. The nature of the descending signal (SDI) which is supposedly generated by shoot receptor-like kinases (RLKs) is still unknown. The action of SDI in the root consists in the inhibition of nodule development beyond the initial stage of meristematic divisions (Reid et al., 2011). Surprisingly, this component of the systemic circuit appears to be conserved among the legumes. This view stems from then observed functional regulation in the intergeneric grafts of *M. truncatula* shoot on the *L. japonicus* rootstock (Lohar and VandenBosch, 2005). In contrast to substance Q, SDI should be a low-molecular weight and stable compound (Lin et al., 2010). An ethanolic fraction of shoot-derived extract inhibiting nodulation in soybean has been reported by Kenjo et al. (2010). Another line of evidence is based on transcription profiling, which indicates the participation of jasmonic acid metabolites in downstream signaling (Seo et al., 2007; Kinkema and Gresshoff, 2008), is consistent with these data.

5. Other Regulatory Circuits of Nodulation

In contrast to the shoot-determined supernodulators, the term hypernodulation is usually applied to the lines showing elevated number of nodules that still follow the wild-type distribution along the root system (Park and Buttery, 1988; Gremaud and Harper, 1989; Novák et al., 1997). The distinctness of the locally-acting factors from the major systemic circuit is evidenced by the additivity of the effect of both mutations in double mutant lines

(Novák et al., 1997; Penmetsa et al., 2003). As a rule, the hypernodulating lines preserve nitrate responsiveness, the trait that clearly distinguishes them from the supernodulating ones. This feature also indicates that a circuit different from the major systemic signaling is affected.

It is probable that the hypernodulating mutations might affect non-systemic and short-acting factors regulating nodule number which existence was suggested by Delves et al., (1987). These factors are supposed to be produced by the established nodules and to suppress the initiation of further nodules in their neighborhood, as shown in experiments with mechanical removal of nodules in alfalfa (*Medicago sativa*) (Caetano-Anollés and Gresshoff, 1991b) and soybean (Caetano-Anollés et al., 1991b). Curiously, some abundantly nodulating mutants lost the sensitivity towards the inhibitory action of ethylene. Typical representative of this group are mutants in *SKL* ("sickle") of *M. truncatula* (Penmetsa et al., 2008). In contrast, the sunn supernodulating mutant is still ethylene-sensitive, i.e., reacts to the exogenous ethylene by reduction in nodule number (Penmetsa and Cook, 1997).

Each of the *M. truncatula* circuits releases a different subset of initial nodules. While the typical supernodulation mutation *sunn* acts through extending the bacterial susceptibility zone of the root and increases nodule initiation along the xylem poles of the central cylinder, i.e., in the wild-type pattern, the mutations in *SKL* allow for the formation of extra nodules throughout the root primary core. It has been shown that *SKL* is a *Medicago* ortholog for the *A. thaliana gene EIN2* which codes for the central regulator of ethylene signaling (Penmetsa et al., 2008).

Another characterized gene involved in non-systemic regulation is *EFD* of *M. truncatula*. The gene codes for a transcription factor which affects both nodule number and differentiation (Vernié et al., 2008). The *EFD* product is essential for nodule meristem persistence. In addition, its activity is necessary to inhibit further nodule initiation. This function fits the requirements imposed on the postulated short-range factor (Caetano-Anollés and Gresshoff, 1991b) or its regulatory molecules. EFD functions also point to the rule that normal nodule functioning and the activity of nodule meristem in the indeterminate nodules or the active central zone in the determinate nodules are necessary for the adjustment of an optimal nodule number in a host plant.

The concept of the short range factor inhibiting nodulation (Caetano-Anollés and Gresshoff, 1991b) assumes the existence of latent nodule primordia. The extra pool of nodule primordia that stop the development before the first mitotic division can enter further development only in the absence of a neighboring nodule. The mechanical excision of differentiated nodules can be mimicked by a pronounced defense-like nodule collapse as observed in alfalfa infected with *S. meliloti* mutant 102F15 (Paau et al., 1985) and soybean infected with a *nifA* mutant of *Bradyrhizobium japonicum* (Studer et al., 1987).

Nodule tissue collapse can be caused also by a host mutation, like in *P. sativum* symbiotic mutant RisfixV (Novák et al., 1995). The involvement of a regulatory circuit distinct from the systemic circuit is indicated by the preservation of nitrate sensitivity of nodulation in the RisfixV mutant, as well as by the additivity of mutation action in double-mutant pea line expressing both the RisfixV mutation and the *nod3* allele of the supernodulator RisfixC. Under nodulation-inhibitory conditions of high ambient nitrate concentrations the components of the phenotype conditioned by each of the two loci reacted in an independent way, i.e., the fraction of nodules contributed by the RisfixV mutation almost disappeared while the RisfixC-dependent fraction remained unchanged (Novák et al.,

1997). Ineffectiveness of nodules caused by bacterial or host mutations or reduction in size due to the developmental mutants brings about a side effect of increasing nodule number. The compensated nodule ineffectiveness or underdevelopment can result from bacterial mutations of *nif, fix, ndv*, and exopolysaccharide type (Spaink, 2000) or can be of plant origin in legume Fix⁻ mutants (Novák et al., 1993*a*). The fine tuning of nodule number and development according to the shoot capability and requirements in nitrogenous compounds keeps total nitrogenase activity at a constant level in the wild-type plants (Atkins and Smith, 2007). Although the increase of nodule number ranges within tens of percent in these cases, in contrast to the fold changes in supernodulating and hypernodulating mutants, it still confirms the existence of compensatory mechanisms based on the negative feedback. Mechanisms of nodule-activity dependent control of nodulation are still being entangled (Kiers et al., 2003).

6. Integrating View on the Regulation of Symbiosis

To date, the amount of experimental data on genes involved in the regulation of nodulation is sufficient for a preliminary reconstruction of the systemic circuit and prediction of the nature of yet unknown components.

6.1. Ascending Regulatory Pathway

The ascending root-derived signal of the systemic regulatory circuit (Figure 1) has been shown to be produced already by initial nodules few days after inoculation of alfalfa (Caetano-Anollés and Gresshoff, 1991*a*). Inefficient nodules of alfalfa formed with a bacterial exopolysaccharide mutant turned to be also a producer of the systemic signal (Caetano-Anollés and Gresshoff, 1991*b*) The signal is generated even in the full absence of bacteria as shown for empty nodules spontaneously occurring in certain alfalfa genotypes (Caetano-Anollés et al., 1991*a*). Using a series of symbiotic mutants, the branching point for the systemic regulatory signal has been located downstream of the transcription factor NSP1a of *L. japonicus*, a *M. truncatula* NSP1 homolog, in the symbiotic signal pathway (Okamoto et al., 2009). On the other hand, the systemic signal was shown to be generated before the action of the pea NIN homolog. The intensity of the signal increased with the stage of nodule development (Li et al., 2009).

The candidate gene approach based on the *L. japonicus* genomic database enabled to identify peptides of the CLE family as low-molecular weight messengers. The name of the peptides is derived from the homology with *A. thaliana* CLV3 (CLAVATA3) participating in meristem development and persistence like CLV1. The *CLE* genes were shown to be regulated both by the symbiosis development, Nod factor alone and nitrate in the wild-type *L. japonicus* (Okamoto et al., 2009) and *M. truncatula* (Mortier et al., 2010). On the other hand, the control was abolished by a supernodulation mutation *rdh1* with a root site of action. This observation is a strong evidence for the assumed role of symbiotic messenger (Okamoto et al., 2009; Ishikawa et al., 2008).

The available experimental data (Li et al., 2009; Novák, 2010) suggest that the *NOD3* product is involved in the same beginning of the regulatory circuit, presumably on the stage preceding the synthesis and release of the substance Q transported upwards into the shoot

(Figure 1). In spite of the advantage stemming from its orthology with *M. truncatula RDN1* gene (Schnabel et al., 2011), the action of the NOD3 protein is still not clear. Moreover, the role of *NOD3* should be in some way related to the participation of small peptides of the CLE family in symbiotic signaling (Oelkers et al., 2008). NOD3 might either directly regulate their synthesis or can directly interact with CLE peptides to sequester their biological activity. Since a protein-protein interaction was described for CLV1 and CLV3 of *A. thaliana* (DeYoung and Clark, 2008), an analogous interaction can be expected between the legume CLE peptides on one side (Okamoto et al., 2009) and LjHAR1 (MtSUNN, GmNARK, PsSYM29) protein kinases on the other side. This notion is based on the structural homologies between AtCLV3 and the CLE peptides (Okamoto et al., 2009) and between AtCLV1 and LjHAR1 (Krusell et al., 2002; Schnabel et al., 2005). In this model, the role of NOD3 might consist in the interaction with the CLE peptides sequestering their level accessible to the LjHAR1 homologs. The alternative of suppressing CLE protein production might occur via the regulation of CLE genes. However, the recent report places the CLE action upstream of NOD3 step (Osipova et al., 2012). The exact role of *NOD3* on early stages of nodulation still requires additional studies oriented at the protein interactions of the *NOD3* product.

6.2. Middle Stage - Signal Conversion in the Shoot and Phytohormone Interference

At the middle stage of systemic signaling, the localization of the *NARK* product to phloem also suggests participation in signal molecule distant transportation. The disturbance of the long-range transport of phytohormones is conceivable as well (Guinel and Geil, 2002). This notion is consistent with discontinuous leaf veins in Klavier hypernodulating mutant of *L. japonicus* (Oka-Kira et al., 2005) and phloem localization of *HAR1/NARK* gene activity (Nontachalyapoom et al., 2007). The search of NARK downstream genes using transcriptomic approach has revealed genes with role in jasmonic acid metabolism (Kinkema and Gresshoff, 2008). This finding, as well as changes in jasmonic acid metabolism in the soybean supernodulation mutant (Seo et al., 2007), points to the jasmonic acid as a relevant molecule in the signaling downstream of HAR1-like proteins.

The role of phytohormones in the systemic signaling is still taking shape. The pleiotropic trait of internode shortening observed in the early blocked (RisfixC) and shoot-acting (*HAR1* orthologs) mutants mimics the deficiency in gibberellic acid synthesis. Nevertheless, a panel of gibberellin mutants of pea (*ls, lh, le, na* and *sln*) did not demonstrate symbiotic changes resembling supernodulation (Ferguson et al., 2005).

Traditionally, the role of auxins was evidenced by nodule initiation upon treatment with auxin transport inhibitors in the absence of rhizobia (Hirsch et al., 1989). Moreover, auxin local accumulation leading to nodule establishment was shown to be linked to the symbiotic flavonoid response (Wasson et al., 2006). Since the metabolism of gibberellin is controlled by auxin in pea (O'Neill and Ross, 2002), the internode length changes in the supernodulators might reflect changes in auxin production or trafficking. However, increased auxin production in the shoot of the supernodulation mutant *sunn* of *M. truncatula* (Van Noorden et al., 2006) disagrees with the observed shoot growth depression.

The role of cytokinins seems to be firmly established to date. They mediate the transfer of rhizobial signal to the symbiotic cytokinin receptor (Gonzales-Rizzo et al., 2006). The persuasive evidence of cytokinin role in the symbiotic signal transfer has been provided by recurrent mutagenesis in the background of *L. japonicus har1* mutant that led to the discovery of supernodulation suppressors. One of them turned to be a legume cytokinin receptor, the mutation in which lead to the restricted symbiotic signal relay and apparent normalization of nodule number (Murray et al., 2006; Tirichine et al., 2007).

In contrast to the active role of auxin and cytokinin in symbiotic signal transfer, abscisic acid (Ding et al., 2008) and ethylene (Penmetsa and Cook, 1997) block nodule development. The effect of ethylene on nodule initiation can be abolished by specific mutations like skl (Penmetsa and Cook, 1997). The identity of the *SKL* gene was determined as an ortholog of *Arabidopsis* ethylene signaling gene *EIN2* (Penmetsa et al., 2008). Although the phenotypic effect of disrupted ethylene perception leads to enhanced nodulation (hypernodulation), the site of ethylene action lies outside of the systemic regulatory circuit (Gresshoff et al., 2009).

6.3. Descending Regulatory Pathway and Cell Cycle Control

In the terminal, descending part of the systemic circuit, two major unclear points remain: what is the nature of the SDI substance and which stage of nodules is the target stage for SDI. Although the supernodulation mutations are revealed by the number of mature nodules formed per root, their primary action should affect the early stages of nodule establishment. The affected stage has long been supposed to be that of latent nodule primordia before nodule meristem formation in indeterminate nodules (Caetano-Anollés and Gresshoff, 1991a; Sagan and Gresshoff, 1996; Kinkema et al., 2006; Li et al., 2009). However, according to the latest view, the stage of "cell divisions in the cortex" is the main target for nodule number regulation (Li et al., 2009). An indirect observation based on the root flavonoid response to rhizobia in RisfixC (*nod3*) supernodulator places the target stage of the feedback circuit even earlier than the stage of nodule initiation (Novák et al., 2004).

Regardless of the final localization of the branching points for the outcoming and incoming signals, the relation between the cell cycle control in initial nodules and symbiotic systemic signals deserves more attention. In the simplified view, the nodule number formed is the result of the conversion of the rhizobial trigger to the cell-cycle regulation. The treatment with Nod factor leads to dedifferentiation of the established root cortical cells that are at the end of the endoreduplication cycle. They then re-enter the mitotic cycle (Inzé and De Veylder, 2006). Also the cells of the established symbiotic nodules of *M. sativa* are at the end of the endoreduplication cycle. The cell cycle progression is stopped by the mitotic inhibitor ccs52, a product of *CCS52A* gene (Cebolla et al., 1999).

The molecular markers of cell cycle might contribute to a better characterization of nodule initiation and development. It has been shown that the auxin-responsive histone H3 gene of *M. truncatula* is more active in the abundantly nodulating sunn and sickle mutants (Penmetsa et al., 2003). Similarly, the early nodulins ENOD40 (Veereshlingam et al., 2004) and ENOD20 (Kuppusamy et al., 2004) were used as markers for nodule primordia. However, histological techniques determining in situ expression are essential in view of the complex patterning of the mitotic zones in the root cortex. It is clear that the transition from the G2 to the mitotic phase in the root cortex cells can be further analyzed using more symbiotic mutants or constructs affecting this stage. Surprisingly, the overexpression of the Ca^{++}/calmodulin-dependent protein kinase coded for by *DMI3* of *M. truncatula* was sufficient alone to induce nodule formation (Gleason et al., 2006). Similarly, the constitutive expression of the *L. japonicus* symbiotic cytokinin receptor LHK1 leads to the determinate nodule induction (Murray et al., 2007; Tirichine et al., 2007). It must be held in mind that the interactions between the initial nodules and the initial infections will distort the frequencies of both structures in the course of symbiosis development. This interdependence will result in the persistence of only a limited fraction of the initial structures. The relations between the epidermal and cortical programs can be analyzed mutationally as well (Oldroyd and Downie, 2008).

Nodule primordia of indeterminate-type nodules that had not been colonized by rhizobia in the form of infection threads did not form nodule apical meristems as demonstrated for *P. sativum* mutant line RisnodA (Voroshilova et al., 2009). On the other hand, the nodule meristems are essential for the persisting activity of indeterminate nodules. Failure in their formation reduces final nodule number, as shown in *M. truncatula* mutant LIN (Kuppusamy et al., 2004). This correlation has been independently confirmed for the *M. truncatula* api and EFD mutants (Teillet et al., 2008; Vernié et al., 2008).

6.4. Future Contribution of High-Throughput Profiling

The possibility to selectively block symbiosis development at different stages enhances the potential of genomic and proteomic approaches available to date. It must be noted that the wide-scale proteomic and transcriptomic profiling has been applied to the resolution of supernodulation phenomenon in soybean already in late nineties (Appel et al., 1999). Subsequently, the subsets of Nod factor- and nitrate-induced genes of soybean have been described (Kinkema and Gresshoff, 2008).

In parallel, the transcriptomic approach based on soybean expression arrays has been applied to the profiling in soybean plants that lost responsiveness to the systemic symbiotic and nitrate stimuli as consequence of a supernodulation mutation (Kinkema and Gresshoff, 2008). Profiling of supernodulating soybean revealed jasmonic acid metabolism-related transcripts as candidates for downstream signaling (Seo et al., 2007). Mutant har1 of *L. japonicus* has recently been included into the thorough transcriptome profiling with 50K hybridization chips under symbiotic/asymbiotic conditions (Hogslund et al., 2009). The resulting database represents a valuable key to decipher the nodulation regulatory mechanism.

7. Where and How Nitrate Signal Joins?

Until this point, only the purely symbiotic aspects of the systemic signaling have been considered. However, as shown in the early classical works (Day et al., 1989), the symbiotic signal is tightly connected to the effect of nitrate. The precise localization of the supernodulation/Nts mutations in the systemic circuit should identify the site where the nitrate signal joins the symbiotic pathway. The efforts to dissociate the effect of the two factors, symbiotic and nitrate, were not successful. The extent of supernodulation can be reduced by leaky alleles in the soybean locus *nts1* (*NARK*) and pea *SYM29*, both coding for

HAR1-homologous protein kinases expressed in the shoot (Delves et al., 1988; Sagan and Duc, 1996). Although these mutants showed improvement in plant performance due to the reduction of overnodulation load (Sagan and Duc, 1996), uncoupling of supernodulation from nitrate resistance has not been shown. Similarly, separation of both traits by genetic recombination has not been achieved (Novák et al., 1997).

In contrast to the systemic spread of the root symbiotic signal, the nitrate signal acts locally. Only that part of the root system that is exposed to a high ambient nitrate level demonstrates inhibition of nodule initiation and growth (Day et al., 1989). Up to date, the localization of the nitrate signal integration point in the signaling circuit is not known for certain (Figure 1). The local action of nitrate in the root implies the integration point at the end of the symbiotic signaling circuit (Day et al., 1989). On the other hand, the regulation of the early CLE peptide messengers by the symbiotic stimulus and nitrate (Okamoto et al., 2009) requires that the integration point is placed in advance of the ascending signal. The CLE peptides are supposed to transduce symbiotic and nitrate signals from the root upwards (Okamoto et al., 2009; Ferguson et al., 2010). Accordingly, the inhibition of nodule formation by the ectopically expressed *M. truncatula* CLE12 and CLE13 inhibited nodule formation only when the systemic circuit was functional. On the other hand, the pleiotropic effect of CLE overexpression consisting in elongated petioles was independent of the sunn mutation. This observation can be interpreted that the pleiotropic signal branches off before the SUNN action. The discrepancies in the localization of CLE action led to the model suggesting two forms of CLE peptides, the locally and distantly acting one (Ferguson et al., 2010; Reid et al., 2011), however, a sufficient experimental evidence still has to be obtained.

Alternatively, the nitrate signal transduction is placed as far downstream as to the transcriptional activator NIN of *L. japonicus* which regulates expression of genes for nodule-specific proteins (nodulins). This function of legume NIN is deduced from the behavior of the *A. thaliana* homolog NLP7 (Castaings et al., 2009). The search of modifier genes interacting with the main molecules comprising the systemic regulatory circuit might be another productive approach (Krusell et al., 2002; Schnabel et al., 2005). The main interactor revealed by mutagenesis is the symbiotic cytokinin receptor which mutational inactivation reduces the frequency of nodule initiation in the *L. japonicus* har1 mutant (Murray et al., 2006; Tirichine et al., 2007). Alternatively, the supernodulation suppressors can recruit shortcut pathways for systemic regulation. This might be the case of *RAE* suppressor of *sunn* mutation in *M. truncatula* (Frugoli et al., 2008).

Similarly to the use of the markers for symbiosis-associated transcription, markers for the nitrate-regulated genes of legumes can be useful in the definition of the mutational blocks in nitrate perception. Today, they can be derived from the inventory of transcripts affected by N status of *M. truncatula* (Ruffel et al., 2008). The data are included in the *Medicago truncatula* Gene Expression Atlas (He et al. 2009) published on the *Medicago* project server (www.medicago.org).

8. The Promise of New Regulatory Mutants

A number of new symbiotic regulatory mutants have been reported during last years (Table 1) that joined the set of yet uncharacterized mutants and affected loci. It appears that the number of mutationally detectable loci taking part in nodule number regulation has not been exhausted to date. Their ongoing characterization is expected to identify a series of new genes and products involved in the systemic signal generation, transduction and perception.

A significant contribution to our understanding of the key phase of symbiotic signal generation in the root might bring further mutants affected in the root-localized processes, like the mutant in the *RDN2* locus of *M. truncatula* (Schnabel et al., 2011). In *L. japonicus*, the root determined loci *RDH1* (*ROOT DETERMINED HYPERNODULATION*) and *TML* (*TOO MUCH LOVE*) were reported to be localized in the range of centimorgan intervals and are close to identification (Ishikawa et al., 2008; Magori et al., 2009).

Also the sets of abundantly nodulating mutants accumulated in traditional legume crops, like pea (Engvild, 1987; Duc and Messager, 1989; Novák et al., 2005; Borisov et al., 2007), soybean (Delves et al., 1988), common bean (*Phaseolus vulgaris*; Park and Buttery, 1988) as well as faba bean (*Vicia faba*), chicken pea (*Cicer arietinum*) and groundnut (*Arachis hypogea*; Bhatia et al., 2001) will certainly be subject of additional studies. In particular, pea loci *NOD4*, *NOD5* and *NOD6* (Sidorova and Shumnyi, 2003) might contribute to our understanding of nodule number regulation, in spite of a more complicated cloning procedure (Sidorova and Shumnyi, 2003).

Nevertheless, the orthology between pea and the fully sequenced model legumes is an efficient tool for the identification of new important genes (Gualtieri.et al., 2002; Schnabel et al., 2011; Novák et al., 2012).

Acknowledgments

The work was supported by the Ministry of Agriculture of the Czech Republic (Institutional Research Concept MZE0002701404).

References

Ané J.M., Kiss G.B., Riely B.K., Penmetsa R.V., Oldroyd G.E.D., Ayax C., Lévy J., Debellé F., Baek J.M., Kaló P., Rosenberg C., Roe B.A., Long S.R., Dénarié J., Cook D.R. (2004) Medicago truncatula dmi1 required for bacterial and fungal symbioses in legumes. *Science* 303, 1364-1367.

Appel M., Bellstedt D.U., Gresshoff P.M. (1999) Differential display of eukaryotic mRNA: Meeting the demands of the new millennium? *J. Plant Physiol.* 154, 561-570.

Arai M., Hayashi M., Takahashi M., Shimada S., Harada K. (2005) Expression and sequence analysis of systemic regulation gene for symbiosis, *NTS1/GmNARK* in supernodulating soybean cultivar, Sakukei 4. *Breeding Sci.* 55, 147-152.

Arrighi J.F., Barré A., Ben Amor B., Bersoult A., Soriano L.C., Mirabella R., De Carvalho-Niebel F., Journet E.-P., Gherardi M., Huguet T., Geurts R., Dénarié J., Rougé P., Gough C. (2006) The Medicago truncatula lysine motif-receptor-like kinase gene family includes NFP and new nodule-expressed genes. *Plant Physiol.* 142, 265–279.

Atkins C.A., Smith P.M.C. (2007) Translocation in legumes: assimilates, nutrients, and signaling molecules. *Plant Physiol.* 144, 550-561.

Bek A.S., Sauer J., Thygesen M.B., Duus J.O., Petersen B.O., Thirup S., James E., Jensen K.J., Stougaard J., Radutoiu S. (2010) Improved characterization of Nod factors and genetically based variation in LysM receptor domains identify amino acids expendable for Nod factor recognition in *Lotus* spp. *Mol. Plant-Microbe In.* 23, 58-66.

Bhatia C.R., Nichterlein K., Maluszynski M. (2001) Mutations affecting nodulation in grain legumes and their potential in sustainable cropping systems. *Euphytica* 120, 415-432.

Bohlool B.B., Ladha J.K., Garrity D.P., George T. (1992) Biological nitrogen fixation for sustainable agriculture: A perspective. *Plant Soil* 141, 1-11.

Borisov A.Y., Danilova T.N., Koroleva T.A., Kuznetsova E.V., Madsen L., Mofett M., Naumkina T.S., Nemankin T.A., Ovchinnikova E.S., Pavlova Z.B., Petrova N.E., Pinaev A.G., Radutoiu S., Rozov S.M., Rychagova T.S., Shtark O.Y., Solovov I.I., Stougaard J., Tikhonovich I.A., Topunov A.F., Tsyganov V.E., Vasilchikov A.G., Voroshilova V.A., Weeden N.F., Zhernakov A.I., Zhukov V.A. (2007) Regulatory genes of garden pea (*Pisum sativum* L.) controlling the development of nitrogen-fixing nodules and arbuscular mycorrhiza: a review of basic and applied aspects. *Appl. Biochem. Micro.* 43, 237-243.

Brewin N.J. (1991) Development of the legume root nodule. *Annu. Rev. Cell Biol.* 7, 191-226.

Caetano-Anollés G., Gresshoff P.M. (1991a) Alfalfa controls nodulation during the onset of *Rhizobium*-induced cortical cell division. *Plant Physiol.* 95, 366-373.

Caetano-Anollés G., Gresshoff P.M. (1991b) Excision of nodules induced by *Rhizobium meliloti* exopolysaccharide mutants releases autoregulation in alfalfa. *J. Plant Physiol.* 138, 765-767.

Caetano-Anollés G., Joshi P.A., Gresshoff P.M. (1991a) Spontaneous nodules induce feedback suppression of nodulation in alfalfa. *Planta* 183, 77-82.

Caetano-Anollés G., Paparozzi E.T., Gresshoff P.M. (1991b) Mature nodules and root tips control nodulation in soybean. *J. Plant Physiol.* 137, 389-396.

Carroll B.J., McNeil D.L., Gresshoff P.M. (1985) Isolation and properties of soybean (*Glycine max* (L.) Merr.) mutants that nodulate in the presence of high nitrate concentrations. *P. Natl. Acad. Sci. USA.* 82, 4162-4166.

Castaings L., Camargo A., Pocholle D., Gaudon V., Texier Y., Boutet-Mercey S., Taconnat L., Renou J.-P., Daniel-Vedele F., Fernandez E., Meyer C., Krapp A. (2009) The nodule inception-like protein 7 modulates nitrate sensing and metabolism in *Arabidopsis*. *Plant J.* 57, 426-435.

Cebolla A., Vinardell E., Kiss G., Olláh B., Roudier F., Kondorosi A., Kondorosi E. (1999) The mitotic inhibitor ccs52 is required for endoreduplication and ploidy-dependent cell enlargement in plants. *EMBO J.* 18, 4476-4484.

Copyright Journal compilation © 2009 Blackwell Publishing Ltd and the Society for Experimental BiologyChen S.K., Kurdyukov S., Kereszt A., Wang X.D., Gresshoff P.M., Rose R.J. (2009) The association of homeobox gene expression with stem cell formation and morphogenesis in cultured Medicago truncatula. *Planta* 230, 827-840.

Chen W.-M., Moulin L., Bontemps C., Vandamme P., Béna G., Boivin-Masson C. (2003) Legume symbiotic nitrogen fixation by β-proteobacteria is widespread in nature. *J. Bacteriol.* 185, 7266-7272.

Cros-Arteil S., Pfaff T., Barker D.G., Journet E.P. (2006) Cuttings and grafts. In: Mathesius U., Journet E.P., Sumner L.W., editors. Medicago truncatula *handbook*. ISBN 0-9754303-1-9, http://www.noble.org/MedicagoHandbook/.

Day D.A., Carroll B.J., Delves A.C., Gresshoff P.M. (1989) Relationship between autoregulation and nitrate inhibition of nodulation in soybeans. *Physiol. Plantarum* 75, 37-42.

Delves A.C., Carroll B.J., Gresshoff P.M. (1988) Genetic analysis and complementation studies on a number of mutant supernodulating soybean lines. *J. Genet.* 67, 1-8.

Delves A.C., Higgins A.V., Gresshoff P.M. (1987) Shoot control of supernodulation in a number of mutant soybeans, *Glycine max* (L.) Merr. *Austral. J. Plant Physiol.* 14, 689-694.

DeYoung B.J., Clark S.E. (2008) BAM receptors regulate stem cell specification and organ development through complex interactions with CLAVATA signaling. *Genetics* 180, 895-904.

Ding Y., Kalo P., Yendrek C., Sun J., Liang Y., Marsh J.F., Harris J.M., Oldroyd G.E.D. (2008) Abscisic acid coordinates Nod factor and cytokinin signaling during the regulation of nodulation in *Medicago truncatula*. *Plant Cell* 20, 2681–2695.

Downie J.A., Parniske M. (2002) Plant biology - fixation with regulation. *Nature* 420, 369-370.

Downie J.A., Walker S.A. (1999) Plant responses to nodulation factors. *Curr. Opinions Plant Biol.* 2, 483-489.

Duc G., Messager A. (1989) Mutagenesis of pea (*Pisum sativum* L.) and the isolation of mutants for nodulation and nitrogen fixation. *Plant Sci.* 60, 207-214.

Edwards A., Heckmann A.B., Yousafzai F., Duc G., Downie J.A. (2007) Structural implications of mutations in the pea *SYM8* symbiosis gene, the *DMI1* ortholog, encoding a predicted ion channel. *Mol. Plant-Microbe In.* 20, 1183-1191.

Endre G., Kereszt A., Kevei Z., Mihacea S., Kalo P., Kiss G.B. (2002) A receptor kinase gene regulating symbiotic nodule development. *Nature* 417, 962-966.

Engvild K.C. (1987) Nodulation and nitrogen fixation mutants of pea, *Pisum sativum. Theor. Appl. Genet.* 74, 711-713.

Ferguson B.J., Indrasumunar A., Hayashi S., Lin M.H., Lin Y.H., Reid D.E., Gresshoff P.M. (2010) Molecular analysis of legume nodule development and autoregulation. *J. Integrative Plant Biol.* 52, 61-76.

Ferguson B.J., Ross J.J., Reid J.B. (2005) Nodulation phenotypes of gibberellin and brassinosteroid mutants of pea. *Plant Physiol.* 138, 2396-2405.

Floss D.S., Hause B., Lange P.R., Küster H., Strack D., Walter M.H. (2008) Knock-down of the MEP pathway isogene 1-deoxy-D-xylulose 5-phosphate synthase 2 inhibits formation of arbuscular mycorrhiza-induced apocarotenoids, and abolishes normal expression of mycorrhiza-specific plant marker genes. *Plant J.* 56, 86-100.

Forde B.G. (2002) The role of long-distance signalling in plant responses to nitrate and other nutrients. *J. Exp. Bot.* 53, 39-43.

Frugoli J., Smith L., Schnabel E., Mukherjee A., Long S. (2008) Regulation of nodule number by multiple genes in *Medicago truncatula*. In: *4th International Congress on Legume Genomics and Genetics, Book of Abstracts.* Cuernavaca, Mexico: UNAM; p. 42.

Fujikake H., Tamura Y., Ohtake N., Sueyoshi K., Ohyama T. (2003) Photoassimilate partitioning in hypernodulation mutant of soybean (*Glycine max* (L.) Merr.) NOD1-3 and its parent Williams in relation to nitrate inhibition of nodule growth. *Soil Sci. Plant Nutr.* 49, 583-590.

Garrity G.M., Boone D.R., Castenholz R.W. (Eds.) (2001) Bergey's Manual of Systematic Bacteriology. Springer, 2nd Edition.

Gelin O., Blixt S. (1964) Root nodulation in peas. *Agri Hortique Genetica* 22, 149-159.

Gleason C., Chaudhuri S., Yang T.B., Munoz A., Poovaiah B.W., Oldroyd G.E.D. (2006) Nodulation independent of rhizobia is induced by a calcium activated kinase lacking autoinhibition. *Nature* 441, 1149-1152.

Gonzalez-Rizzo S., Crespi M., Frugier F. (2006) The *Medicago truncatula* CRE1 cytokinin receptor regulates lateral root development and early symbiotic interaction with *Sinorhizobium meliloti*. *Plant Cell* 18, 2680-2693.

Gremaud M.F., Harper J.E. (1989) Selection and initial characterization of partially nitrate tolerant nodulation mutants of soybean. *Plant Physiol.* 89, 169-173.

Gresshoff P.M., Lohar D., Chan P.K., Biswas B., Jiang Q., Reid D., Ferguson B., Stacey G. (2009) Genetic analysis of ethylene regulation of legume nodulation. *Plant Signal. Behav.* 4, 818-823

Gualtieri G., Kulikova O., Limpens E., Kim D.J., Cook D.R., Bisseling T., Geurts R. (2002) Microsynteny between pea and Medicago truncatula in the *SYM2* region. *Plant Mol. Biol.* 50, 225-235.

Guinel, F.C., Geil, R.D. (2002) A model for the development of the rhizobial and arbuscular mycorrhizal symbioses in legumes and its use to understand the roles of ethylene in the establishment of these two symbioses. *Can. J. Bot.* 80, 695-720.

He J., Benedito V.A., Wang M., Murray J.D., Zhao P.X., Tang Y., Udvardi M.K. (2009) The Medicago truncatula gene expression atlas web server. *BMC Bioinformatics* 10, 441.

Hirsch A.M., Bhuvaneswari T.V., Torrey J.G., Bisseling T. (1989) Early nodulin genes are induced in alfalfa root outgrowths elicited by auxin transport inhibitors. *P. Natl. Acad. Sci. USA* 86, 1244–1248.

Hogslund N., Radutoiu S., Krusell L., Voroshilova V., Hannah M.A., Goffard N., Sanchez D.H., Lippold F., Ott T., Sato S., Tabata S., Liboriussen P., Lohmann G.V., Schauser L., Weiller G.F., Udvardi M.K., Stougaard J. (2009) Dissection of symbiosis and organ development by integrated transcriptome analysis of Lotus japonicus mutant and wild-type plants. *PLOS ONE* 4, e6556, DOI: 10.1371.

Imaizumi-Anraku H., Takeda N., Charpentier M., Perry J., Miwa H., Umehara Y., Kouchi H., Murakami Y., Mulder L., Vickers K., Pike J., Downie J.A., Wang T., Sato S., Asamizu E., Tabata S., Yoshikawa M., Murooka Y., Wu G.J., Kawaguchi M., Kawasaki S., Parniske M., Hayashi M. (2005) Plastid proteins crucial for symbiotic fungal and bacterial entry into plant roots. *Nature* 433, 527-531.

Inzé D., De Veylder L. (2006) Cell cycle regulation in plant development. *Annu. Rev. Genet.* 40, 77-105.

Ishikawa K., Yokota K., Li Y.Y., Wang Y.X., Liu C.T., Suzuki S., Aono T., Oyaizu H. (2008) Isolation of a novel root-determined hypernodulation mutant rdh1 of *Lotus japonicus*. *Soil Sci. Plant Nutr.* 54, 259-263.

Jacobsen E., Feenstra W.J. (1984) A new pea mutant with efficient nodulation in the presence of nitrate. *Plant Sci. Lett.* 33, 337-344.

Kenjo T., Yamaya H., Arima Y. (2010) Shoot-synthesized nodulation-restricting substances of wild-type soybean present in two different high performance liquid chromatography peaks of the ethanol-soluble medium-polarity fraction. *Soil Sci. Plant Nutr.* 56, 399-406.

Kévei Z., Lougnon G., Mergaert P., Horvath G.V., Kereszt A., Jayaraman D., Zaman N., Marcel F., Regulski K., Kiss G.B., Kondorosi A., Endre G., Kondorosi E., Ané J.M. (2007) 3-Hydroxy-3-methylglutaryl coenzyme A reductase1 interacts with NORK and is crucial for nodulation in *Medicago truncatula*. *Plant Cell* 19, 3974-3989.

Kiers E.T., Rousseau R.A., West S.A., Denison R.F. (2003) Host sanctions and the legume–rhizobium mutualism. *Nature* 425, 78-81.

Kinkema M., Gresshoff P.M. (2008) Investigation of downstream signals of the soybean autoregulation of nodulation receptor kinase GmNARK. *Mol. Plant-Microbe In.* 21, 1337-1348.

Kinkema M., Scott P.T., Gresshoff P.M. (2006) Legume nodulation: successful symbiosis through short- and long-distance signalling. *Funct. Plant Biol.* 33, 707-721.

Kokubun M., Akao S. (1994) Inheritance of supernodulation in soybean mutant En6500. *Soil Sci. Plant Nutr.* 40, 715-718.

Krusell L., Madsen L.H., Sato S., Aubert G., Genua A., Szczyglowski K., Duc G., Kaneko T., Tabata S., De Bruijn F., Pajuelo E., Sandal N., Stougaard J. (2002) Shoot control of root development and nodulation is mediated by a receptor-like kinase. *Nature* 420, 422-426.

Krusell L., Sato N., Fukuhara I., Koch B.E.V., Grossmann C., Oka-Kira E., Otsubo Y., Aubert G., Nakagawa T., Sato S., Okamoto S., Tabata S., Parniske M., Wang T.L., Kawaguchi M., Stougaard J., Duc G. (2011) The *Clavata2* genes of pea and *Lotus japonicus* affect autoregulation of nodulation. *Plant J.* 65, 861–871.

Kuppusamy K.T., Endre G., Prabhu R., Penmetsa R.V., Veereshlingam H., Cook D.R., Dickstein R., VandenBosch K.A. (2004) *LIN*, a Medicago truncatula gene required for nodule differentiation and persistence of rhizobial infections. *Plant Physiol.* 136, 3682-3691.

Landau-Ellis D., Angermuller S., Shoemaker R., Gresshoff P.M. (1991) The genetic locus controlling supernodulation in soybean (*Glycine max* L.) co-segregates tightly with a cloned molecular marker. *Mol. Gen. Genet.* 228, 221-226.

Lee H.S., Chae Y.A., Park E.H., Kim Y.W., Yun K.I., Lee S.H. (1997) Introduction, development, and characterization of supernodulating soybean mutant. I. Mutagenesis of soybean and selection of supernodulating mutant. *Korean J. Crop Sci.* 42, 247-253.

Lestari P., Van K., Kim M.Y., Lee S.-H. (2006) Nodulation and growth of a supernodulating soybean mutant SS2-2 symbiotically associated with Bradyrhizobium japonicum. *Jurnal AgroBiogen* 2, 8-15.

Lévy J., Bres C., Geurts R., Chalhoub B., Kulikova O., Duc G., Journet E.P., Ané J.M., Lauber E., Bisseling T., Dénarié J., Rosenberg C., Debellé F. (2004) A putative Ca^{2+} and calmodulin-dependent protein kinase required for bacterial and fungal symbioses. *Science* 303, 1361-1364.

Li D.X., Kinkema M., Gresshoff P.M. (2009) Autoregulation of nodulation (AON) in Pisum sativum (pea) involves signalling events associated with both nodule primordia development and nitrogen fixation. *J. Plant Physiol.* 166, 955-967.

Limpens E., Franken C., Smit P., Willemse J., Bisseling T., Geurts R. (2003) LysM domain receptor kinases regulating rhizobial Nod factor-induced infection. *Science* 302, 630-633.

Limpens E., Ramos J., Franken C., Raz V., Compaan B., Franssen H., Bisseling T., Geurts R. (2004) RNA interference in *Agrobacterium rhizogenes*-transformed roots of Arabidopsis and Medicago truncatula. *J. Exp. Bot.* 55, 983-992.

Lin Y.H., Ferguson B.J., Kereszt A., Gresshoff P.M. (2010) Suppression of hypernodulation in soybean by a leaf-extracted, NARK- and Nod factor-dependent, low molecular mass fraction. *New Phytol.* 185, 1074-1086.

Lohar D.P., VandenBosch K.A. (2005) Grafting between model legumes demonstrates roles for roots and shoots in determining nodule type and host/rhizobia specificity. *J. Exp. Bot.* 56, 1643-1650.

Magori S., Oka-Kira E., Shibata S., Umehara Y., Kouchi H., Hase Y., Tanaka A., Sato S., Tabata S., Kawaguchi M. (2009) *TOO MUCH LOVE*, a root regulator associated with the long-distance control of nodulation in Lotus japonicus. *Mol. Plant-Microbe In.* 22, 259-268.

McComb A.J., McComb J.A. (1970) Growth substances and relation between phenotype and genotype in Pisum sativum. *Planta* 91, 235-345.

Miyazawa H., Oka-Kira E., Sato N., Takahashi H., Wu G.-J., Sato S., Hayashi M., Betsuyaku S., Nakazono M., Tabata S., Harada K., Sawa S., Fukuda H., Kawaguchi M. (2010) The receptor-like kinase KLAVIER mediates systemic regulation of nodulation and non-symbiotic shoot development in Lotus japonicus. *Development* 137, 4317-4325.

Mortier V., Deen Herder G., Whitford R., Van de Velde W., Rombauts S., D'haeseleer K., Holsters M., Goormachtig S. (2010) CLE peptides control Medicago truncatula nodulation locally and systemically. *Plant Physiol.* 153, 222-237.

Murray J., Karas B., Ross L., Brachmann A., Wagg C., Geil R., Perry J., Nowakowski K., MacGillivary M., Held M., Stougaard J., Peterson L., Parniske M., Szczyglowski K. (2006) Genetic suppressors of the Lotus japonicus har1-1 hypernodulation phenotype. *Mol. Plant-Microbe In.* 19, 1082-1091.

Murray J.D., Karas B.J., Sato S., Tabata S., Amyot L., Szczyglowski K. (2007) A cytokinin perception mutant colonized by *Rhizobium* in the absence of nodule organogenesis. *Science* 315, 101-104.

Murray J.D., Muni R.R., Torres-Jerez I., Tang Y., Allen S., Andriankaja M., Li G., Laxmi A., Cheng X., Wen J., Vaughan D., Schultze M., Sun J., Charpentier M., Oldroyd G., Tadege M., Ratet P., Mysore K.S., Chen R., Udvardi M.K. (2011) Vapyrin, a gene essential for intracellular progression of arbuscular mycorrhizal symbiosis, is also essential for infection by rhizobia in the nodule symbiosis of Medicago truncatula. *Plant J.* 65, 244-252.

Nazaryuk V.M., Sidorova K.K., Shumny V.K., Kalimullina F.R., Klenova M.I. (2006) Physiological and agrochemical properties of different symbiotic genotypes of pea (*Pisum sativum* L.). *Izv. Akad. Nauk Biol.* 2006, 688-697.

Nelson L.M. (1987) Response of Rhizobium leguminosarum isolates to different forms of inorganic nitrogen during nodule development in pea (*Pisum sativum* L.). *Soil Biol. Biochem.* 19, 759-763.

Nishimura R., Ohmori M., Fujita H., Kawaguchi M. (2002a) A *Lotus* basic leucine zipper protein with a RING-finger motif negatively regulates the developmental program of nodulation. *P. Natl. Acad. Sci. USA* 99, 15206–15210.

Nishimura R., Ohmori M., Kawaguchi M. (2002b) The novel symbiotic phenotype of enhanced-nodulating mutant of Lotus japonicus: astray mutant is an early nodulating mutant with wider nodulation zone. *Plant Cell Physiol.* 43, 853-859.

Nontachalyapoom S., Scott P.T., Men A.E., Kinkema M., Schenk P.M., Gresshoff P.M. (2007) Promoters of orthologous Glycine max and *Lotus japonicus* nodulation autoregulation genes interchangeably drive phloem-specific expression in transgenic plants. *Mol. Plant-Microbe In.* 20, 769-780.

Novák K. (2010) Early action of pea symbiotic gene *NOD3* is confirmed by adventitious root phenotype. *Plant Sci.* 179, 472-478.

Novák K., Biedermannová E., Vondrys J. (2009) Symbiotic and growth performance of supernodulating forage pea lines. *Crop Sci.* 49, 1227-1234.

Novák K., Biedermannová E., Vondrys J. (2012) Functional markers delimiting a Medicago orthologue of pea symbiotic gene *NOD3*. *Euphytica*, DOI: 10.1007/s10681-011-0586-8.

Novák K., Lisá L., Škrdleta V. (2004) Rhizobial *nod* gene-inducing activity in pea nodulation mutants: Dissociation of nodulation and flavonoid response. *Physiol. Plantarum* 120, 546-555.

Novák K., Lisá L., Škrdleta V. (2011) Pleiotropy of pea RisfixC supernodulation mutation is symbiosis-independent. *Plant Soil* 342, 173-182.

Novák K., Pešina K., Nebesářová J., Škrdleta V., Lisá L., Našinec V. (1995) Symbiotic tissue degradation pattern in the ineffective nodules of three nodulation mutants of pea (*Pisum sativum* L.). *Ann. Bot.* 76, 303-313.

Novák K., Škrdleta V., Kropáčová M., Lisá L., Němcová M. (1997) Interaction of two genes controlling symbiotic nodule number in pea (*Pisum sativum* L.). *Symbiosis* 23, 43-62.

Novák K., Škrdleta V., Němcová M., Lisá L. (1993a) Symbiotic traits, growth, and classification of pea nodulation mutants. *Rost. Výroba* 39, 157-170.

Novák K., Škrdleta V., Němcová M., Lisá L. (1993b) Behavior of pea nodulation mutants as affected by increasing nitrate level. *Symbiosis* 15, 195-206.

Novák K., Šlajs M., Biedermannová E., Vondrys J. (2005) Development of an asymbiotic reference line for pea cv. Bohatýr by de novo mutagenesis. *Crop Sci.* 45, 1837–1843.

Oelkers K., Goffard N., Weiller G.F., Gresshoff P.M., Mathesius U., Frickey T. (2008) Bioinformatic analysis of the CLE signaling peptide family. *BMC Plant Biology 8, 1, DOI:* 10.1186/1471-2229-8-1.

Oka-Kira E., Kawaguchi M. (2006) Long-distance signaling to control root nodule number. *Curr. Opinions Plant Biol.* 9, 496-502.

Oka-Kira E., Tateno K., Miura K., Haga T., Hayashi M., Harada K., Sato S., Tabata S., Shikazono N., Tanaka A., Watanabe Y., Fukuhara I., Nagata T., Kawaguchi M. (2005) klavier (klv), a novel hypernodulation mutant of Lotus japonicus affected in vascular tissue organization and floral induction. *Plant J.* 44, 505-515.

Okamoto S., Ohnishi E., Sato S., Takahashi H., Nakazono M., Tabata S., Kawaguchi M. (2009) Nod factor/nitrate-induced *CLE* genes that drive HAR1-mediated systemic regulation of nodulation. *Plant Cell Physiol.* 50, 67-77.

Oldroyd G.E., Downie J.A. (2008) Coordinating nodule morphogenesis with rhizobial infection in legumes. *Annu. Rev. Plant Biol.* 59, 519-546.

O'Neill D.P., Ross J.J. (2002) Auxin regulation of the gibberellin pathway in pea. *Plant Physiol.* 130, 1974-1982.

Op den Kamp R., Streng A., De Mita S., Cao Q.Q., Polone E., Liu W., Ammiraju J.S.S., Kudrna D., Wing R., Untergasser A., Bisseling T., Geurts R. (2011) LysM-type mycorrhizal receptor recruited for *Rhizobium* symbiosis in nonlegume *Parasponia*. *Science* 331, 909-912.

Osipova M.A., Mortier V., Demchenko K.N., Tsyganov V.E., Tikhonovich I.A., Lutova L.A., Dolgikh E.A., Goormachtig S. (2012) WUSCHEL-RELATED HOMEOBOX5 gene expression and interaction of CLE peptides with components of the systemic control add two pieces to the puzzle of autoregulation of nodulation. *Plant Physiol.* 158, 1329-1341.

Paau A.S., Leps W.T., Brill W.J. (1985) Regulation of nodulation by Rhizobium meliloti 102F15 on its mutant which forms an unusually high number of nodules on alfalfa. *Appl. Environ. Microb.* 50, 1118-1122.

Park S.J., Buttery B.R. (1988) Nodulation mutants of white bean (*Phaseolus vulgaris* L.) induced by ethyl-methane sulphonate. *Can. J. Plant Sci.* 68, 199-202.

Penmetsa R.V., Cook D.R. (1997) A legume ethylene-insensitive mutant hyperinfected by its rhizobial symbiont. *Science* 275, 527-530.

Penmetsa R.V., Frugoli J.A., Smith L.S., Long S.R., Cook D.R. (2003) Dual genetic pathways controlling nodule number in Medicago truncatula. *Plant Physiol.* 131, 998-1008.

Penmetsa R.V., Uribe P., Anderson J., Lichtenzveig J., Gish J.C., Nam Y.W., Engstrom E., Xu K., Sckisel G., Pereira M., Baek J.M., Lopez-Meyer M., Long S.R., Harrison M.J., Singh K.B., Kiss G.B., Cook D.R. (2008) The *Medicago truncatula* ortholog of Arabidopsis EIN2, sickle, is a negative regulator of symbiotic and pathogenic microbial associations. *Plant J.* 55, 580-595.

Peters N.K., Frost J.W., Long S.R. (1986) A plant flavone, luteolin, induces expression of Rhizobium meliloti nodulation gene. *Science* 233, 977-980.

Postma J.G., Jacobsen E., Feenstra W.J. (1988) Three pea mutants with an altered nodulation studied by genetic analysis and grafting. *J. Plant Physiol.* 132, 424-430.

Prell J., White J.P., Bourdes A., Bunnewell S., Bongaerts R.J., Poole P.S. (2009) Legumes regulate Rhizobium bacteroid development and persistence by the supply of branched-chain amino acids. *P. Natl. Acad. Sci. USA* 106, 12477-12482.

Quandt H.J., Pühler A., Broer I. (1993) Transgenic root nodules of Vicia hirsuta: a fast and efficient system for the study of gene expression in indeterminate-type nodules. *Mol. Plant-Microbe In.* 6, 699-706.

Radutoiu S., Madsen L.H., Madsen E.B., Felle H.H., Umehara Y., Gronlund M., Sato S., Nakamura Y., Tabata S., Sandal N., Stougaard J. (2003) Plant recognition of symbiotic bacteria requires two LysM receptor-like kinases. *Nature* 425, 585-592.

Reid D.E., Ferguson B.J., Hayashi S., Lin Y.H., Gresshoff P.M. (2011) Molecular mechanisms controlling legume autoregulation of nodulation. *Ann. Bot.* 108, 789-795.

Riely B.K., Lougnon G., Ané J.M., Cook D.R. (2007) The symbiotic ion channel homolog DMI1 is localized in the nuclear membrane of Medicago truncatula roots. *Plant J.* 49, 208-216.

Ruffel S., Freixes S., Balzergue S., Tillard P., Jeudy C., Martin-Magniette M.L., Van der Merwe M.J., Kakar K., Gouzy J., Fernie A.R., Udvardi M., Salon C., Gojon A., Lepetit M. (2008) Systemic signaling of the plant nitrogen status triggers specific transcriptome responses depending on the nitrogen source in Medicago truncatula. *Plant Physiol.* 146, 2020–2035.

Sagan M., Duc G. (1996) *Sym28* and *Sym29*, two new genes involved in regulation of nodulation in pea (Pisum sativum L). *Symbiosis* 20, 229-245.

Sagan M., Gresshoff P.M. (1996) Developmental mapping of nodulation events in pea (Pisum sativum L.) using supernodulating plant genotypes and bacterial variability reveals both plant and Rhizobium control of nodulation regulation. *Plant Sci.* 117, 167-179.

Schauser L., Roussis A., Stiller J., Stougaard J. (1999) A plant regulator controlling development of symbiotic root nodules. *Nature* 402, 191-195.

Schnabel E., Journet E.P., De Carvalho-Niebel F., Duc G., Frugoli J. (2005) The Medicago truncatula SUNN gene encodes a CLV1-like leucine-rich repeat receptor kinase that regulates nodule number and root length. *Plant Mol. Biol.* 58, 809-822.

Schnabel E., Kassaw T., Smith L., Marsh J., Oldroyd G., Long S., Frugoli J. (2011) *ROOT DETERMINED NODULATION 1* regulates nodule number in *M. truncatula* and defines a highly conserved, uncharacterized plant gene family. *Plant Physiol.* 157, 328-340.

Schnabel E., Mukherjee A., Smith L., Kassaw T., Long S.R., Frugoli J. (2010) The lss supernodulation mutant of Medicago truncatula reduces expression of the *SUNN* gene. *Plant Physiol.* 154, 1390-1402.

Searle I.R., Men A.E., Laniya T.S., Buzas D.M., Iturbe-Ormaetxe I., Carroll B.J., Gresshoff P.M. (2003) Long-distance signaling in nodulation directed by a CLAVATA1-like receptor kinase. *Science* 299, 109-112.

Seo H.S., Li J., Lee S.Y., Yu J.W., Kim K.H., Lee S.H., Lee I.J., Paek N.C. (2007) The hypernodulating nts mutation induces jasmonate synthetic pathway in soybean leaves. *Mol. Cells* 31, 185-193.

Sidorova K.K., Shumnyi V.K. (1998) Analysis of pea (*Pisum sativum* L.) supernodulating mutants. *Genetika* 34, 1452-1454.

Sidorova K.K., Shumnyi V.K. (2003) A collection of symbiotic mutants in pea Pisum sativum L.: Creation and genetic study. *Russian J. Genet.* 39, 406-413.

Sinjushin A.A., Konovalov F.A., Gostimskii S.A. (2008) *Sym28*, a gene controlling stem architecture and nodule number, is localized on linkage group V. *Pisum Genet.* 40, 15-18.

Smit P., Raedts J., Portyanko V., Debelle F., Gough C., Bisseling T., Geurts R. (2005) NSP1 of the GRAS protein family is essential for rhizobial Nod factor-induced transcription. *Science* 308, 1789-1791.

Spaink H.P. (2000) Root nodulation and infection factors produced by rhizobial bacteria. *Annu. Rev. Microbiol.* 54, 257-288.

Sprent J.I., James E.K. (2007) Legume evolution: where do nodules and mycorrhizas fit in? *Plant Physiol.* 144, 575-581.

Staehelin C., Xie Z.-P., Illana A., Vierheilig H. (2011) Long-distance transport of signals during symbiosis. Are nodule formation and mycorrhization autoregulated in a similar way? *Plant Signaling Behav.* 6, 372-377.

Stracke S., Kistner C., Yoshida S., Mulder L., Sato S., Kaneko T., Tabata S, Sandal N., Stougaard J., Szczyglowski K., Parniske M. (2002) A plant receptor-like kinase required for both bacterial and fungal symbiosis. *Nature* 417, 959-962.

Streeter J. (1988) Inhibition of legume nodule formation and N_2 fixation by nitrate. *CRC Cr. Rev. Plant Sci.* 7, 1-23.

Studer D., Gloudemans T., Franssen H.J., Fischer H.-M., Bisseling T., Hennecke H. (1987) Involvement of the bacterial nitrogen fixation regulatory gene (*nifA*) in control of nodule-specific host-plant gene expression. *Eur. J. Cell Biol.* 45, 177-184.

Teillet A., Garcia J., De Billy F., Gherardi M., Huguet T., Barker D.G., De Carvalho-Niebel F., Journet E.-P. (2008) api, a novel *Medicago truncatula* symbiotic mutant impaired in nodule primordium invasion. *Mol. Plant-Microbe In.* 21, 535-546.

Tirichine L., Sandal N., Madsen L.H., Radutoiu S., Albrektsen A.S., Sato S., Asamizu E., Tabata S., Stougaard J. (2007) A gain-of-function mutation in a cytokinin receptor triggers spontaneous root nodule organogenesis. *Science* 315, 104-107.

Tsyganov V.E., Voroshilova V.A., Priefer U.B., Borisov A.Y., Tikhonovich I.A. (2002) Genetic dissection of the initiation of the infection process and nodule tissue development in *the Rhizobium-pea (Pisum sativum L.)* symbiosis. *Ann. Bot.* 89, 357-366.

Turnbull C.G.N., Booker J.P., Leyser H.M.O. (2002) Micrografting techniques for testing long-distance signalling in *Arabidopsis. Plant J.* 32, 255-262.

Van de Wiel C., Scheres B., Franssen H., Van Lierop M.-J., Van Lammeren A., Van Kammen A., Bisseling T. (1990) The early nodulin transcript ENOD2 is located in the nodule parenchyma (inner cortex) of pea and soybean root nodules. *EMBO Journal* 9, 1-7.

Van Noorden G.E., Ross J.J., Reid J.B., Rolfe B.G., Mathesius U. (2006) Defective long-distance auxin transport regulation in the Medicago truncatula super numeric nodules mutant. *Plant Physiol.* 140, 1494–1506.

Vasse J., De Billy F., Truchet G. (1993) Abortion of infection during the Rhizobium meliloti-alfalfa symbiotic interaction is accompanied by a hypersensitive reaction. *Plant J.* 4, 555-566.

Veereshlingam H., Haynes J.G., Penmetsa R.V., Cook D.R., Sherrier D.J., Dickstein R. (2004) nip, a symbiotic Medicago truncatula mutant that forms root nodules with aberrant infection threads and plant defense-like response. *Plant Physiol.* 136, 3692-3702.

Vernié T., Moreau S., De Billy F., Plet J., Combier J.P., Rogers C., Oldroyd G., Frugier F., Niebel A., Gamas P. (2008) EFD is an ERF transcription factor involved in the control of nodule number and differentiation in *Medicago truncatula. Plant Cell* 20, 2696-2713.

Voroshilova V.A., Demchenko K.N., Brewin N.J., Borisov A.Y., Tikhonovich I.A. (2009) Initiation of a legume nodule with an indeterminate meristem involves proliferating host cells that harbour infection threads. *New Phytol.* 181, 913-923.

Wasson A.P., Pellerone F.I., Mathesius U. (2006) Silencing the flavonoid pathway in Medicago truncatula inhibits root nodule formation and prevents auxin transport regulation by rhizobia. *Plant Cell* 18, 1617-1629.

White F.F., Taylor B.H., Huffman G.A., Gordon M.P., Nester E.W. (1985) Molecular and genetic analysis of the transferred DNA regions of the root-inducing plasmid of *Agrobacterium rhizogenes. J. Bacteriol.* 164, 33-44.

Willems A. (2006) The taxonomy of rhizobia: an overview. *Plant Soil* 287, 3-14.

Wopereis J., Pajuelo E., Dazzo F.B., Jiang Q.Y., Gresshoff P.M., de Bruijn F.J., Stougaard J., Szczyglowski K. (2000) Short root mutant of Lotus japonicus with a dramatically altered symbiotic phenotype. *Plant J.* 23, 97-114.

Yokota K., Li Y.Y., Hisatomi M., Wang Y.X., Ishikawa K., Liu C.T., Suzuki S., Aonuma K., Aono T., Nakamoto T., Oyaizu H. (2009) Root-determined hypernodulation mutant of Lotus japonicus shows high-yielding characteristics. *Biosci. Biotech. Bioch.* 73, 1690-1692.

Yoshida C., Funayama-Noguchi S., Kawaguchi M. (2010) plenty, a novel hypernodulation mutant in *Lotus japonicus. Plant Cell Physiol.* 51, 1425-1435.

In: Symbiosis: Evolution, Biology and Ecological Effects   ISBN: 978-1-62257-211-3
Editors: A. F. Camisão and C. C. Pedroso   © 2013 Nova Science Publishers, Inc

Chapter 3

LEGUMES: PROPERTIES AND SYMBIOSIS

Corina Carranca[*]

Instituto Nacional de Investigação Agrária, Oeiras, Portugal
CEER, Instituto Superior de Agronomia, Lisboa, Portugal

Abstract

Improved nitrogen (N) management is needed to optimize economic returns to farmers and minimize environmental concerns associated with N use. Symbiotically N2 fixed is of particular significance in sustainable agriculture as it allows reducing the use of chemical N in the production of field crops. The legume-bacteria symbiosis is the most important N2 fixing system. This relationship is between bacteria collectively called Rhizobium and annual or perennial leguminous plants. In general, the rhizobia infect the plant via the root hair and invade the tissues. The cells are infected intracellular, divided, and enlarged to produce the characteristic nodules (determinate or indeterminate). The importance of this relationship relies in the ability of the rhizobia to fix (reduce) the atmospheric N2 to the host-plant, which in turn provides soluble carbohydrates to the microorganisms. Nitrogen fixation depends on legume performance, environmental conditions, soil nutrients availability, contaminants, bacteria abundance and diversity in soil, and bacteria specificity and infectivity for the legume. Under favourable conditions, legumes fix more than 50% of plant N.

Keywords: determinate nodules and legumes, factors affecting fixation, indeterminate nodules and legumes, pattern of fixation, *Rhizobium*

1. Introduction

Though nitrogen (N) is abundant in nature it is only directly available to plants when converted through biological or industrial processes to certain forms, primarily as ammonium (NH_4^+), followed by nitrate (NO_3^-). Some N is made available to plants through the decay of existing soil organic matter (endogenous OM), but generally this process is not sufficient to replenish N required by the crops. Adequate replacement of plant available N in the soil is

---

[*] E-mail address: c.carranca.ean©clix.pt.

mostly accomplished by applying fertilizers (inorganic or organic fertilizers, including manure, industrial and urban waste, crop residues, etc.).

The goal of reducing chemical fertilizers usage will be to this century what the goal of reducing pesticides was to the last century. Legume cover crops or crop residues are important tools for this purpose supplementing the soil with N. Symbiotically fixed N2 by legumes is of particular significance in sustainable agriculture. Sustainability is defined here as the agricultural production where natural resources are sustained and restored, and agricultural productivity is more than just the basic needs of subsistence, although what is an acceptable or optimal degree of productivity is difficult to generalise across environments and farming systems. Symbiotic N2 fixation contributes to productivity both directly, where the fixed N2 is harvested in grain or other plant material, or indirectly by contributing to the maintenance or enhancement of soil fertility in the agricultural system.

Since atmospheric N2 is stable in atmosphere and is a renewable resource, symbiotic fixation in agricultural systems is a sustainable N source. In contrast to the large amounts of fossil energy required to produce mineral N fertilizers, the energy derived from N2 fixation is free and derived from photosynthesis. Symbiosis is then considered as an economically friendly approach to supplying N to the agro-ecosystems, and is particularly recommended for organic farming systems, where chemical fertilizers (especially N) are forbidden. N2 fixers are then called biofertilizers.

There are several benefits a farmer can get from N_2 fixation by legumes. The most obvious is the reduced need for fertilizer application (specifically inorganic N). Also, fixed N_2 by bacteria is not as quickly leached from the soil as is the applied mineral N, resulting in a more stable soil N pool. By reducing mineral fertilizer N inputs, legumes reduce the costs of production and the potential for environmental N contamination.

1.1. Legumes

Leguminous plants are of the *Leguminosae* Family or, more recently *Fabaceae* (Table 1). Include annual or permanent crops, either as herbaceous or forest trees [e.g. acacia (*Acacia*), false acacia (*Robinia pseudoacacia* L.), *Dalbergia* L., *Sophora* L.]. They are included in three sub-families: *Faboideae,* which includes four hundred and thirty species [e.g. soybean (*Glycine max* L.), pea (*Pisum sativum* L.), fababean (*Vicia faba* L.), bean (*Phaseolus* sp.), alfalfa (*Medicago sativa* L.), lentil (*Lens culinaris* Medikus), lupine (*Lupinus* sp.), chickpea (*Cicer arietinum* L.)], *Mimosoideae* [e.g. mimosa (*Mimosa* L.), acacia], and *Caesalpinioideae* [e.g. carob tree (*Ceratonia siliqua* L.)].

The more important cultivated species are included in the *Fabaceae* Family. For consumption, legumes are mainly produced as arable crops by their protein content and oilseed and cultivated as pasture or forage crops (Almeida 2006). Edible seeds of legumes are known as pulses (Latin word *puls* means thick soup). Grain has a high nutritional value (protein, minerals and vitamins). The same quality is true for forage and pasture legumes.

**Table 1. Taxonomic framework for leguminous plants
(Sources: Almeida 2006; USDA classification)**

Kingdom	*Plantae*
Subkingdom	*Tracheobionta* (vascular plants)
Superdivision	*Spermatophyta* (seed plants)
Division	*Magnoliophyta*
Class	*Magnoliopsida*
Subclass	*Rosidae*
Order	*Fabales*
Family	*Fabaceae*
Sub-family	*Faboideae, Mimosoideae, Caesalpinoideae*

Pulses are annual leguminous crops (peas, lentils, fababeans, soybeans, chickpeas, beans, lupine) yielding from one to twelve grains or seeds of variable size, shape and color within a pod. These pods may have dehiscent or indehiscent fruits.

Peas and beans can produce 4-5 t grain ha^{-1}. Green beans (*P. vulgaris* L.) are podded legumes that can yield 8-11 t pod ha^{-1} (larger podded varieties can yield 18-20 t pod ha^{-1}) (Davies 1997).

2. Symbiotic N$_2$ Fixation

Leguminous plants are grown as pulses for grain or as pastures or forage in agro-forestry or in natural ecosystems, and provide the major N input into the soil as a result of their ability to convert atmospheric N$_2$ to an N form that can be assimilated by the plant. According to Almeida (2006), 97% of *Faboideae*, 90% of *Mimosoideae* and 23% of *Caesalpinioideae* sub-families are supposed to establish symbiotic relationships with a N$_2$ fixing soil bacteria. The growth of legumes is then partially free of the constraint of sufficient amounts of N in the soil due to their general ability to live in association with the soil rhizobia, capable of fixing N$_2$ from the air. This process, called symbiotic N$_2$ fixation is a mutually beneficial association between a prokaryotic N$_2$ fixing microorganism (e.g. *Rhizobium*, *Nostoc* and *Frankia*) and a eukaryotic, usually a photosynthetic host-plant (legume) (Table 2).

Rhizobia are saprophytic bacteria of the *Rhizobiacae* Family, mostly of the Genera *Rhizobium*, *Mesorhizobium*, *Sinorhizobium* and *Bradyrhzobium* and are capable of fixing atmospheric N$_2$ in symbiosis with the *Fabaceae* legumes. They represent only a small fraction of the soil microflora.

The rhizobia strains ability to form nodules with a wide range of hosts (legumes) may contribute to their persistence in soil. Nodulation in legumes is mostly a highly specific and complex process, but legumes can also be very promiscuous [e.g. beans, peanuts (*Arachis hypogaea* L.)], i.e. they can form nodules with several bacteria strains (Júnior and Reis 2008). Fixation by promiscuous legumes or bacteria normally is less efficient.

Nitrogen fixation follows several pathways, but generally rhizobia infect the plant via the root hair. Few plants can be infected via the stem, most in the *Papilionoideae*, such as *Aeschynomene*, *Discolobium* and *Sesbania* species.

Table 2. Some examples of *Fabaceae* legumes and the respective specific rhizobia strains (Sources: Amarger 2001; Almeida 2006)

Principal host-legumes	*Rhizobium* strain	biovar
Trifolium (clover)	*R. leguminosarum*	trifolii
Pisum, Vicia, Lens, Lathyrus (peas, faba, lentil)	*R. leguminosarum*	viciae
Phaseolus (bean)	*R. leguminosarum*	phaseoli
Lupinus sp. (lupine)	*R. lupini, Bradyrhizobium*	-
Glycine max L. (soybean)	*Bradyrhizobium*	japonicum
Medicago, Melilotus, Trigonella (alfalfa, clover)	*Sinorhizobium*	meliloti
Cicer arietinum L. (chickpea)	*Mesorhizobium*	ciceri
Lotus (lotus)	*Mesorhizobium*	loti
Acacia (acacia)	*Sinorhizobium*	acacia
Sesbania	*Sinorhizobium*	sesbaniae

These plants grow in tropical wetlands such as the Pantanal in Brazil, as well as in other seasonally-flooded regions of the tropics (e.g. Amazon and Orinoco basins) and flooded rice fields. Nodulation of their stems in a flooded environment allows these legumes to continue symbiotic N2 fixation in wet conditions that would normally prevent it in non-hydrophytic legumes, as both nodulation and nodule functioning are very much dependent on a continued supply of oxygen (O2) to support the aerobic respiration required by this highly energetically costly process. The nodules formed on the aerial stems of these plants are slightly unusual in developmental and structural terms and do not fit easily into the determinate nodules.

A sustained molecular dialogue between both partners during the infection process is a prerequisite for a successful symbiosis. In general, rhizobia bind to immature root hairs. During the initial step of the interaction, the presence of plant root-secreted flavonoids (specific phenolic compounds) attracts and stimulates bacteria growth around the rhizosphere and induces the expression of rhizobia nodulation (Nod genes) (symplasmid gene). The flavonoids can be recognized by several bacteria, or by very few. The Nod genes encode enzymes involved in the synthesis and excretion of lipochitooligosaccharide molecules (signals), also termed Nod factors, which are powerful growth hormones. These phytohormones act like auxins, signal molecules that activate several physiological and morphological alterations in the roots (hair curling and deformation) at the point of attachment (Xi et al. 2000; Trevaskis et al. 2002; Júnior and Reis 2008). Successful infections may be visible in about five days afterwards. Enclosed within the plant-derived infection thread, bacteria move down the root hair in the direction of the root cortex. A rapid multiplication of rhizobia within the infected cells occurs and the cells enlarge, divide and produce the characteristic nodules (Figure 1).

All Rhizobium species analyzed so far produce a family of structurally related lipochitooligosaccharide derivative. These Nod factors are major host-range determinants (Schultze et al. 1995; Reddy et al. 1998; Xi et al. 2000; Júnior and Reis 2008). Alone they can induce many of the changes associated with nodulation. They differ substantially in their activity towards host or non-host plants, and this depends on the length of the oligosaccharide chain, the nature of the fatty acyl substituent, or the presence or absence of the 0-acetyl group (Schultze et al. 1995). It is possible that *Nod* factors of low activity compete with active ones and may thus influence the efficiency of nodule induction. Therefore, it is important to know which parameters determine, for example, the length of the oligosaccharide chain and the efficiency of each type of chemical substitution.

These *Nod* signals are major determinants of host specificity, i.e., each *Rhizobium* species produces characteristic *Nod* factors carrying a combination of structural modifications not found in other species or strains. The key for a specific or promiscuous symbiosis results from the recognition of both molecules (*Nod* genes by legumes and *Nod* factors by bacteria) and the expression of genes that regulate its synthesis (Júnior and Reis 2008). Rhizobia can thus be characterized by the range of hosts they can nodulate (Amarger 2001). Some bacteria appear highly specific and nodulate only plants belonging to a single Genus or to some species within a Genus. For instance, rhizobia isolated from European species of clovers will not nodulate clover species native from Africa. Other rhizobia are symbionts of multiple species (promiscuous).

Some bradyrhizobia do not produce *Nod* factors, and a different dialog with the host-legume is probably established (Reddy et al. 1998).

During the root infection, bacteria are released from the unwalled root hair of the infection thread in the plant cell cytoplasm by endocytosis of the membrane surrounding the infection thread. As a result, bacteria are separated from the host cell cytoplasm by being enclosed in a membrane, and mediate nutrient and signal exchanges between the partners. This new sub-cellular compartment is named a symbiosome or peribacteroid (Xi et al. 2000).

a)

b)

Figure 1. Part of *a)* fababean (*Vicia faba* L.) and *b)* soybean (*Glycine max* L.) roots systems bearing nodules of *Rhizobium* strains, respectively.

The generation and maintenance of this organelle-like compartment is an essential process of the coexistence between rhizobia and living plant cells, protecting the bacteria from the defence response of the host. Within these compartments bacteria divide and differentiate into N_2 fixing bacteroids, which are about forty-two times larger than the original bacteria, change their form into rod and may fill the cell (Trevaskis et al. 2002). Only at this moment the N_2 fixation can begin (in general, eight to fifteen days after nodules are visible). Isolated bacteroids contain all the metabolic machinery, including the enzyme nitrogenase, which is required for fixation. Nitrogenase consists of two proteins [an iron (Fe) protein and molybdenum (Mo)-Fe protein].

The host-plant supplies the nutrients, such as carbohydrates (malate, succinate) to the bacteroids with which they synthesize the large amounts of ATP needed to convert the atmospheric N_2 into ammonia (NH_3). Sucrose formed by photosynthesis in the leaves (cytosol) is translocated to the root system where is converted to malate and succinate to provide carbon (C) skeletons and energy for symbiosis (N_2 fixation) and assimilation of NH_3 within the infected nodule tissue through the glutamine synthetase to form amides (such as aspargine in beans or soybeans) or purines. These are converted to ureides, allantoin (in lupine, pea, clover, alfalfa, fababaean) and allantoic acid in nodules and exported in the xylem sap to the shoots to be degraded and produce NH_4^+. This NH_4^+ is assimilated to produce aminoacids and other organic compounds (De Varennes 2003; Hardarson and Atkins 2003; Fotolli et al. 2011).

To produce ATP (by cellular respiration), the bacteroids need O_2. However, nitrogenase is strongly inhibited by O_2. To help the bacteroids, nodules are filled with hemoglobin (a freshly-cut nodule has a red colour when is active). The hemoglobin in the nodule (leghemoglobin, which is a protein containing Fe in the legume) continuously transports just the right amount of O_2 from the peribacteroid to the bacteroids to satisfy their conflicting requirements (Bergersen and Turner 1975; De Varennes 2003). If the nodule is white or greenish brown, either the symbiosis is ineffective or the nodule is undergoing breakdown, and is said to be senescing.

Symbiotic fixation can thus be represented by the following equation (eq. 1), in which one mole of N_2 produces two moles of NH_3 at the expense of sixteen moles of ATP and a supply of electrons and protons, and is mediated by the nitrogenase enzyme:

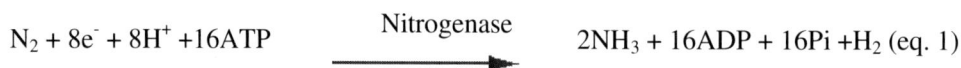

$$N_2 + 8e^- + 8H^+ + 16ATP \xrightarrow{\text{Nitrogenase}} 2NH_3 + 16ADP + 16Pi + H_2 \text{ (eq. 1)}$$

A minimum of 10 moles of ATP can be used in highly efficient symbiosis, and under stress conditions the number of ATP moles required to reduce one mole of N2 can vary from 20 to 30.

This reaction (eq. 1) occurs when N_2 is bound to the nitrogenase enzyme complex. The Fe protein is first reduced by electrons donated by ferredoxin. The reduced Fe protein then binds ATP and reduces the Mo-Fe protein, which donates electrons to N_2 producing HN=NH. In two further cycles of this process, HN=NH is reduced to H_2N-NH_2, and this in turn is reduced to $2NH_3$. The reduced ferredoxin which supplies electrons for this process is generated by photosynthesis.

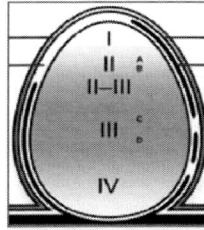

Figure 2. Diagram illustrating the different zones of an indeterminate root nodule (adapted from Trevaskis et al. 2002).

Fixation has high costs for the plant: direct costs for the energy production, costs for reduction of N_2 into NH_3, and for protein synthesis and transport to the aerial part, and indirect costs for nodules formation and maintaining. Estimated costs are of the order of 2.4–7.0 g C g^{-1} of N_2 fixed (Saari and Ludden 1987).

Few *Papilionoideae* tropical legumes, such as soybean, *Phaseolus* or *Lotus japonicus* form determinate nodules, which lose meristematic activity shortly after initiation and are spherical in shape when mature. The cells containing the newly-released rhizobia divide to form the functional N_2 fixing nodule. The infected region of determinate nodule contains infected host cells that are all at a similar developmental stage (Trevaskis et al. 2002).

The indeterminate nodules are found in temperate legumes as pea, clovers and medic (*Medicago truncatula* Gaertn.) plants and have an elongated shape, with a pronounced meristematic region. These nodules increase in length over the growing season, maintaining active the apical meristem. According to Trevaskis et al. (2002), the central tissue of indeterminate nodules is divided into several zones, differing in their degree of cell differentiation because these nodules maintain an active growth (Figure 2): (I) the meristematic zone or active meristem; (II) the pre-fixation or invasion zone, where cell division is arrested and cells differentiate and one invaded; (III) the N_2 fixation zone where plant cells and bacteria are terminally differentiated for N_2 fixation; and (IV) the senescent zone, closest to the root, which contains degenerated plant cells and bacteroids. The breakdown of the heme component of leghemoglobin results in a visible greening at the base of the nodule.

When the legume plant matures and dies, nodules on the root system decompose and release the rhizobia into the soil, particularly from indeterminate nodules (the senescence of determinate nodules occurs earlier in the growing cycle). If the same legume species is planted again the following year, sufficient number of rhizobia is usually present to provide good nodulation.

The extent to which legumes fix N_2 and factors that limit this fixation capacity along with the availability of fixed N_2 to enhance soil fertility are critical components in exploiting the utility of legumes in sustainable farming systems. Rates of symbiotic fixation in legumes vary with host, microsymbiont and environmental factors, but rates as high as 650 kg N ha^{-1} $year^{-1}$ have been reported in temperate clover pastures and alfalfa in Argentina. Under favourable conditions, grain legumes may fix from 65 to 450 kg N ha^{-1} ($year^{-1}$), corresponding to 60-98% of plant N (Table 3).

In a Mediterranean area, fixed N_2 at pod-filling by white and yellow lupine (*Lupinus albus* L. and *L. luteus* L., respectively) with native rizobia was greater than 100 kg N ha^{-1}

(year^{-1}) and corresponded to 100% fixation (Carranca et al. 2009a,b). At maturity, fixation by these lupines almost replenished the crop N requirements (93-98%). Inoculated soybean in Brazil also depends about 100% on N_2 fixation.

Table 3. Rates of fixation (%) by different legumes under rainfed conditions in Portugal, in sub-humid climate in Chile, and in rainfed and irrigated areas in eastern Australia

Legume	Location	Soil type	Rate of fixation (%)	Sources
Pea[a] (*Pisum sativum* L.)	Central Portugal	Ortic Luvisols[d, e]	20*-65	Carranca et al. (1999a)
Pea (*P. sativum* L.)	Eastern Australia	Grey cracking clay; Red brown earth	98	Carranca et al. (2008, 2009d)
Fababean (*Vicia faba* L.)	Central Portugal	Ortic Luvisols[d, e]	60-70	Carranca et al. (1999a)
Fababean (*V. faba* L.)	Eastern Australia	Grey craking clay; Red duplex; Red clay	46-78	Peoples et al. (2001)
Green bean (*Phaseolus vulgaris* L.)	Central Portugal	Fluvisol[d, e]	ns	Carranca (1996)
Chickpea[a] (*Cicer arietinum* L.)	Central Portugal	Ortic Luvisols[d, e]	45*-80	Carranca et al. (1999a)
Chickpea (*C. arietinum* L.)	Eastern Australia	Grey craking clay; Red duplex; Red clay	19-53	Peoples et al. (2001)
Soybean[b] (*Glycine max* L.)	Central Portugal	Ortic Luvisols[d, e]	25	Carranca (1996)
White lupine (*Lupinus albus* L.)	Central Portugal	Ortic Podzol[d]	93	Carranca et al. (2009a)
Yellow lupine (*L. luteus* L.)	Western Portugal	Eutric Regosols[e]	98	Carranca et al. (2008, 2009b)
Narrow-leafed lupine (*L. angustifolius* L.)	Eastern Australia	Yellow duplex	60-80	Peoples et al. (2001)
Sub-clover (*Trifolium subterraneum* L.)	Chile	Fluvisol[d, e]	74-91	Ovalle et al. (2006)
Sub-clover (*T. subterraneum* L.)	Chile	Fluvisol[d, e]	74-91	Ovalle et al. (2006)
Subclover[c] (*T. subterraneum* L.) (mixed sward)	South Portugal	Litosols[d]	35-50	Carranca et al. (1999b)
Sub-clover[c] (*T. subterraneum* L.) (mixed sward)	Eastern Australia	Grey craking clay; Red brown earth	87	Peoples et al. (2001)
Perennial medics (*Medicago sativa* L.)	Eastern Australia	Grey cracking clay; Red brown earth; Red clay	25-68	Peoples et al. (2001)
Annual medics (*M. sativa* L.)	Eastern Australia	Grey craking clay	85	Peoples et al. (2001)
Annual medics (*M. polymorpha* L.)	Chile	Fluvisol[d, e]	79	Ovalle et al. (2006)
Serradella (*Ornythopus compressus*)	Chile	Fluvisol[d, e]	79	Ovalle et al. (2006)
Serradella (*O. sativus* Brot.)	Chile	Fluvisol[d, e]	87	Ovalle et al. (2006)

[a]= inoculum from ICARDA;

[b]= inoculation with *Bradyrhizobium japonicum* from ICARDA;

[c]= inoculation with *Rhizobium leguminosarum* L. bv. trifolii;

ns= not significant;

[d] FAO (1971-1981);

[e]= IUSS Working Group WRB (2006);

*= under drought stress.

Rates of N_2 fixation are directly related to legume growth rate. Nitrogen fixation is closely linked to legume biomass production in pulse, pasture and forage crops. On average, 20-25 kg of shoot N was fixed for average ton of legume shoot dry matter (DM) accumulated in Eastern Australia (Peoples et al. 2001), whereas 16 and 5-11 kg N_2 fixed t^{-1} DM were estimated for pulses and subclover, respectively produced under rainfed conditions in Mediterranean regions (Carranca et al. 1999a,b). A recommended generic value of 20 kg N_2 fixed t^{-1} legume DM was proposed as a means of predicting the amount of shoot N_2 fixed in different legume-based rotations around the world, particularly pulses (Peoples et al. 2001). As a consequence, any factor that reduces plant growth has the potential to alter the pattern and frequency of nodulation and fixation. Generally, unfavourable climatic or edaphic factors that restrict plant growth are likely to have negative impacts on the potential to fix N_2. The factors involved are diverse and include mostly an inadequate soil fertility, temperature, light intensity, drought or water-logging, soil contaminants, and plant pests or diseases as indirect factors.

Environmental factors will often influence rhizobia survival and diversity in soil. A favourable rhizosphere is vital for the symbiosis, but the magnitude of stress effects and the rate of inhibition of this association usually depend on the plant growth phase and development, as well as the severity of the stress.

2.1. Pattern of Symbiotic N_2 Fixation

Before *rhizobia* form nodules in legume roots, bacteria must be attracted to the roots through a bidirectional host–bacteria communication (signaling), as described in a previous section. Not all legumes show high fixing capacity for the entire growth period. Some plants form nodules during the vegetative period and have a maximum N_2 fixing rate at flowering or pod-filling stage, declining sharply thereafter [pea, chickpea, soybean, clover (*Trifolium* sp.)]. Others, such as white and yellow lupine and fababean form nodules immediately after plant germination and during the growth cycle, and show a high N_2 fixing rate for the whole growth cycle (Carranca et al. 1999a, 2008, 2009a,b).

In determinate peas, Jensen (1987) reported that more than 50% of pea nodules can be destroyed and died, and roots can stop to grow before flowering, showing that N was mobilized from vegetative organs to seeds, with a reduction in the amount of photosynthates translocated to nodules.

Nitrogen fixation in chickpea declines later than peas, with the onset of pod-filling (Kyei-Boahen et al. 2002). This apparent decline in N_2 fixation has been attributed partly to nodule senescence after flowering, despite the increase in nodule weight, acetylene reduction activity, leghemoglobin and plant protein content. This suggests that although nodules continued to grow due to indeterminate growth habit of chickpea nodules, the growing zones could not sustain the N_2 fixation levels indicating the decline in N_2 fixation during the pod-fill (Kyei-Boahen et al. 2002). Some researchers have also linked this decline to the carbohydrate deprivation hypothesis, which attributes the decrease in nodule function to a diminished supply of photosynthate to the nodules.

The decline in N_2 fixation during the reproductive phase has been associated with the development of pods as a competing sink, therefore limiting carbohydrate availability to the root nodules. But this explanation does not seem generally feasible since it is not applied to fababean or lupine.

Fababean and lupine also appear to have an excellent cool-temperature tolerance for nodule growth and function (Boote et al. 2001; Carranca et al. 2009a). The adaptation of both host and bacteria genotypes to stress during the early symbiotic interaction is a possible explanation for the nodule growth and function from the early stages (Ampomah et al. 2008).

2.2. Constraints for Symbiotic N_2 Fixation

Nitrogen fixation depends on bacteria abundance and diversity in soil, bacteria specificity and infectivity for the host-plant, legume performance, environmental conditions (soil type and climate), cultural practices and nutrients availability.

2.2.1. Rhizobium Abundance and Diversity, and Specificity and Infectivity

Not all legumes form nodules and those that can form may only be infected by specific groups of rhizobia. No rhizobia nodulate more than 1% of the known leguminous species, though some rhizobia are promiscuous and nodulate hosts from a number of different species and Genera of legumes (e.g. *R. tropici* strain which nodulates *Leucaena*, *Phaseolus*, etc.). Because of the specificity of the interaction between Nod factor and the receptor on the legume, some rhizobia strains will infect only pea, some only clover, others only lucerne (alfalfa or medic), etc. The determinants of competitiveness for nodule occupancy by rhizobia are not fully understood. However, some factors such as genotype of both the host and the competing bacteria and the adaptation to stress during the early symbiotic interactions have been shown to influence the attachment (Ampomah et al. 2008).

A legume growing in its native habitat is likely to have the appropriate bacteria present in the soil. If nodules are present on the roots, and if the nodules are pink or red inside, then an appropriate bacteria strain is living in the soil and actively fixing N_2 in symbiosis with this legume. If the legume to be planted has not been grown previously in the area, if the legume is to be planted on land currently planted with non-leguminous crops, if the legume to be planted is a species widely different from those which were grown previously on the land or if the legume grown recently failed a good nodulation, the efficient N_2 fixing bacteria are not present in the soil and it is necessary to introduce them into the soil where legumes are to be grown, or in the seeds by the inoculation process.

The success of the symbiosis depends on the presence of highly infective (ability of a rhizobia to form nodules with a particular legume) and effective (ability of a rhizobia to fix N_2) bacteria in the soil to the specific host-plant. When the indigenous bacteria in the soil are not highly efficient with the present host, inoculation with a suitable strain of the *Rhizobium* species at sowing is the most useful agronomic practice to ensure the maximum legume yield with a high rate of plant N accumulation.

In the inoculation process, rhizobia can be provided in the seeds (coating) or directly into the soil. Rhizobia inoculants remain relatively immobile in the soil and, as a consequence, application on or with the seed at sowing results predominantly in crown and taproot nodulation. Mixing bacteria with the soil after application of inoculants in the seeding row leads to more profuse nodulation on most parts of the root system, including lateral roots (Hardarson and Atkins 2003). I

In case of inoculation, poor nodulation may occur even if good inoculation practices were used. The bacteria introduced must be able to adapt to the actual soil conditions, multiply in the rhizosphere, and compete with the native rhizobia species for the infection sites. Such an introduction depends on local conditions, but bacteria often become persistent (Amarger 2001).

Kyei-Boahen et al. (2002) found that the application of inoculants to the soil in the granular form (or spread in the row in the liquid form) produce more nodules in the lateral roots and in the lower part of the root system. Such nodules die latterly. The inoculants applied to seeds in the liquid or peat form produce mostly nodules in the crown region. These nodules are formed first during the vegetative growth than the nodules in the lateral roots and lower root system. Nodules in the crown region are in a layer of soil which is subject to great fluctuation in both temperature and moisture during pod-filling which can accelerate the rate of nodule senescence in the crown region relative to nodules located on lateral roots. Lateral root nodules are significant contributors to N_2 fixation, particularly during the later stages of plant growth especially the pod-fill. According to Hardarson et al. (1989), the position of the nodules in the soybean root system rather than the number or fresh weight (FW) of nodules plays a greater role in influencing the amount of N_2 fixed.

Criteria to choose the proper inoculants include survival and mobility of bacteria in soil, colonization of the rhizosphere, the ability to compete for rapid infection of the host root system, as well as to provide sufficient effective nodules to improve the fixation capacity of the legumes. Inoculation should be done in shade, within one hour before seeding, and inoculated seeds should not be exposed directly to sunlight to avoid killing the N_2 fixing bacteria. Rhizobia bacteria begin dying as soon as the inoculated seeds are sown. The longer the seed lies in the soil before germination, the fewer viable rhizobia bacteria are present. Inoculant containing a sticker or that is coated on the seed provides more protection for the bacteria (namely for low soil pH) and improves its survival to about three weeks.

To maximize symbiotic fixation one must select for both the best *Rhizobium* strain (very effective exotic strain) and the best host-cultivar. Although the arguments to suggest that inoculation is the approach most likely to result in maximising N_2 fixation, despite years of research on inoculation the only example where there is widespread adoption of rhizobia inoculants in developing countries is that of soybean. For small farmers, the use of the promiscuous soybean varieties seems the most appropriate. In developed countries, inoculation of clovers, particularly the subclover has also been widely adopted by using coated seeds with *R. trifolii*.

2.2.2. Nutrients Availability

Several nutrients are required in greater amounts for nodulated legumes than the same plants supplied with fertilizer N. They include phosphorus (P) for energy supply, Mo for the nitrogenase activity, Fe for infection, leghemoglobin and nitrogenase, sulphur (S) for protein synthesis, calcium (Ca) for the infection, and cobalt (Co) for cyanocobalamine. Deficient legumes in these elements may show N deficiency symptoms in the absence of mineral N.

Unlike the deficient level in soil, the excess of available N, particularly the NO_3^- can depress the fixation since symbiosis requires a higher energy level compared to the plant N uptake, and plant absorbs NO_3^- preferentially, although this is also a highly energetic process when compared to NH_4^+ uptake.

The more important nutrients will be reported below.

2.2.2.1. Nitrogen

Mineral N is known to inhibit N_2 fixation when present at high levels in the soil. On the other hand, a certain level of supplied N may be needed during early crop development in order to overcome N deficiency during the period from when the cotyledon N source is exhausted until nodules are formed and capable of supplying the plant with symbiotically N_2 fixed (Jensen 1987). These plants cropped in soils low in N require a starter dose of N (5-10 kg N ha^{-1}) to stimulate seedling growth and early nodulation such that both N_2 fixation and yield are enhanced (Carranca 1996; Hardarson and Atkins 2003). Nitrogen "starter" is not useful for peas, among other pulses since nodules are formed before the exhaustion of cotyledons to N (Jensen 1987). At flowering, N fertilization is often used for peas, soybeans and beans when nodules are senescing and pods are significant competitors for carbohydrates. Forage and pasture legumes are often supplied with a starter N due to the small size of seeds.

The extent to which soil N restricts N_2 fixation depends on genetic diversity between legume species and bacteria strains. The nitrogenase activity can be strongly reduced in presence of high soil N. The mechanism of inhibition of N_2 fixation (nitrogenase activity) by NO_3^- implies various factors. The inhibitory effect of exogenous NO_3^- on N_2 fixation has been attributed to a direct competition between nitrate reductase and nitrogenase for reducing power or to the fact that nitrite (NO_2^-, a by-product of nitrate reductase) inhibits the function of nitrogenase and leghemoglobin (Zahran 1999). Oliveira et al. (2004) observed a depressing effect of 450 kg mineral N ha^{-1} on nodulation and nitrogenase activity in alfalfa inoculated with *S. meliloti*, but no effect was observed on yield, total N concentration, crude protein, non-protein N and digestibility, in relation to the treatment without mineral N application.

Legumes prefer to absorb soil N instead of fixing it from the atmosphere since the energetic cost is lower in case of plant N uptake. However, *Lupinus* sp., for instance, maintains a high rate of N_2 fixation during the whole growth stage and obtains more than 90% of required N derived from the atmosphere even in soils of moderate N status (Hardarson and Atkins 2003; Carranca et al. 2008, 2009a). Fababean shows a similar feature (Carranca et al. 1999a; Hardarson and Atkins 2003). Some supernodulating mutants of soybeans also show high fixing rates in presence of high levels of soil NO_3- (Eskew et al. 1989). Camargos and Sodik (2010) reported that nodule growth (a function of photoassimilates) and N_2 fixation by *Calopogonium mucunoides* L., a tropical South American perennial legume were not affected by high levels of soil NO_3^-, although nodule number was reduced (higher energetic process).

Organic residues (agricultural or industrial) are often added to agricultural soils as an economical way of disposal, to improve the soil physical characteristics and to increase OM and plant nutrients. Castro and Ferreira (2006) observed a reduction on symbiotic N_2 fixation by subclover (*T. subterraneum* L.) one year after addition of 20-50 t biosolid ha^{-1} to the soil, comparing to the addition of 0.5-10 t ha^{-1}. Fixation decreased from 80% to 35% due to the addition of high levels of available N in the biosolid. Five years later, authors did not find any significant difference on fixation rate among treatments due to the stability of the organic residue in soil. Teixeira et al. (2006) and Araújo et al. (2007) reported no significant effects on nodulation of leguminous trees such as leucaena species (*Leucaena* sp.) and algarobe

[*Prosopis juliflora* (Sw)] after addition of low amounts of an industrial biosolid to improve soil fertility.

Aiming at estimating the amount of N_2 fixed in a grazed mixed pasture, Vinther (1998) observed that urine and faeces deposited by dairy cows promoted the growth of grass but reduced the proportion of clover and the rate of fixation by the legume. Considering the stocking density of 4–6 cows ha^{-1}, the length of the grazing period, the frequency of excretion, and the area covered by individual patches, Vinther (1998) estimated that the N_2 fixed in the grass-clover pasture was reduced by 10–15% compared to the ungrazed sward. This reduction was explained by the presence of available soil N from the excreta. The distribution of faeces and urine on the pasture is a problem. With continuous grazing at low stocking rates, much of the animal excreta are concentrated around local areas. Animal excreta distribution in the sward is improved with rotational grazing systems where stock density is higher.

2.2.2.2. Phosphorus

Phosphorus deficiency in soil is especially important under acidic conditions where soluble P may be precipitated in the presence of aluminium (Al), and in weathered soils of the tropics and subtropics.

Leguminous plants usually require more P than similar plants supplied with fertilizer N. Strains of rhizobia differ markedly in tolerance to P deficiency. Slow-growing bacteria (e.g. *R. leguminosarum*, *R. meliloti*, *R. phaseoli*, *R. trifolii*) appear more tolerant to low P levels than do fast-growing (e.g. *R. japonicum*, *R. lupini*) *Rhizobium* strains (Zahran 1999; Amarger 2001).

The high requirement for P in symbiotic legumes is consistent with the involvement of P in the high rate of energy transfer that takes place in the nodule (Sulieman et al. 2008). Nodules are important P sinks, and commonly have the highest concentration of this element in the plant. This is because of the high energy cost of N_2 fixation (eq. 1) and the cost of building and maintaining functioning nodules. Under low P supply, fixation is strongly impaired, both in respect to nodule initiation and functioning. A starter P supply results in enhanced nodule number and mass and greater N_2 fixation per plant and per gram of nodules (Davies 1997).

In sustainable agriculture, which has recently become important, research has been aimed at improving soil and crop quality as well as the effectiveness and persistence of the two symbionts (rhizobia and mycorrhiza) and other beneficial microorganisms within the soil microflora. In developed countries, the exploration of more efficient mechanisms to gather soil P reserves, for example the use of vesicular-arbuscular mycorrhiza or improved cultivars is viable due to often excessive use of P fertilizers in the past. Rock phosphates have been depleted and mineral P is a new facing problem. Goss and De Varennes (2002) reported a positive effect of dual symbiosis with rhizobia or bradyrhizobia and mycorrhizal fungi in leguminous plants. Both microorganisms, particularly the arbuscular mycorrhiza (AM) are active in root cortical cells, but they do not seem to compete for infection sites by rhizobia, except when the photosynthetic rate is limiting. Since AM symbiosis improves the legume P nutrition, especially in low soil available P, it has been assumed that the beneficial effect on N_2 fixation is mainly due to increased supply of P to the nodules, probably by changes in the signalling process between plants and bacteria. Improved nodulation and N_2 fixation in mycorrhized legumes can also be the result of enhanced uptake of essential micronutrients. As the fungus does not enter the nodule tissue, the supply of P has to be a plant-mediated process. This dual symbiosis has an added energy cost to the host, but can result in a striking growth improvement. Arbuscular mycorrhiza and *Rhizobium* in combination significantly increase nodulation, root colonization, the nutrient content of seeds, yield and yield components compared to the control in irrigated and rainfed chickpea (Erman et al. 2011).

In tropical countries, where soils have a poor capacity to supply P for crop growth due to a strong P adsorption, the N_2 fixing systems are well suited to utilize rock phosphates due to their net root proton excretion (soil acidification effect) and a greater sink for Ca than non-legumes which help to solubilise the rock-P. This effect is particularly relevant in mycorrhized legumes.

2.2.2.3. Molybdenum

The important trace element, Mo, becomes increasingly less available as soil pH decreases (pH<5.0). Molybdenum deficiency reduces symbiotic fixation since it is a component of nitrogenase enzyme. Carranca et al. (1999a) found a significant decrease in fixation rate of uninoculated and inoculated peas in a Mo deficient soil at Elvas (central Portugal) compared to Casas Velhas, in the same region, but without Mo deficiency in the soil.

Molybdenum salts have sometimes been incorporated into the inoculants, and commonly reduce the survival of rhizobia in the inoculants. For this reason, this is not a recommended practice. An alternate approach is to apply Mo by foliar fertilization.

2.2.2.4. Iron

Most of a third of the world population suffers from iron (Fe) deficiency, but the part of population most affected is women of reproductive age and children. Recent studies that connect Fe deficiency with deficient cognitive development emphasize the impact of the problem. Plants are the main source of Fe in the majority of diets, so assuring consumption of vegetables including legumes with an adequate level of Fe, constitutes an essential part of the strategy for improving the level of human nutrition.

Iron is frequently one of the most limiting nutrients for plants grown in calcareous soils. Iron chlorosis is a consequence of extreme Fe deficiency and is characterized by interveinal chlorosis. This metal is essential for many physiological and biochemical processes such as: photosynthesis, respiration, DNA synthesis, and early nodule development and N_2 fixation (Slatni et al. 2008). Fabaceae-Rhizobium symbiosis is particularly sensitive to Fe deficiency with respect to NO_3^--dependent plants (Krouma and Abdelly 2003). Iron is required for some key proteins involved in N_2 fixation like leghemoglobin and nitrogenase, important for the regulation of O_2 supply to bacteroids, and N assimilation like glutamate reductase, and NO_3^- and NO_2^- reductase. Iron deficiency can limit root nodule bacteria survival and multiplication, as well as host-plant growth, nodule initiation, development and function. Iron deficient plants can develop many nodules, but few are functioning nodules.

Some genotypes of lentil and chickpea differ in their ability to nodulate and fix N_2 on calcareous soils. Tolerant genotypes show a better Fe uptake efficiency and a preferential allocation of this nutrient towards the nodules. Nodule initiation is less sensitive to Fe deficiency in peanuts than in *L. angustifolius* L. (Slatni et al. 2008), whereas beans are very sensitive to Fe deficiency (Davies 1997). Under Fe deprivation, Slatni et al. (2008) observed that the symbiotic N_2 fixation by common bean 'Flamingo' with colored seeds is linked to a better Fe allocation to nodules and a significant Fe use efficiency for nodules growth and N_2 fixation, compared to 'Coco blanc' beans with white seeds. Iron use efficiency was 1.5 times greater in 'Flamingo' than 'Coco blanc' subjected to Fe deficiency. This parameter and symbiotic fixation discriminated clearly the two common bean cultivars and 'Flamingo' appeared as the most efficient cultivar (Krouma and Abdelly 2003; Slatni et al. 2008). The current study demonstrated that the most tolerant cultivar to Fe deficiency showed the following characteristics: (i) a better ability to protect its photosynthetic organs from Fe chlorosis, (ii) a capacity to preserve nodular development (number and growth of nodules), and (iii) a preferential allocation of Fe to nodules. Results suggested that Fe use efficiency for symbiotic N_2 fixation could be used to screen tolerant bean lines to Fe deficiency in condition of symbiotic N_2 fixation (Krouma and Abdelly 2003; Krouma et al. 2006; Slatni et al. 2008).

2.2.3. Environmental Factors

2.2.3.1. Temperature

The optimum temperature for most legumes and bacteria is within 25-30 ℃. Higher temperatures can cause flowers abortion (Corks and Vanorle 1988). Early sowing is necessary so that flowering begins before the warmest period. The negative effects of high temperatures for both symbionts are normally associated with water deficit. Bean (a summer crop) is very sensitive to temperatures above 30 ℃ at flowering time, showing increased abscission of flower buds, flowers, young pods, and reduced seed development. Genotypes selected for high temperature tolerance show less reduction of pollen viability at high temperature than susceptible ones. Beans are also very sensitive to low temperatures, which can limit their growth in the early season. Breeding of green beans for cold tolerance was reported (Davies 1997). Unlike, peas are grown successfully in cool but not excessively cold climates. Optimum growing temperature for this legume is between 13 ℃ and 18 ℃ (Muehlbauer and McPhee 1997).

Rhizobia are mesophile bacteria and most do not grow below 10 ℃ or above 37 ℃. Exceptions are the bradyrhizobia collected from some Artic legumes and in the hot and dry Sahel savannah of Africa (Júnior and Reis 2008). These bacteria may show different cell morphology, with thicker cell walls to reduce water loss. High and low temperatures influence growth and survival of rhizobia in soil, bacterial infection, bacteroids differentiation, nodules growth and functioning, and the time period when nodules are active reducing the N_2 fixed (Carranca 1996; Davies 1997; Zahran 1999; Júnior and Reis 2008). Rhizobia survival in soil exposed to high temperatures is greater in soil aggregates than in light texture soils and is favoured by dry rather than moist conditions (Zahran 1999). While 4-5% of a *B. japonicum* inoculum was recovered from a soil 24 h after sowing the legume at 28 ℃, lower than 0.2% survived when sowing took place at 38 ℃ (Chalk et al. 2010). Exposure of legumes to temperatures greater than 40 ℃, even for short periods, can cause irreparable loss of nodule function. The rate of symbiotic fixation by white lupine was reduced by about 20% when the soil temperature was lower than 0 ℃ in January, at central Portugal, affecting nodules growth and functioning. Indeterminate nodules were produced once the temperature increased above 8 °C (Carranca et al. 2009a).

Temperature during the transport and storage of bacteria inoculants and after seed inoculation and planting is also very critical.

Table 4. Pastures composition by the end of winter (February 2011) and spring (April 2011) in Mediterranean conditions, as influenced by the pasture age and diversity, and tree (*Quercus ilex* L.) canopy (Carranca, unpublished data)

Source of variation	Legume aerial biomass (kg DM ha^{-1})	Legume root biomass (kg DM ha^{-1})	Non-legume aerial biomass (kg DM ha^{-1})	Non-legume root biomass (kg DM ha^{-1})
		Vegetative period		
Pasture type				
improved>30 years-old	784 a	389 a	299 a	216 b
improved 12 years-old	646 a	122 a	497 a	532 ab
improved 5 years-old	240 a	53 a	370 a	436 b
natural>25 years-old	254 a	99 a	549 a	1239 a
Tree canopy influence				
under	88 b	21 b	541 a	861 a
out	875 a	311 a	313 b	350 a
ANOVA				
Pasture type	ns	ns	ns	3.27[*]
Tree canopy influence	15.40[***]	8.18[**]	7.16[*]	ns
Interaction	ns	ns	ns	ns
	Legume aerial biomass (kg DM ha^{-1})	Legume root biomass (kg DM ha^{-1})	Non-legume aerial biomass (kg DM ha^{-1})	Non-legume root biomass (kg DM ha^{-1})
		Bloom period		
Pasture type				
improved>30 years-old	710 b	52 c	6199 a	415 b
improved 12 years-old	2102 a	244 b	5557 a	1056 ab
improved 5 years-old	1059 ab	513 a	8769 a	750 ab
natural>25 years-old	907 b	120 c	7071 a	1252 a
Tree canopy influence				
under	933 a	267 a	7689 a	1100 a
out	1456 a	198 b	6109 a	636 b
ANOVA				
Pasture type	5.30[**]	53.39[***]	ns	4.45[*]
Tree canopy influence	ns	6.19[*]	ns	7.16[*]
Interaction	ns	30.95[***]	ns	ns

ANOVA=Analysis of variance; DM=dry matter; ns, [*], [**], [***] = *F*-values not significant (*P*≥0.05), and significant for *P*<0.05, 0.01 and 0.001, respectively; in each column and for each characteristic, means with the same letter are not significantly different (*P*<0.05), according to Bonferroni's test.

2.2.3.2. Light Intensity

Light intensity affects the symbiosis since light improves the photosynthetic capacity of legumes, improving the plant nutritional status. Rhizobia need a continuous supply of carbohydrates to produce the required energy to capture the N_2. Maintaining sufficient leaf area in a legume stand to intercept most of the sunlight is critical to maintain a high growth rate and support N_2 fixation. In a mixed pasture, the capacity of legumes to intercept the solar radiation is thus an important factor affecting the competitiveness of legumes to survive. The cut of leaves in pastures for animal consumption reduces the *Rhizobium* activity in the nodules by reduction of the photosynthetic capacity of the host-plant.

In Mediterranean regions, grasses often dominate the natural pastures in the autumn and winter period, but legumes generally dominate in spring and summer. This was observed in Portugal in a natural pasture greater than 25 years-old (Table 4), but was not the case for improved (biodiverse) pastures, with different ages (Carranca, unpublished data). In average, the aboveground biomass produced by the subclover during the autumn-winter under different rotational grazing pastures (natural and biodiverse), varying from 5, 12 and >25 years-old was 482 kg DW ha^{-1}, i.e. 11% greater than associated non-legumes (429 kg DW ha^{-1}). In spring, subclover aerial material (1194 kg DW ha^{-1}) was 51% lower than non-legumes (2340 kg DW ha^{-1}). In general, these pastures, in agro-forest systems, with different age and composition did not differ in growth response. In the autumn-winter, legumes growth was sinificantly greater out of the tree (*Quercus ilex* L.) influence, compared to the under tree canopy (Table 4) by the light interception. Pastures composition in the improved stands was based on *Lolium* and *Phalaris* as non-legumes, and composite plants and *Plantago* in the natural sward. The above values agreed with Rochon et al. (2004) for Mediterranean environments. According to them, annual legume-based pastures need to be grazed in order to maintain a total biomass averaging 1000 and 1500 kg DW ha^{-1}. They found higher forage availability under rotational grazing than continuous stocking, but swards were richer in annual ryegrass and poorer in leguminous species, compared to continuous stocking.

2.2.3.3. Atmospheric CO_2 Concentration

Legumes are grouped into cool (C_3) season category. These plants are less efficient to carry out photosynthesis as the temperature increases, comparing to C_4 plants. This explains the greater richness of grasses in tropical pastures. Nevertheless, it is supposed that the actual elevated atmospheric carbon dioxide [CO_2] may increase both the amount and percentage of host-plant N derived via symbiotic N_2 fixation, and benefit the overall N economy of terrestrial ecosystems. This increase is expected since (C_3) plants can have a greater photosynthetic capacity which in turn can supply the bacteroids in the nodules with greater amounts of photoassimilates. Ainsworth et al. (2002) reported that either the photosynthetic capacity or yield of soybean increased by increasing [CO_2], but this increase only represented 7%, less than the theoretical change needed to maximize the response. Garten Jr. et al. (2008) did not find any significant response of legumes [perennial red clover (*T. pratense* L.) and shrubs (*Lespedeza cuneata* (Dum. Cours.) G. Don.)] under elevated [CO_2].

2.2.3.4. Water Availability

The majority of legumes are sensitive to drought stress. This is particularly important in light soils, with low water retention. The timing of the stress is particularly harmful to legume biomass and fixation rate if it is experienced especially during germination, flowering or grain filling (Chalk et al. 2010). The most critical period is flowering and early pod set, when moisture stress can cause flowers and pods to drop, decreasing the yield. Early moisture stress, at the stage of two trifoliate, can have a lasting effect by reducing vegetative growth and affecting floral initiation, so that crop maturity becomes more uneven. The fixation response to soil water availability depends, in part, on how fixed N is transported within the host legume (Garten Jr. et al. 2008).

Beans are very sensitive to both drought and excess water. Approximately 60% of beans production worldwide is limited by drought (Davies 1997). Chickpea is the fourth most important pulse in the world, characterized not only by high protein content in the grains and a good source of carbohydrates, both important in human diets. In Mediterranean systems, chickpea is a spring crop, sown in March-April and is extremely affected by long periods of drought during the reproductive phase (Carranca et al. 1999a; Duarte-Maçãs 2003). Plant breeding should select plant genotypes that could fasten the vegetative cycle to avoid the hydric stress during the reproductive phase.

Perennial lucerne (*M. sativa* L.) has a long growing season and a deep-rooted system allowing a greater access to soil water down the soil profile (Peoples et al. 2001). This ability to grow under conditions unsuitable for annuals shows its greater importance for pastures than annual legumes. Legumes with a deep-rooting characteristic and a perennial or indeterminate growth habit are favoured in their ability to withstand drought and fix N_2 (Chalk et al. 2010). These characteristics should be highly considered by plant breeders. A broad choice of available cultivars is offered in lucerne and subterranean clovers for rainfed conditions (Lelièvre et al. 2008). Legumes with a certain tolerance to water stress usually exhibit an osmotic adjustment, which is partly accounted for by changing turgid cell and by accumulation of some osmotically active solutes (Corks and Vanorle 1988; Zahran 1999; Duarte-Maçãs 2003).

Drought also reduces the fixation capacity of sensitive legumes. Numerous studies show a clear decline in nodulation, nodule mass, N_2 fixation, or legume production under conditions of water scarcity (Carranca et al. 1999a; Garten Jr. et al. 2008). De Varennes (2003) and Júnior and Reis (2008) reported that drought stress causes a reduction on the formation of fine roots, a discontinuity on leghemoglobin production, the abortion of certain nodules, and lower translocation of ureides and aminoacids formed in roots to the aerial plant. Zahran (1999) verified that water stress imposed during the vegetative growth is more detrimental to nodulation and N_2 fixation than that imposed during the reproduction stage. A possible explanation is the little chance for legumes to recover from drought stress imposed during the vegetative period, particularly for determinate legumes.

Fababean and white or yellow lupine have high tolerance to drought and water-logging due to their strong taproots and indeterminate habits (Carranca et al. 1999a; Boote et al. 2001). Carranca et al. (1999a) observed that inoculation increased the fixation capacity of chickpea and pea under drought conditions from 45 to 80% and 20 to 65%, respectively (Table 3), but did not affect fababean and white or yellow lupine, which showed fixation rates above 70% (Carranca et al. 1999a, 2009a). In Ankara (Turkey), Albayrak et al. (2006)

observed increased fixation rates by inoculating common vetch (*Vicia sativa* L.) cultivars grown under rainfed conditions.

The occurrence of rhizobia populations in desert soils and the effective nodulation of legumes growing therein emphasize the fact that rhizobia can exist in soils with limiting moisture levels. However, population density tends to be lowest under the most desiccated conditions and to increase as the moisture stress is relieved (Zahran 1999). The selection for hydric stress *Rhizobium* strains is possible and can markedly affect the symbiosis success.

2.2.3.5. Soil Acidity and Alkalinity

Soil acidity is a significant problem facing agricultural production in many areas of the world and limits legume productivity. In Latin America there is more than 800 million Oxisols and Ultisols having a pH<5.0. Most leguminous plants require a neutral or slightly acidic soil for growth, especially when they depend on symbiotic N_2 fixation. Unfavourable soil pH (<5.0) affects the symbiosis, especially the nodulation success, by the toxic levels of Al and manganese (Mn) in soils, or induced Ca, P and Mo deficiencies. Host species vary in tolerance to Al and Mn, and are generally more affected by these elements than bacteria. Fababean and *Lupinus* sp. have a high tolerance to soil acidity due to their strong taproot (Boote et al. 2001; Carranca et al. 2008, 2009a).

Soil acidity can limit rhizobial growth and persistence in soil. Fast-growing rhizobia are generally more sensitive than are the slow-growing bradyrhizobia, but low pH tolerant strains exist in many species, namely the fast-growing *R. loti*. Failure to nodulate is common in acid soils, partly explained by the lowered number of rhizobia in the soil, but also because acidic soils affect the attachment. This susceptibility can be caused by the alteration on *Nod* factors production (Júnior and Reis 2008).

Isolates of *S. meliloti* (slow-growing bacteria) are particularly acid sensitive (pH<6.0). Lucerne lives in symbiosis with *S. meliloti* (Table 2) which is more sensitive to low pH and Ca deficiency than other strains. Pelleting the seeds has been an effective measure to allow good growth of lucerne in acidic soils in Europe (Mediterranean regions), Australia, New Zealand and Canada. The symbiotic dependence of white clover and subclover is unaffected by moderate acidity (pH 4.5-5.5) (Chalk et al. 2010).

The selection for acid tolerant rhizobia strains is possible and can markedly affect the symbiosis success. The use of acid-tolerant inoculant strains or the inoculation of seeds by pelleting (coating) the seeds with a layer of ground rock phosphate or limestone to reduce the negative effects of low pH are alternative practices. As many tropical and temperate soils are acidic, soil liming is also a solution for this problem. Liming reduces phytotoxic concentrations of Al and Mn.

Leguminous plants that fix N_2 absorb more cations than anions, because as uncharged N_2 enters and root protons are excreted to balance the internal pH. Thus symbiotic N_2 fixation causes some soil acidification, whereas plants that uptake NO_3^- raise the soil pH. As an example, Jensen and Hauggaard-Nielsen (2003) reported that alfalfa, producing 10 t ha^{-1} of fresh matter under symbiotic N_2 fixation, acidified the soil to such an extent that 600 kg $CaCO_3$ ha^{-1} $year^{-1}$ had to be applied to neutralize the soil pH.

The effect of soil alkalinity is likely to be greater on the host-plant than on rhizobia. Alkaline conditions limit the availability of Fe, zinc (Zn), boron (B) and Mn in the soil, thereby reducing the plant growth and symbiotic fixation. These deficiencies are normally corrected by foliar application.

2.2.3.6. Soil Salinity

Salinity is a serious threat to agriculture in arid and semiarid regions, including the Mediterranean and tropical areas. A soil is classified as saline when the electrical conductivity (EC) of the saturation extract exceeds the threshold value of 4 dS m^{-1}. The effect of salinity is likely to be greater on the host-plant than on rhizobia. The salinity response of legumes varies greatly and depends on factors such as climatic conditions, soil properties and growth stage (Dardanelli et al. 2010). Increase in the salinity of soils by wrong fertilizer practices or bad quality water supply results in decreased productivity of most crops and leads to marked changes in their growth pattern and yield, depending upon plant species, salinity level, and ionic composition of the salts.

Rhizobia species can tolerate higher concentrations than legumes. Increasing salt concentration may have a detrimental effect on soil microbial population as a result of direct toxicity and osmotic stress. Fixation in certain pulses such as fababean, bean, chickpea and soybean is more salt tolerant than others like peas, which are little affected up to 5 dS m^{-1} (Zahran 1999; Chalk et al. 2010). Perennial lucerne is also a salt tolerant legume, but this host is less tolerant than is the specific *Rhizobium*. The response of a white clover-ryegrass pasture to salinity showed that ryegrass was not affected, whereas clover DM declined 32%, but fixation rate was not significantly affected (Chalk et al. 2010). According to these authors, the white clover-*Rhizobium* symbiosis was resilient to moderate salinity, except for legume biomass which was predicted to decline 12% for each unit increase in EC above the threshold of 1.5 dS m^{-1}.

Cells of *Rhizobium* species exposed to high salt concentrations often accumulate osmoregulants such as glutamic acid, trehalose, glycine betaine and proline, which help to maintain turgid in the cell, limit the damage caused by salts, and change their morphology: the cells appear as spiral or filament-like structures and the cell size greatly expands the activity (Zahran 1999). Growth of a number of rhizobia species was inhibited by 100 mM sodium chloride (NaCl), while others, such as *S. meliloti* are tolerant to 300-700 mM NaCl. Rhizobia from woody legumes show substantial salt tolerance: strains from acacia, algarobe and leucaena are tolerant to 500-850 mM NaCl.

Although some rhizobia species can tolerate extremely high levels of salt (up to 1.88 M NaCl) they may show significantly reduced symbiotic efficiency under salt stress. The legume-*Rhizobium* association and nodules formation are more salt sensitive than *Rhizobium*, since salt stress inhibits the initial phases of symbiosis, by little curling or deformation of root hairs, reducing the respiration of nodules by reducing the cytosolic protein production, specifically leghemoglobin, and a lower photosynthetic activity (Zahran 1999). Rao et al. (2002) concluded that host-symbiont interaction under salt stress is not governed by the salt tolerance of bacteria. They suggested that the best strategy for improving legume symbiotic performance under salt stress is to screen and select legume cultivars that have the best nodulation and grain yield, followed by matching the most effective *Rhizobium* strains, irrespective of their relative performance under saline and non-saline conditions.

2.2.3.7. Soil Tillage

Tillage can destroy the rooting system (and nodules formed), decreasing the fixation rate (Table 5). Leguminous fixation capacity often increases under no-tillage, when compared to conventional practice also due to the absence of soil disturbance, without disruption of soil aggregates and disturbance of soil microbial population. Carranca et al. (2009a) found no significant response on fixation rate by lupine (*L. albus* L. 'Estoril') either with the native fixing bacteria or by soil inoculation, under both conventional and no-tillage practice in a podzolized soil at central Portugal (93%). Tillage also did not affect the fixation capacity (>95%) of uninoculated *L. luteus* L. grown in a podzolized soil at west Portugal (Carranca et al. 2008).

2.2.3.8. Contaminants

The association legume-*Rhizobium* is sometimes used as an index of soil contamination since nodulation is reduced before the host-plant shows toxicity symptoms. Contaminants, e.g. heavy metals and/or polycyclic aromatic hydrocarbons (PAHs) associated with certain organic fertilizers such as biosolids may affect negatively the survival of various soil microorganisms. A decline in rhizobia populations in amended soils may be due to the presence of heavy metals which become available during the mineralization of biosolids in soils.

Table 5. Effect of soil disturbance on nodules formation and N_2 fixing capacity of soybean (Source: Goss and De Varennes 2002)

Soil disturbance	Nodule weight (mg dry matter plant^{-1})	N_2 fixed (mg N plant^{-1})
+	12b	8b
-	32a	28a

Means in the same column with different letters are significantly different at $P<0.05$ for *t*-test.

One of the mechanisms adopted by microbes to cope up with variations caused by the heavy metal stress is induction of specific proteins (Satyanaryama and Johri 2005). Exposure of *R. leguminosarum* cells to 100 mM each of nickel (Ni), copper (Cu) and Zn induced several proteins in periplasmatic fraction.

Sewage sludge application to agricultural soils is an economical way of disposal. It improves the physical characteristics of the soil and increases OM content and essential plant nutrients, in particular N and P. Concern about the use of sewage sludge contaminated by heavy metals has been increased because heavy metals can persist in the soil over long periods and have ecotoxicological effects on plants and soil microorganisms. European Community legislation for sewage sludge application into agricultural soils is very strict as to the heavy metal content, which must not be above certain toxic limits for plants, but for microbes the legislation is missing to include microbial protection.

Studies on the impact of heavy metals on rhizobia population and infection, and on the rate of fixation are rare. Adverse effects of heavy metals on nodulation and N_2 fixation of legumes were reported for clover and chickpea. Castro et al. (1997) observed a reduction to 15% of nitrogenase activity in nodules of subclover 'Clare' inoculated with *R.*

leguminosarum bv. trifolii. Obbard et al. (1994) reported important effects of sludge type and rate, and concentration of heavy metals on reduction of the size of effective rhizobia populations, and on the symbiosis by white clover. Zahran (1999) suggested two possibilities to explain the mechanism by which the elevated metal concentrations reduced N_2 fixation: (i) one or more of the metals present can prevent the formation of N_2-fixing nodules by effective *Rhizobium* strains present in the soil, or (ii) the metal contamination can result in elimination of the effective *Rhizobium* strains from the soil.

Castro et al. (2008) did not find any significant effect of heavy metals contamination [mercury (Hg) and arsenic (As)] on the symbiosis of lotus with the specific rhizobia strain. These authors recommended this crop for soils remediation. Successful establishment of pioneer species improve soil characteristics by enriching its OM content and possibly reducing soil toxicity by heavy metals, so that more sensitive plants can develop and thus a healthy ecosystem can eventually be organized.

The polycyclic aromatic hydrocarbon reduces nodulation before visible damage of the plant can occur. Wetzel and Werner (1995) found a slight decrease in nodulation of alfalfa after application of cadmium chloride ($CdCl_2$), sodium arsenite ($NaAsO_2$), fluoranthene ($C_{16}H_{10}$) and other PAHs to the soil.

Crop residues left in the soil also have been found to cause phytotoxicity (Mallik and Tesfai 1988). Rice straw residue in soil reduced significantly the N_2 fixation by beans (Mallik and Tesfai 1988; Lovett and Ryuntyn 1992). In Taiwan, yields of grain legumes decreased when grown more than once a year in the same area. This was the case of mungbean (*Vigna. radiate* L.) planted in amended soils with mungbean root residues (Lovett and Ryuntyn 1992). Nodule number, nodule weight and N_2 fixation were significantly reduced.

Root exudates of allelopathic species, leaf leachates, and decomposing residues can also be effective in reducing nodule number and leghemoglobin content. Allelopathic effects of weeds on legume-rhizobia symbiosis are poorly understood. Weeds can cause greater loss in soybean production than all other pests combined (Mallik and Tesfai 1988). These authors (1988) found that aqueous extracts of some weeds, root exudates, leaf leachates and decaying plant residues were toxic to rhizobia *in vitro*, and adversely affected nodulation and N_2 fixation in white clover, red kidney beans (*P. vulgaris* L.) and Korean lespedeza (*Lespedeza stipulacea*).

Plant growth and symbiotic N_2 fixing microorganisms can be inhibited by the allelophatic chemicals contained in the residues of tree species, though Heckman and Kluchinski (1995) did not find any indication of an allelophatic inhibition on nodulation or N_2 fixation by soybean from heavy application to soil of oak, maple, sycamore or walnut leaves. *Eucalyptus* residues have substances that cause injures on the development of some leguminous plants, as well as the infection by the *Rhizobium*. Investigations to identify the allelopathic compounds in leachates of bark, fresh leaves and leaf litter of *Eucalyptus tereticornis* L., *E. camadulensis* L., *E. polycarpa* L. and *E. microtheca* L. showed the negative influence of phenolics and leachates on the germination, DM yield and nitrogenase activity of redgram [*Cajanus cajan* (L.) Mills], an important pulse in India (Sasikumar et al. 2001). Carranca et al. (2008) observed that the incorporation of *E. globulus* L. residues in a light Portuguese soil as an organic amendment did not affect the yellow lupine productivity and N_2 fixation, compared to the unamended control plots.

Conclusion

By the increase of the world's population and as the natural resources that supply fertilizer-N diminish the objective is achieved through the development of superior legume cultivars, improvements in agronomic practices, and increase efficiency of the N_2 fixing process itself by a better management of the symbiotic relationship between plants and bacteria.

The populations of native bacteria collectively called by *Rhizobium* should be adequately characterized and their diversity and performance on nodulating different legume species (promiscuous bacteria) must be evaluated. Very often, these native bacteria are very infective and effective to the grown legumes, particularly the local varieties. Inoculants should contain two or more isolates in order to get success with the environmental adaptation and competition for infection and formation of nodules on the intended host-legume.

The acclimation of legumes to grow under elevated $[CO_2]$ in the field is also a priority to adapt plants to rising $[CO_2]$ and to maximize fixation under this condition.

References

Ainsworth, E.A.; Davey, P.A.; Bernacchi, C.J.; Dermody, O.C.; Heaton, E.A.; Moore, D.J.; Morga, P.B.; Naidu, S.L.; Ra, H-SY., Zhu, X-G.; Curtis, P.S. and Long, S.P. (2009). A meta-analysis of elevated $[CO_2]$ effects on soybean (*Glycine max*) physiology, growth and yield. *Global Change Biology* 8:695–709.

Albayrak, S.; Sevimay, C.S. and Çöçü, S. (2006). Effect of Rhizobium inoculation on forage and seed yield and yield components of common vetch (*Vicia sativa* L.) under rainfed. *Acta Agriculturae Scandinavica Section B-Soil and Plant Science* 56:235-240.

Almeida, D. (2006). Manual de culturas hortícolas II. Editoral Presença, Queluz de Baixo, Portugal.

Amarger, N. (2001). Rhizobia in the field. *Adv Agron* 73:109-168.

Ampomah, O,Y.; Ofori-Ayeh, E.; Solheim, B. and Svenning, M.M. (2008). Host range, symbiotic effectiveness and nodulation competitiveness of some indigenous cowpea bradyrhizobia isolates from the transitionsl savanna zone of Ghana. *African Journal of Biotechnology* 7:988-996.

Araújo, A.S.F., Monteiro, R.T.R. and Carvalho, E.M.S. (2007). Effect of composted textile sludge on growth, nodulation and nitrogen fixation of soybean and cowpea. *Bioresource Technology* 97:1028-1032.

Bergersen, F.J. and Turner, G.L. (1975). Leghaemoglobin and the supply of O_2 to nitrogen-fixing root nodule bacteroids: Studies of an experimental system with no gas phase. *Journal of General Microbiology* 89:31-47.

Boote, K.J.; Minguez, M.I. and Sau, F. (2001). Adapting the CROPGRO legume model to simulate growth of faba bean. *Agron J* 94:743-756.

Camargos, L.S. and Sodik, L. (2010). Nodule growth and nitrogen fixation of *Calopogonium mucunoides* L. show low sensitivity to nitrate. *Symbiosis* 51:167-174.

Carranca, C. (1996). Nitrogen cycling in Portuguese soils and its assessment by [15]N. Ph.D. Thesis in Agricultural Engineering, University of Lisbon, Portugal.

Carranca, C.; De Varennes, A. and Rolston, D.E. (1999a). Biological nitrogen fixation by fababean, pea, and chickpea under field conditions estimated by the ^{15}N isotope dilution technique. *Eur J Agron* 10**:**49-56.

Carranca, C.; De Varennes, A. and Rolston, D.E. (1999b). Biological nitrogen fixation estimated by ^{15}N dilution, natural ^{15}N abundance, and N difference techniques in a subterranean clover-grass sward under Mediterranean conditions. *Eur J Agron* 10:81-89.

Carranca, C.; Madeira, M.; Torres, M.O.; Pina, J.P. and Marques, P. (2008). Variação sazonal da fixação simbiótica em duas espécies de *Lupinus* sujeitas a diferentes práticas de preparação do solo. Resumo das Comunicações do 1 Congresso Luso-Espanhol de Fixação de Azoto. *Estoril*: 14. (1-4 de Junho).

Carranca, C., Torres, M.O. and Baeta, J. (2009a). White lupine as a beneficial crop in Southern Europe. I - Potential for N mineralization in lupine amended soil and yield and N_2 fixation by white lupine. *Eur J Agron* 31:183-189.

Carranca, C., Rocha, I, De Varennes, A.; Oliveira, A.; Pampulha, M.E. and Torres, M.O. (2009b). Effect of tillage and temperature on potential nitrogen mineralization and microbial activity and microbial numbers of lupine amended soil. *Agrochimica LIII* (3), May-June:183-195.

Castro, I.V. and Ferreira, E.M. (2006). Fertilización y contaminación: Metales pesados y lodos de depuradoras. In: E.J. Bedms; C.L. González and B. Rodelas (Eds.), *Fijación de nitrógeno: Fundamentos y aplicaciones* (pp. 298-310). Granada.

Castro, I.V.; Ferreira, E.M. and MacGrath, S.P. (1997). Effectiveness and genetic deversity of Rhizobium leguminosarum bv. trifolii isolates in Portuguese soils polluted by industrial effluents. *Soil Biol Biochem* 29:1209-1213.

Castro, I.V.; Sá-Ferreira, P.; Simões, F.; Matos, J.A. and Ferreira, E.M. (2008). Use of Lotus/Rhizobium symbiosis in regeneration of potential soils. *Lotus Newsletter* 37:87-88.

Chalk, P.M.; Alves, B.J.R.; Boddey, R.M. and Urquiaga, S. (2010). Integrated effects of abiotic stresses on inoculants performance, legume growth and symbiotic dependence estimated by ^{15}N dilution. *Plant Soil* 328:1-16.

Corks, F. and Vanorle, L. (1988). Technical and economical aspects of the introduction of the proteagineous pea crop in Belgium. In: P. Plancquart and R. Haggar (Eds.), *Legumes in farming systems* (pp. 117-125). Kluwer Acad. Publish., Commission of the European Communities.

Dardanelli, M.; González, P.S.; Medeot, D.B.; Paulucci, N.S.; Bueno, M.A. and Garcia, M.B. (2010). Effects of peanut rhizobia on the growth and symbiotic performance of Arachis hypogea under abiotic stress. *Symbiosis* 47:175-180.

Davies, J.H.C. (1997). Phaseolus beans. In: H.C. Wien (ed.), *The physiology of vegetable crops* (pp. 409-428). CAB International, USA.

De Varennes, A. (2003). Produtividade dos solos e ambiente. Escolar Editora, Lisboa, Portugal.

Duarte-Maçãs, I. (2003). Selecção de Linhas de Grão-de-bico (*Cicer arietinum* L.) Adaptadas ao Ambiente Mediterrânico–Critérios Morfológicos e Fisiológicos. Ph.D. Thesis, University of Évora (Portugal).

Erman, M., Demir, S.; Ocak, E.; Tufenkc, S.; Oguz, F. and Akkopru, A. (2011). Effects of Rhizobium, arbuscular mycorrhiza and whey applications on some properties in chickpea (Cicer arietinum L.) under irrigated and rainfed conditions 1 - Yield, yield components, nodulation and AMF colonization. *Field Crops Research* 122:14–24.

Eskew, D.L.; Kapuya, J. and Danso, S.K.A. (1989). Nitrate inhibition of nodulation and nitrogen fixation by supernodulating nitrate-tolerant symbiosis mutants of soybean. *Crop Sci* 29:1491-1496.

FAO. (1971-1981). FAO/UNESCO Soil Map of the World (1:5 000 000).

Fotolli, M.N.; Tsikou, D.; Kolliopoulou, A.; Aivalakis, G.; Katinakis, P.; Udvardi, M.K.; Rennenberg, H. and Flemetakis, E. (2011). Nodulation enhances dark CO_2 fixation and recycling in the model legume Lotus japonicus. *J Exp Bot* 62:2959-2971..

Garten, Jr. C.T.; Classen, A.T.; Norby, R.J.; Brice, D.J.; Weltzin, J.F. and Souza, L. (2008). Role of N_2-fixation in constructed old-field communities under different regimes of [CO_2], temperature, and water availability. *Ecosystems* 11:125-137.

Goss, M. and De Varennes, A. (2002). Soil disturbance reduces the efficacy of mycorrhizal associations for early soybean growth and N_2 fixation. *Soil Biol Biochem* 34:1167-1173.

Hardarson, G. and Atkins, C. (2003). Optimising biological N_2 fixation by legumes in farming systems. *Plant Soil* 252:41-54.

Hardarson, G., Golbs, M. and Danso, S.K.A. (1989). Effect of nodulation patterns on nitrogen fixation by soybean (*Glycine max* (L.) Merrill). *Soil Biol Biochem* 21:783-787.

Heckman, J.R. and Kluchinski, D. (1995). Soybean nodulation and nitrogen fixation on soil amended with plant residues. *Biol Fertil Soils* 20:284-288.

IUSS (International Union of Soil Science), Working Group WRB (World Reference Base). (2006). World reference base for soil resources 2006. World Soil resources Reports (103), 2nd ed. FAO, Rome (Italy).

Jensen, E.S. (1987). Seasonal patterns of growth and nitrogen fixation in field-grown pea. *Plant Soil* 101:29-37.

Jensen, E.S. and Haugggaard-Nielsen, H. (2003). How can increased use of biological N_2 fixation in agriculture benefit the environment? *Plant Soil* 252:177-186.

Júnior, P.I.F. and Reis, V.M. (2008). Algumas limitações à fixação biológica de nitrogénio em leguminosas. *Embrapa, Documentos* 252:1-40.

Krouma, A. and Abdelly, C. (2003). Importance of Fe-use efficiency of nodules in common bean (Phaseolus vulgaris L.) for iron deficiency chlorosis resistance. *J Plant Nutr Soil Sci* 166:525-528.

Krouma, A.; Drevon, J.-J. and Abdelly, C. (2006). Genotypic variation of N_2-fixing common bean (Phaseolus vulgaris L.) in response to iron deficiency. *J Plant Physiol* 163:1094-1100.

Kyei-Boahen, S.; Slinkard, A.E. and Walley, F.L. (2002). Time course of N_2 fixation and growth of chickpea. *Biol Fertil Soils* 35:441-447.

Lelièvre, F.; Norton, M.R. and Volaire, F. (2008). Perennial grasses in rainfed Mediterranean farming systems-Current and potential role. *Options Méditerranéennes, Series A* (79):137-146.

Lovett, J. and Ryuntyn, M. (1992). Allelopathy: Broadening the context. In: S.J.H. Rizvi and V. Rizvi (Eds.), *Allelopathy: Basic and applied aspects* (1st ed, pp. 11-20). Chapman and Hall, London.

Mallik, M.A.B. and Tesfai, K. (1988). Allelopathic effect of common weeds on soybean growth and soybean-Bradyrhizobium symbiosis. *Plant Soil* 112:177-182.

Muehlbauer, F.J. and McPhee, K.E. (1997). Peas. In: H.C. Wien (ed.), *The physiology of vegetable crops* (pp. 409-428). CAB International, USA.

Obbard, J.P.; Sauerbeck, D.R. and Jones, K.C. (1994). The effect of heavy metal-contaminated sewage sludge on the rhizobial soil population of an agricultural field trial. In: M.H. Donker; H. Eijsackers and F. Heimback (Eds.), *Ecotoxicology of soil organisms* (pp. 127–161). Lewis Publishers, London.

Oliveira, W.S.; Oliveira, P.P.A.; Corsi, M.; Duarte, F.R.S. and Tsai, S.M. (2004). Alfalfa yield and quality as function of nitrogen fertilization and symbiosis with *Sinorhizobium meliloti. Scientia Agrícola* 61. (doi: 10.1590/S0103-90162004000400013).

Ovalle, C.; Urquiaga, S.; Del Pozo, A.; Zagal, E, and Arredondo, S. (2006). Nitrogen fixation in six forage legumes in Mediterranean central Chile. *Acta Agric. Scand., Section B-Plant Soil Science* 56:277-283.

Peoples, M.B.; Bowman, A.M.; Gault, R.R.; Herridge, D.F.; McCallum, M.H.; McCormick, K.M.; Norton, R.M.; Rochester, I.J.; Scammell, G.J. and Schwenke, G.D. (2001). Factors regulating the contributions of fixed nitrogen by pasture and crop legumes to different farming systems of eastern Australia. *Plant Soil* 228:29-41.

Rao, D.L.N.; Giller, K.E.; Yeo, A.R. and Flowers, T.J. (2002). The effects of salinity and sodicity upon nodulation and nitrogen fixation in chickpea (*Cicer arietinum*). *Ann Bot* 89:563-570.

Reddy, P.M., Ladha, J.K.; Ramos, M.C.; Maillet, F.; Hernandez, R.J.; Torrizo, L.B.; Oliva, N.P.; Datta, S.K. and Datta, K. (1998). Rhizobial lipochitooligosaccharide nodulation factors activate expression of the legume early nodulin gene ENOD12 in rice. *The Plant Journal* 14:693-702.

Rochon, J.J.; Dayle, C.J.; Greef, J.M.; Hopkins, A.; Molle, G.; Sitzia, M.; Scholefield, D.; Smith, C.J. (2004). Grazing legumes in Europe: A review of their status, management, benefits and future prospects. *Grass and Forage Science* 59:197-214.

Saari, L.L. and Ludden, P.W. (1987). The energetic and energy cost of symbiotic nitrogen fixation. In: T. Kosuge and E.W. Nester (Eds.), *Plant microbe interactions* (2) . MacMillan Publ. Co., New York, USA.

Sasikumar, K.; Vijayalakshmi, C. and Parthiban, K.T. (2001). Allelophatic effects of four eucalytptus species on redgram (Cajanus cajan L.). *Journal of Tropical Agriculture* 39:134-138.

Satyanaryama, T. and Johri, B.N. (Eds.). (2005). Microbial Diversity: Current Perspectives and Potential Applications, I.K. Int. Publish. House Ltd., New Delhi, India.

Schultze, M.; Staehelin, C.; Rohrig, H.; John, M.; Scmidt, J.; Kondorosi, E. and Schell, J. (1995). *In vitro* sulfotransferase activity of Rhizobium meliloti NodH protein: Lipochitooligosaccharide nodulation signals are sulfated after synthesis of the core structure. *Proc Natl Acad Sci USA* 92:2706-2709.

Slatni, T.; Krouma, A.; Aydi, S.; Chaiffi, C.; Gouia, H. and Abdelly, C. (2008). Growth, nitrogen fixation and ammonium assimilation in common bean (Phaseolus vulgaris L.) subjected to iron deficiency. *Plant Soil* 312:49–57.

Sulieman, S.; Fischinger, S. and Schulze, J. (2008). N-feedback regulation of N_2 fixation in Medicago Truncatula under P-deficiency. Genn. Apply. *Plant Physiology Special* 34:33-54.

Teixeira, K.R.G.; Gonçalves Filho, L.A.R.; Carvalho, E.M.S.; Araújo, A.S.F. and Santos, V.B. (2006). Efeito da adição de lodo de curtume na fertilidade do solo, nodulação e rendimento de matéria seca do caupi. *Ciência e Agrotecnologia* 30:1071-1076.

Trevaskis, B.; Colebatch, G.; Desbrosses, G.; Wandrey, M. and Wienkoop, S. (2002). Differentiation of plant cells during symbiotic nitrogen fixation. *Comp Funct Genom* 3:151-157.

Vinther, F.P. (1998). Biological nitrogen fixation in grass–clover affected by animal excreta. *Plant Soil* 203:207–215.

Wetzel, A. and Werner, D. (1995). Ecotoxicological evaluation of contaminated soil using the legume root nodule symbiosis as effect parameter. *Environ Toxicol Water Qual* 10:127-134.

Xi, C.; Schoeters, E.; Vanderleyden, J. and Michiels, J. (2000). Symbiosis-specific expression of Rhizobium etli casA encoding a secreted calmodulin-related protein. *Proc Natl Acad Sci USA* 97:11114–11119.

Zahran, H.H. (1999). Rhizobium-legume symbiosis and nitrogen fixation under severe conditions and in an arid climate. *Microbiol Mol Biol Rev* 63:968–989.

In: Symbiosis: Evolution, Biology and Ecological Effects   ISBN: 978-1-62257-211-3
Editors: A. F. Camisão and C. C. Pedroso   © 2013 Nova Science Publishers, Inc

Chapter 4

SYMBIOSIS OF SEA ANEMONES AND HERMIT CRABS IN TEMPERATE SEAS

Chryssanthi Antoniadou, Anna-Maria Vafeiadou and Chariton Chintiroglou*

School of Biology, Department of Zoology, Aristotle University,
Thessaloniki, Greece

Abstract

Symbiosis, according to its initial meaning, refers to the biological interaction between two organisms living in close association. However, this definition is rather controversial, with the term being often used generically, since the outcome can vary across a continuum from negative to positive interactions. Symbiosis is a widespread phenomenon in temperate marine communities, and the association between sea anemones and hermit crabs belongs to the most common cases, being a familiar example of mutualism. In these latter specific cases of interactions gastropod shells are involved as prerequisite, since they provide both refuge for hermit crabs and substratum for the settlement of sea anemones; thus, shell resource availability is crucial for the establishment of this particular type of symbiosis. Within this context the present study aims to integrate the results of various studies to provide a general review about the symbiotic interactions of sea anemones and hermit crabs in temperate seas, addressing the following issues: (1) clarify the relevant terminology, which is differently interpreted by various authors; (2) provide a general description of the sea anemone - hermit crab association, as most studies examine separately the species involved and not the symbiosis as a whole; (3) assess the diversity and distribution of sea anemone - hermit crab associations in temperate seas, also incorporating gastropod shells and their availability, which although crucial, has been only little investigated; (4) address the behavioural patterns of both symbionts for the establishment of the symbiosis, including as well the behavioural plasticity of hermit crab related to shell resource utilization, and (5) report relevant information about co-evolution of the participant species, referring to the existing hypotheses on the evolution of the symbiosis, underlining its importance.

---

* E-mail address: antonch@bio.auth.gr, Tel. +302310998901, Fax. +302310998269, Address: Chryssanthi Antoniadou Aristotle University, School of Biology, Department of Zoology, Thessaloniki, Greece, Gr - 54124, (Corresponding Author)

Symbiosis: Meaning and Relevant Terminology Considering the Specific Case of Sea Anemones - Hermit Crabs

In nature very few species, if any, live separated; almost all species depend on other to gain vital resources, such as habitat, food and protection. This dependency among species has been very early recognized from biologists under the concepts of biotic interactions and symbiosis. The term symbiosis was coined originally by Anton de Bary in 1879 in his study about lichens, to mean any association between different species, with the implication that the organisms are in persistent contact, but that the relationship does not need to be advantageous to all participants (see Douglas, 2010). Thus, according to its initial meaning, symbiosis refers to the biotic interaction between two organisms living in close association; the latter phrase differentiates symbionts from simply interacting species. However, this definition is rather controversial, since the outcome of interactions can vary across a continuum from negative to positive results, and among participant species. Moreover, the term has often been used generically and its meaning has frequently been deviated from the original definition. The subsequent proposition of additional definitions and the lack of agreement for a specific one within the scientific community have further complicated the strength of this term; similar problems can be found for many other, widely applicable, terms in the field of marine ecology (see Dauvin et al., 2008).

After the first definition of symbiosis, very little awareness about the term, as defined subject, existed between biologists and up to 1950s; the phenomenon has been encountered as scattered among organisms and very little research was in progress, almost exclusively covering terrestrial associations (see Smith, 2001 for a thorough review of symbiosis research trends over the last century). Thereafter, and especially after 1970s, symbiosis research advanced incorporating many topics and including major marine taxa, such as sponges, corals and sea anemones. Symbiotic procedures are thought to be less diverse and widespread in aquatic domain (Smith, 2001), despite the recognition of their prominent role in particular marine ecosystems, such as coral reefs in tropics, and shallow benthic communities in temperate seas (Grutter and Irving, 2007).

Considering all the above the first task of this study is to thoroughly revise and clarify the relevant terminology, which is differently interpreted by various authors, focusing on the marine domain and the specific case of interactions between closely associated sea anemones and hermit crabs.

As Smith (2001) clearly pointed "there is still no clear and universally agreed definition of symbiosis, even though it is 130 years after de Bary devised the term". Currently symbiosis is used under a wide range (Martin and Britayev, 1998) referring to all cases in which two species live in close association (Henry, 1966), although many researchers attempted to restrict the term to associations where partners mutually benefit (Rhode, 1981) or alternatively and more sophisticated defined, symbiosis refers to intimate mutualism involving direct supply of nutrients or other resources between physiologically integrated species (Grutter and Irving, 2007). Proving benefit existence is highly problematic since, at least in some associations, the partner's cost surpasses any hypothetical benefit. Douglas (1994) rejected mutual gains and suggested the acquisition of a novel metabolic capability from one partner as the basis of symbiosis. However, this concept is complex as the gain is strictly connected with metabolism, although practically applicable to some specific cases of interactions between bacteria and plants or metazoans (e.g. symbiotic zooxanthellae and corals or sea anemones). This fact, together with the largely unknown nature of species interactions, hinders the general acceptance of the latter idea and enhances the generic sense of the term.

As mentioned earlier, symbiosis constitutes a rather loose term up to date, which includes a wide range of interactions that cover the specific cases of: (i) parasitism, i.e. when the symbiosis is advantageous to one partner at the expense of the other, (ii) commensalism, i.e. when the symbiosis is advantageous to one partner without harming the other, and (iii) mutualism, i.e. when the symbiosis is reciprocally advantageous to both partners; these cases are symbolized as follows +/-, +/0, +/+, respectively (Martin and Britayev, 1998; Bruno et al., 2003; Patzner, 2004). Apart from parasitism which is interpreted as a negative interaction (at least for one partnership) symbiosis is also described under the terms of facilitative or positive interactions (Stachowicz, 2001). The latter terms are becoming of increasing applicability in the scientific audience as they give a more precise description about the nature of species interactions and thus, a trend to replace the more generically defined symbiosis is evolving (Stachowicz, 2001; Grutter and Irving, 2007). A clear distinction, however, among the above cases of symbiosis is not always evident, because many factors define the nature of these interactions, such as the degree of association between the species and their specialization, its necessity for the species survival, the temporal pattern, and the life stage at which interaction occurs (Martin and Britayev, 1998). Considering all the above it seems rather reasonable to adopt the latter authors' opinion suggesting the use of the term symbiosis as "stepping stone in helping to understand the real relationships in any particular association".

Symbiosis appears to be more common in tropical marine communities (Grutter and Irving, 2007); nevertheless, the phenomenon is widespread also in temperate seas with the association between sea anemones and hermit crabs belonging to the most common and widely acknowledged cases of mutualism (Williams and McDermott, 2004; Vafeiadou et al., 2011). More specifically, each case of symbiosis, including mutualism, can be categorized as: (i) obligate or facultative, in the first case partners may survive only in association and in the second, while benefiting from the presence of each other, they may also survive in absence of their partner (Boucher et al., 1982), (ii) direct or indirect, in the first case partners interact physically and in the second they benefit from the each other's presence without direct contact (Boucher et al., 1982), (iii) permanent or temporary, in the first case partners are living together during their whole life and in the second only in some phase of their life cycle (Martin and Britayev, 1998), and (iv) monoxenic, oligoxenic or polyxenic, in the first case the symbiont is associated with only one host, whereas in the other two cases few or several different host species are involved, respectively (Lom, 2001); the latter category is used only for parasitism.

Considering the particular case of symbiosis between sea anemones and hermit crabs, its development requires the involvement of a third part, i.e. gastropod shells, which provide both refuge for hermit crabs and substratum for the settlement of sea anemones. These tripartite associations were assigned as ecological triangles by Ross and Sutton (1963). Nevertheless, the term has been expanded and is currently used in the broad fields of ecology and environmental biology to describe interactions among three biotic or abiotic parameters (Styron, 1977; Kareiva, 1982; Xu et al., 2006). Taking into account its original description, the limited implementation from other authors (Chintiroglou, et al. 1992; Christidis et al., 1997), or even from the ones who suggested it (Ross 1974a, 1974b, 1979), and the doubt concerning its validity, since gastropods do not actively participate in the association although their shells are vital for the development of the sea anemones - hermit crabs symbiosis (Vafeiadou et al., 2011), the term ecological triangle is abandoned at the present review.

Sea Anemones - Hermit Crabs Symbiosis: A General Description

The interaction of sea anemones and hermit crabs is one of the most familiar examples of symbiosis in temperate seas, interpreted as a typical case of mutualism. Considering symbiosis terminology (see above), this specific case can further be described as a clear paradigm of indirect, permanent, facultative, in most cases, mutualism. If we can expand the use of the terms monoxenic/oligoxenic/polyxenic which so far is used for parasitism, we assume most sea anemones as polyxenous symbionts, as they can be hosted by several different hermit crab species; however, this term is rather species-specific (see for example the case of the sea anemone *Adamsia obvolva* which associates only with the hermit crab *Sympagurus pictus* as a monoxenous symbionts).

Nevertheless, much discussion around this aspect has followed due to confusion through terms and suggestions by several authors; although sea anemone - hermit crab symbiosis had been considered as mutualism from early studies (Roughgarden, 1975; Hazlett, 1981; Ross, 1984; Brooks, 1989), it has only recently been characterized as facultative mutualism (Patzner, 2004; Williams and McDermott, 2004; Vafeiadou et al., 2011). With older studies using contradictory terminology, given that an exact description of the symbiotic relationship was missing, the kind of interaction should be re-examined, at least for some particular species. The interaction between the sea anemone *Adamsia palliata* and the hermit crab *Pagurus prideaux* for example had long been interpreted as a case of obligate commensalism, before the anemone species was proved first to live alone, without any association with hermit crabs, and second to live in association with other hermit crab species too (Ates, 1995). Even further, hermit crabs of some species may prey on their symbiotic sea anemones under starvation, or under increased sea anemone densities (Imafuku et al., 2000). Williams and McDermott (2004) in their review study on hermit crab symbiosis stress the difficulties of such categorization. There are some examples of species among cnidarians in association with hermit crabs that happen to feed on the eggs of hosts but the relationship had been previously described as commensalism, or other cases of temporal changes in the symbiotic nature of the relationship, i.e. switching from commensalism to mutualism or parasitism, depending on different environmental and biological factors.

In this aspect, a general description of the sea anemone - hermit crab symbiosis is presented below, encompassing all the relevant information included in the literature, as such to underline the importance of symbiosis for both participant species, and for marine ecosystems, respectively.

In the particular case of sea anemones - hermit crabs symbiosis, though, the presence of a third, indirect participant is required: gastropod shells. They constitute the linking part of the symbiosis, providing refuges for hermit crabs (to protect their abdomen part) and suitable substratum for the settlement of sea anemones (Conover, 1978; Brooks, 1989). Thus, shell availability is a crucial factor for the establishment of the symbiosis.

The development of the symbiotic interaction initiates by the detachment of sea anemones from the substratum and their placement on gastropod shells inhabited by hermit crabs. A cooperation of both symbionts is necessary for the well-establishment of the symbiosis; however, some cases where symbiosis initiates by only one of the symbionts have also been reported.As such, hermit crabs detach sea anemones, using tactile stimulation, and actively transfer them on their shells (Brunelli, 1910; Cowles, 1919; Ross, 1970); in some cases with the cooperation of the sea anemones, which loosen their connection with the substratum to enhance their transfer (Ross, 1974a, 1974b; Lawn, 1976; McFarlane, 1976).

In particular, sea anemones are the only symbionts among cnidarians associated with hermit crabs which are actively hosted by them and not haphazardly fixed on the shells during larval settlement (Gusmão and Daly, 2010). In other cases, sea anemones do also transfer themselves on shells inhabited by hermit crabs, without aid of the latter, to establish a symbiotic relationship with them (Davenport et al., 1961; Ross, 1959, 1965; Ross and Sutton, 1961; see also section 4 for details in behavioural patterns).

The importance of symbiosis for both partners is diverse (Table 1). The hermit crab enforces its defence to predators, gaining protection via the sea anemone nematocysts (Brooks, 1989). As known, the main predators of hermit crabs are cephalopod molluscs (e.g. octopus) which are not resistant to the toxins excreted by the nematocysts of cnidarians (Ross, 1967, 1971; Brooks, 1991). As a result, hermit crabs actively host sea anemones on the gastropod shells they inhabit (Gusmão and Daly, 2010), evolving a whole behaviour towards the establishment of the symbiosis, including gathering increased number of anemones under predator pressure, or stealing anemones from other crabs (Ross and Boletzky, 1979; see also section 4 for details in behavioural patterns).

Table 1. Overview of the advantages and disadvantages of symbiosis for sea anemones and hermit crabs

		Hermit crabs	Sea anemones
Advantages		Protection from predators	Protection from predators
		Increased shell strength	Substratum availability
		Decreased energetic costs of changing/searching for shells	Increased feeding capacity (increased food resource exploitation)
		Prey on symbionts in case of starvation (only some species)	Increased dispersal Direct feeding by their host
Disadvantages		Increased energetic costs of carrying heavier shells	Predation by the host (only in specific cases)
		Increased intra- and inter-specific competition	

Additional benefits for the hermit crab may also derive from expansion of the anemone over the shell, forming a so called "living cloak" inhabited by the hermit crab, strengthening the shell in this way and thus, the crab's structural defence (Faurot, 1910; Doumenc, 1975; Ross, 1984).Furthermore, sea anemones of the genus *Adamsia* form a chitin shell-like structure, known as carcinoecium, which probably gives further protection to the hermit crab while it grows, without the need of switching shells (Dunn et al., 1980; Gusmão and Daly

2010), as it has also been reported for the genus *Stylobates* in tropical seas (Dunn and Liberman, 1983; Fautin, 1987, 1992).

Protection against predators is a benefit for the sea anemones too, since symbiosis with hermit crabs ensures their mobility, in addition with their active defence by hermit crabs against animals which endeavour to prey on their symbiotic sea anemones (Brooks and Gwaltney, 1993). Moreover, sea anemones increase their dispersal capability via hermit crab mobility (Balss, 1924), gaining suitable substrata for their settlement (Nyblade, 1966; Riemann-Zürneck, 1994).

Increased exploitation of food sources by sea anemones has also been reported as a consequence of hermit crab mobility. For example, the sea anemones of the species *Calliactis parasitica* when settled on stable substrata (e.g. rocky) are able to exploit food supplies from only a limited area (ca. 0.5 m^2/day), whereas they are able to move up to 20 m^2/day due to symbiosis, thus, increasing their feeding potential (Stachowitsch 1979, 1980). Increased food supplies for the sea anemones can also derive from the food residuals of hermit crabs (Ross, 1960; Stachowitsch, 1979, 1980; Chintiroglou and Koukouras, 1991; Fautin, 1992). The exact position the sea anemones are placed on the shell has also proved to be important, as the closer they are to the shell aperture, and thus to the hermit crab, the more they benefit during its feeding (Balasch et al., 1977; Brooks, 1989); however, the sea anemone is often placed on the top of the shell, which may potentially increase their accessibility to suspended particulate organic matter from the water column. The anemones are usually oriented with their mouth below the shell aperture, to increase protection and allow their host to avoid changing shells when it grows (Ross, 1974b). Direct feeding of the sea anemones by their associated hermit crabs has also been mentioned in the literature (e.g. Wortley, 1863; Fox, 1965), being though a rather controversial possibility (Ross, 1974a).

Apart from the positive outcomes for both hermit crabs and sea anemones, the symbiosis has a great importance for biodiversity in marine benthic ecosystems, too. It is broadly known that gastropod shells that are inhabited by hermit crabs host also a variety of other organisms (epibiotic and endolithic), thus, formatting small biotic communities (Conover, 1979; Stachowitsch, 1980; Hazlett, 1984; McClintock, 1985; Caruso et al., 2003; Turra, 2003; Williams and McDermott, 2004).

Although gastropod abundance and distribution are important for the establishment of such micro-communities (McLean, 1983), hermit crabs have also a key-role. They prolong the presence of empty gastropod shells on the sea bottom by occupying them, avoiding their burial in soft sediments in the opposite situation (Conover, 1975, 1979), and thus, the shells can be available as substrata and colonized by a great diversity of organisms (McLean, 1983; Williams and McDermott, 2004). As a result, the abundance and distribution of hermit crabs, and the selection of shells, affect the abundance and distribution of a variety of organisms, which use the shells as micro-habitats. With respect to this function of hermit crabs, they had been characterized as allogenic ecosystem engineers, which are defined as these organisms able to transform biotic or abiotic substances from one physical situation to another (Jones et al., 1997; Gutiérrez et al., 2003; Jones and Gutiérrez, 2007).

Sea Anemones - Hermit Crabs Symbiosis: Diversity and Distribution Patterns

In the comprehensive review of hermit crab associated species, Williams and Mc Dermott (2004) reported 37 species of sea anemones living as symbionts with hermit crabs, whereas Gusmão (2010) reduced the number of associate sea anemone species to 32. According to our revision a total of 35 valid sea anemone species belonging to 14 genera (*Adamsia, Aiptasia, Antholoba, Calliactis, Carcinactis, Gonactinia, Hormathia, Neoaiptasia, Paracalliactis, Paranthus, Sagartiogeton, Sagartiomorphe, Stylobates, Verrillactis*) and seven families (Actiniidae, Actinostolidae, Aiptasiidae, Gonactiniidae, Hormathiidae, Sagartiidae, Sagartiomorphidae) have been reported as hermit crab symbionts (see Table 2). The vast majority of those species belong to Hormathiidae family (22 valid species), whereas other three sea anemone species are under uncertain taxonomic status (i.e. *Paracalliactis mediterranea, P. japonica* and *Verrillactis guttata*). Hermit crabs of 41 species hosted sea anemones (Table 2); those species belong to 15 genera (*Anapagurus, Catapaguroides, Catapagurus, Clibanarius, Dardanus, Diacanthurus, Diogenes, Lophopagurus, Micropagurus, Oncopagurus, Paguristes, Pagurus, Parapagurus, Petrochirus, Sympagurus*) and three families (Diogenidae, Paguridae, Parapaguridae).

Overall, 68 different types of sea anemones - hermit crabs symbiosis, have been reported in the literature up to date. The hermit crab *Dardanus arrosor* appeared to host the larger diversity of sea anemones, i.e. seven species, followed by *Pagurus alatus, P. bernhardus, P. cuanensis* and *Paguristes eremita* that were found in symbiosis, each, with three different anemone species. The sea anemone *Calliactis polypus* is involved in symbiosis with eight hermit crab species, followed by *C. parasitica* that has been found on the shells of seven hermit crabs; *C. tricolor* and *Adamsia palliata* are associated with six hermit crabs, and *Verrillactis paguri* with five. The rest hermit crab and sea anemone species appeared to be more specialized as they have been reported associated with one or two different species.

Considering diversity of shell utilization, whether hermit crabs prefer the shells of specific gastropod species remains unknown (see also Ates et al., 2007), and in most cases the abundance of shells seems to be the major factor influencing shell utilization (Kellogg, 1976; Barnes, 1999). Vafeiadou et al. (2011) studying shell resource utilization of hermit crab species in symbiosis with *Calliactis parasitica* in the Mediterranean, reported that 53 different shells are occupied by the four hermit crabs: *Dardanus arrosor, D. calidus, Pagurus excavatus* and *Paguristes eremita*, associated with *C. parasitica* (Figure 1). All crabs utilized a large variety of discarded shells, although a preference for specific gastropods has also been suggested, at least for some species.For example *Pagurus excavatus* inhabits 17 different species, but in most cases it was found in *Bolinus brandaris* and *Galeodea echinophora* shells, while *Paguristes eremita* most frequently occupied *Hexaplex trunculus* and *B. brandaris* shells, although it is occasionally found in the shells of other 33 gastropod species (Vafeiadou et al., 2011). A selective behaviour of hermit crabs towards the size of shells has been suggested (Childress, 1972; Chintiroglou et al., 1992; Wada et al., 1997; Côté et al., 1998; Caruso et al., 2003); nevertheless, selectivity to shells of certain gastropod species remains doubtful and further research is necessary to elucidate relevant patterns.

Table 2. Taxonomic list and temperate zone distribution of sea anemone and hermit crab species reported to live in symbiosis; ? Refers to species under uncertain taxonomic status (participant species data based on Williams and Mc Dermott, 2004; taxonomic status checked with World Register of Marine Species; distribution data based on Fautin, 2008 and Ocean Biogeographic Information System)

Sea anemone species	Temperate zone distribution
Actiniidae	
Stylobates aeneus Dall, 1903	
Stylobates cancrisocia (Carlgren, 1928)	Indian
Stylobates loisetteae Fautin, 1987	
Actinostolidae	
Antholoba achates (Drayton in Dana, 1846)	SW Atlantic, SE SW Pacific
Paranthus rapiformis (Le Sueur, 1817)	NW SW Atlantic
Aiptasiidae	
Aiptasia sp.	
Neoaiptasia commensali Parulekar, 1969	
Gonactiniidae	
Gonactinia prolifera (Sars, 1835)	NE NW Atlantic, SE Pacific
Hormathiidae	
Adamsia obvolva Dally et al., 2004	NW Atlantic
Adamsia palliata (Muller 1776)	NE Atlantic, Mediterranean
Adamsia sociabilis Verrill, 1882	NW Atlantic
Calliactis algoaensis Carlgren 1938	Indian
Calliactis argentacolorata Pei, 1996	
Calliactis conchiola Parry 1952	SW Pacific
Calliactis japonica Carlgren, 1928	NW Pacific
Calliactis parasitica (Couch, 1842)	NE Atlantic, Mediterranean
Calliactis polypores Pei, 1996	NW Pacific
Calliactis polypus (Forskal, 1775)	NW Atlantic, NW SW NE SE Pacific, Indian
Calliactis reticulata Stephenson, 1918	SW Atlantic
Calliactis tricolor (Le Sueur 1817)	NW SW Atlantic
Calliactis variegata Verrill, 1869	SE Pacific
Calliactis xishaensis Pei, 1996	
Hormathia coronata (Gosse, 1858)	NE Atlantic, Mediterranean, Indian
Paracalliactis consors (Verrill, 1882)	N Atlantic
Paracalliactis lacazei Dechance and Dufaure, 1959	Mediterranean
Paracalliactis mediterranea Ross and Zamponi, 1982?	Mediterranean
Paracalliactis michaelsarsi Carlgren 1928	NE NW Atlantic
Paracalliactis japonica Carlgren 1928 ?	NW Pacific
Paracalliactis rosea Hand 1976	SW Pacific
Paracalliactis sinica Pei, 1982	NW Pacific
Paracalliactis stephensoni Carlgren 1928	NE Atlantic
Paracalliactis valdiviae Carlgren 1928	Indian

Sea anemone species	Temperate zone distribution
Sagartiidae	
Carcinactis dolosa Riemann-Zurneck, 1975	SW Atlantic
Carcinactis ichikawai Uchida, 1960	NW Pacific
Sagartiogeton undatus (Muller, 1788)	NE Atlantic, Mediterranean
Verrillactis guttata (Agassiz in Verrill, 1864)?	N Atlantic
Verrillactis paguri (Verrill, 1869)	NW SE Pacific, Indian
Sagartiomorphidae	
Sagartiomorphe carlgreni Kwietniewski, 1898	SW NW Pacific
Hermit crab species	
Diogenidae	
Clibanarius erythropus (Latreillei, 1818)	NE Atlantic, Mediterranean
Clibanarius padavensis De Mann, 1888	
Clibanarius vittatus (Bosc, 1802)	NW SW Atlantic
Dardanus arrosor Herbst, 1796	NE SE Atlantic, Indian, Mediterranean, NW SW Pacific
Dardanus calidus (Risso, 1827)	NE Atlantic, Mediterranean
Dardanus deformis (H. Milne Edwards, 1836)	SE SW Pacific, Indian
Dardanus impressus (De Haan, 1849)	NW Atlantic
Dardanus lagopodes (Forskal, 1775)	
Dardanus pedunculatus (Herbst, 1804)	NW SW SE Pacific, Indian
Dardanus tinctor (Forskal, 1775)	
Dardanus venosus (H. Milne Edwards, 1848)	NW SW Atlantic
Diogenes custos (Fabricius, 1798)	
Diogenes edwardsii (De Haan, 1849)	NW Pacific
Diogenes sp.	
Paguristes eremita (Linnaeus, 1767)	Mediterranean
Paguristes subpilosus (Henderson, 1888)	SW Pacific
Petrochirus diogenes (Linnaeus, 1767)	NW SW Atlantic
Paguridae	
Anapagurus chiroacanthus (Lilljeborg, 1856)	NE Atlantic, Mediterranean
Anapagurus laevis (Bell, 1846)	NE Atlantic, Mediterranean
Catapaguroides fragilis (Melin, 1939)	
Catapagurus sharreri A. Milne Edwards, 1880	NW Atlantic
Diacanthurus rubricatus (Henderson, 1888)	SW Pacific
Lophopagurus lacertosus (Henderson, 1888)	SE SW Pacific
Micropagurus polynesiensis (Nobili, 1906)	
Pagurus alatus Fabricius, 1775	NE Atlantic, Mediterranean
Pagurus bernhardus (Linnaeus, 1758)	NW NE SW SE Atlantic
Pagurus cuanensis Bell, 1846	NW NE Atlantic, Mediterranean
Pagurus excavatus (Herbst, 1791)	Mediterranean
Pagurus forbesi Bell, 1846	NE Atlantic, Mediterranean
Pagurus impressus (Benedict, 1892)	NW Atlantic

Table 2. (Continued)

Sea anemone species	Temperate zone distribution
Pagurus longicarpus Say, 1817	NW Atlantic
Pagurus pollicaris Say, 1817	NW Atlantic
Pagurus prideaux Leach, 1815	NE Atlantic, Mediterranean Indian
Parapaguridae	
Oncopagurus bicristatus (A. Milne Edwards, 1880)	NW NE Atlantic
Parapagurus pilosimanus Smith, 1879	NE SE NW Atlantic, Mediterranean, Indian, NE NW Pacific
Parapagurus sp.	
Sympagurus andersoni (Henderson, 1896)	Indian
Sympagurus dimorphus (Studer, 1883)	SW SE Atlantic, Indian, SE SW Pacific
Sympagurus dofleini (Balss, 1912)	SE SW Pacific
Sympagurus pictus Smith, 1883	NW Atlantic
Sympagurus trispinosus (Balss, 1911)	SE SW Pacific, Indian

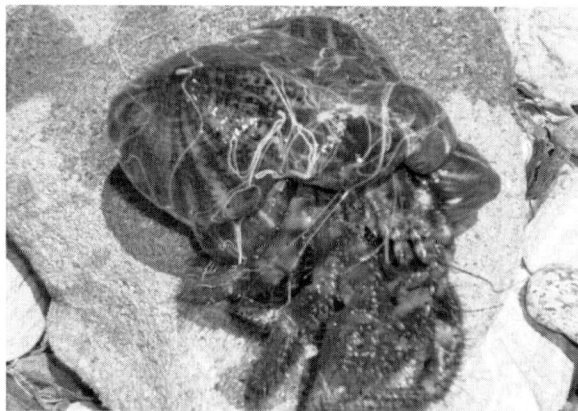

Figure 1. Sea anemone – hermit crab symbiosis: specimens of the sea anemone *Calliactis parasitica* in symbiosis with the hermit crab *Pagurus excavatus*, in Thermaikos Gulf (north Aegean Sea) using a *Bolinus brandaris* shell (above) and with *Dardanus calidus*, in Sifnos Island (Cyclades plateau, South Aegean Sea) using a *Phalium granulatum* shell (below).

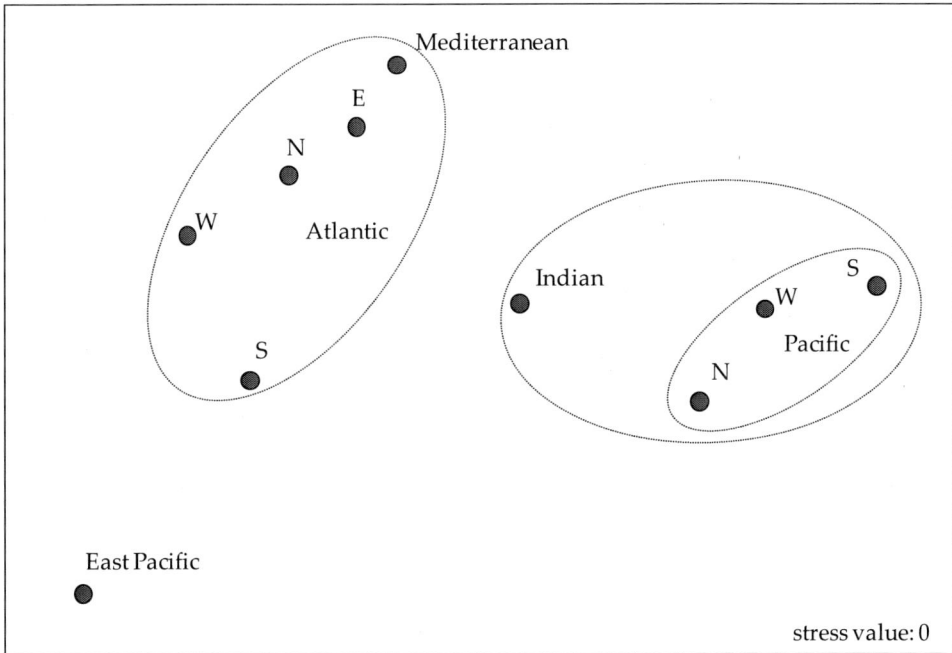

Figure 2. Temperate zone distribution of sea anemone - hermit crab symbiosis types as visualized by applying multi-dimensional scaling ordination via Bray-Curtis distances on presence – absence data (N = north, S = south, E = east, W = west).

Out of the 68 different types of sea anemones - hermit crabs symbiosis, 53 are distributed in temperate zones. Their biogeographic distribution, as visualized applying non-metric multidimensional scaling ordination via Bray-Curtis distances on presence – absence data (Figure 2), revealed the increased affinity in symbiotic types' composition between: (i) the Mediterranean and the Atlantic and (ii) within the Pacific Ocean with the exception of its eastern part, where very few such associations have been reported. The symbiotic types reported from the Indian Ocean (data from tropical zone excluded from the analysis) showed increased similarity with the Pacific group, although to a smaller degree (Figure 2).

These results conform to the findings of Ross (1974a) which documented that sea anemone - hermit crab symbiosis types cover mainly the circumtropical zone, extending also to some warm-temperate areas such as the Mediterranean Sea, and differ in their qualitative structure according to the geographic region in a global scale; accordingly the author claimed the existence of different zoogeographic zones. An analogous pattern has been revealed in a much smaller spatial scale, i.e. over the Aegean Sea (Vafeiadou et al., 2011). According to the latter authors, the symbiotic types, even when only one sea anemone species has been considered, followed a consistent pattern of spatial distribution according to the geographic areas studied. After considering the diversity of shell utilisation as well, a similar pattern emerged, concurring to the recently proposed latitudinal cline of shell resource utilisation by hermit crabs (Barnes, 2003). Therefore, this trend in the distribution of sea anemones - hermit crabs symbiosis types may be useful in biogeography studies.

Sea Anemones - Hermit Crabs Symbiosis: Manifested Behavioural Patterns

How beneficial symbiosis is for both sea anemones and hermit crabs has been already discussed (see section 2, general description of the symbiosis); nevertheless, specific behavioural patterns promoting their association exhibited by both symbionts, and verify once again its importance. Since the development of a symbiotic relationship depends on specific actions driven by particular behaviours of the participant species, behavioural patterns exhibited by sea anemones and hermit crabs (summarized in Table 3) are reported in this part, giving also examples of specific cases.

As aforementioned, the establishment of the sea anemones - hermit crabs symbiotic interaction is primarily based on hermit crabs for they detach sea anemones from their substrata, using tactile stimulation, and place them on the gastropod shells they inhabit (Brunelli, 1910; Cowles, 1919; Ross, 1970). Hermit crabs of the species *Dardanus arrosor*, known to host sea anemones of the species *Calliactis parasitica* in temperate seas, move the anemones by manipulating their base little by little, a behaviour performed, though, only by female individuals (Ross, 1967). Prior species recognition and selectivity by hermit crabs towards particular sea anemone species has been mentioned in the literature (Ross, 1974b; Brooks and Mariscal, 1986); however, such behaviour has not so far been confirmed by experimental results. On the other extreme, the hermit crab species *Pagurus alatus* does not facilitate the anemone transfer and settlement on the shell (Ross and Zamponi, 1982).

Table 3. Synopsis of the main behavioural patterns manifested by sea anemones and hermit crabs to enhance symbiosis; (?) refers to uncertain behaviours

Hermit crabs	Sea anemones
Detachment / transfer of sea anemones (using tactile stimulation)	Facilitation of its detachment by the hermit crab
Preference towards particular sea anemone species (?)	Active transfer on the shell without the participation of the host
Placement of increased number of sea anemones on the shell under predator presence	
Preference towards shells with increased number of sea anemones under predation stress	
Arrangement of sea anemones on the shell (balance, predation stress)	
Symmetric placement of sea anemones on the shell (?)	
Intra- and inter-specific competition for sea anemones	
Plasticity on shell selectivity patterns (depending on various factors and previous experience)	
Intra- and inter-specific competition for gastropod shells	

Active participation of the sea anemones during their detachment from the substratum by the hermit crabs is one of the most interesting behaviours in the symbiosis (Balasch and Mengual, 1974; Ross and Boletzky, 1979; Bach and Herrnkind, 1980). It has been reported that sea anemones are loosening their connection with the substratum during their manipulation by hermit crabs to enhance their transfer on the shells (Ross, 1974b; Lawn, 1976; McFarlane, 1976). According to Ross (1979), although sea anemones move themselves from rocky substrata to attach on gastropod shells, they do not seem to actively change their shell substratum for another, but only when getting transferred by their host crab. Nevertheless, experiments revealed that they may transfer themselves by tentacle adhesion (followed by pedal disc attachment) to a shell inhabited by a hermit crab – without the active participation of it – under conditions of high predation risk for the latter, e.g. upon perception of mollusc presence (Davenport et al., 1961; Ross, 1959, 1965; Ross and Sutton, 1961). This particular behaviour has been observed in four species of the genera *Calliactis* and *Paracalliactis* (Gusmão and Daly, 2010) and has been characterized as one of the most complex behaviours of cnidarians (Ross, 1974b).

The perception of predator presence has as a result the active behaviour of hermit crabs too (Balasch and Mengual, 1974; Ross and Boletzky, 1979), which prefer to inhabit shells with more sea anemones, and/or place more sea anemones on their shell under increased density of predators, in comparison with predator absence circumstances (Balasch and Mengual, 1974; Ross and Boletzky, 1979; Brooks and Mariscal, 1986; Brooks, 1989). The placement of sea anemones on the shell is also influenced by predation stress, with anemones being typically placed close to the aperture of the shell, a key-position for better protection (Cutress and Ross, 1969; Brooks, 1988, 1991); though, the balance of the crab is first and foremost considered, with anemones being arranged in accordance with the center of gravity of the shells (Balasch et al., 1977; Brooks, 1989; Caruso et al., 2003).

Preference towards a symmetric placement of the sea anemones by hermit crabs has also been assumed, in particular for the species *Dardanus pedunculatus* living in symbiosis with *Calliactis tricolor* in reef ecosystems (Giraud, 2011). The author mentioned a consistent pattern, probably related to the balance of the shell, although this specific study does not use anemone weight distribution data. Additionally, this behaviour of non-random but symmetrical anemone placement by the hermit crab could be related to shell cover with sea anemones in a way to maximize protection, without necessarily needing a large number of them (Giraud, 2011) and thus, reducing the energy costs of the crab by carrying a heavier shell. Another remarkable behavioural pattern manifested by hermit crabs is their strategy for gathering more sea anemones, including intra- and inter-specific competition. Accordingly, they steal sea anemones from the shells of other hermit crabs (Mainardi and Rossi, 1969; Ross, 1974b, 1979), which sometimes might even be of the same species (Giraud, 2011). As an example, the hermit crab species *Dardanus arrosor* appears to dominate over *Pagurus excavatus* or *Paguristes eremita* when they occur at the same habitat, stealing their symbiotic anemones as a result of antagonism (Ross, 1979). The size of both the hermit crab (and in particular the size of its cheliped) and its shell are considered as the main factors for its competitive dominance (Giraud, 2011; Yasuda, 2011; Yoshino, 2011), giving the advantage to larger individuals and/or species.

In spite of its indirect benefit, gastropod shell selectivity by hermit crabs is a very important aspect for the well-establishment of the symbiosis, and should not be neglected from the behavioural patterns manifested by hermit crabs. Numerous studies have focused on shell selection behaviours of hermit crabs (e.g. Reese, 1962; Balasch and Cuadras, 1976; Fotheringham, 1976; Hazlett, 1978, 1984, 1992; Abrams, 1982; Dowds and Elwood, 1983, 1985; McClintock, 1985; Liszka and Underwood, 1990; Wada et al., 1997; Côté et al., 1998;

Hahn, 1998; Osorno et al., 1998). Experimental studies suggest gastropod mass, weight, total size (McClintock, 1985) and protective ability (see Reese, 1962), as the main factors that influence the selection by hermit crabs (Buckley and Ebersole, 1994). Intrinsic shell properties (e.g. shape, spines, center of gravity, shell axis) are other important features in shell selection by hermit crabs (Reese, 1963; Caruso and Chemello, 2009), influencing also the placement of sea anemones on the shell (Ross and Boletzky, 1979; Brooks, 1989).

According to Wada et al. (1997), the preferential shell size for a hermit crab depends on the growth rate of the latter. The same authors showed that hermit crabs tend to occupy larger shells in proportion with their size in the following cases: (i) when shell resource availability is restricted, (ii) when they are going to change their exoskeleton, and (iii) when the growth of their body size after the next moulting phase is expected to be large. On the contrary, the size of the gastropod shell may influence the rate of the hermit crab growth (Wada et al., 1997), a fact which illustrates the complexity of these associations. For instance, by selecting an oversized shell in proportion to its size, the hermit crab may on the one hand delay its searching for a larger shell during its growth, as to assure its further growth and reproduction (Childress, 1972; Wada et al., 1997; Côté et al., 1998), and on the other hand gain some advantage over antagonists; however, the energy cost is much higher. Occupying a shell that is too large could negatively affect growth and fecundity of the crab and its ability to protect itself from predators (Vance, 1972; Bertness, 1981; Elwood et al., 1995). Hermit crabs seem to select suitable shells not only with respect to their size but also regarding the environmental conditions, as for instance the strength of marine currents, showing a preference towards stronger/heavier shells under strong current conditions (Hahn, 1998), balancing the energetic constraints of carrying a heavier shell by increasing their protection.

Among the most important factors affecting the choice of an adequate shell by hermit crabs should also be considered their previous experience on shell selection, beginning from the early stage of their life (Gilchrist, 1985; Hazlett, 1992; Hahn, 1998; Gherardi, 2006). The preference hermit crabs show towards shells of specific gastropod species has been also discussed to be related to such previous experience (Reese, 1963; Elwood et al., 1979; Borjesson and Szelistowski, 1989). According to Hazlett (1992), individual hermit crabs can also adjust their preferences on shell size/type depending on recent shell availability experience. In spite of the importance of species-specific selectivity, whether such behaviours are typical, or exhibited only by some hermit crab species, or even only in particular cases related with shell availability, is very hard to be explicitly demonstrated, and thus remain uncertain.

An exceptional behaviour of hermit crabs, under conditions of limited shell resources, includes their fighting for a more suitable (better-fitting) gastropod shell than they already have (Abrams, 1982; Dowds and Elwood, 1983, 1985; Gherardi, 2006). These fights seem to either benefit both antagonists, as at the end they both gain a better shell than what they initially had, or only the stronger crab (Hazlett, 1978; Abrams, 1982). As a result of these competitions, or possibly of the lack of previous experience in shell selection, smaller hermit crabs usually end up carrying less suitable shells. The ability of larger crabs to obtain more suitable shells creates a pressure over smaller individuals to inhabit the remaining ones, without the possibility of selection; this behaviour is considered indicative of the crabs' "high social status", according to Balasch and Cuadras (1976).

Analogous behaviours have been confirmed by examining the biometric relationships between hermit crab weight and shell weight or total sea anemone biomass, for the species *D.*

calidus, *D. arrosor* and *P. excavatus* in symbiosis with *C. parasitica* in the Aegean Sea, SE Mediterranean (Vafeiadou et al., 2011). The results of this study revealed that smaller hermit crabs carry heavier shells and increased anemone biomass in proportion to their weight. Analogous observations have been previously reported by two other studies (Balasch and Cuadras, 1976; Chintiroglou et al., 1992) which examined the biometric relationships between the shell and the symbionts, referred as biometric indicators (e.g. shell weight / crab weight, shell and anemone weight / crab weight). Such biometric relationships are used to describe the ability of the crab to carry its shell and the latter's protective capacity, and their application, at the very end, can give an approximation of the functionality of the symbiosis.

Sea Anemones - Hermit Crabs Symbiosis: Co-Evolution of the Participant Species

Reciprocal altruism is among the first theories proposed to explain the evolution of mutualism; such an interaction can be develop and maintained when individuals interact by providing benefit to another in the expectation of future reciprocation, as in the case of marine cleaning behaviour (Trivers, 1971). Reciprocal altruism has been formalized in the iterated Prisoner's Dilemma game (two individuals that can defect or cooperate, receive a high payoff from defection independently of partner's behaviour but receive higher payoff if they cooperate than if both defect). Thereafter, other approaches also emerged, including by-product mutualism (partners act selfishly but a benefit results from their behaviour), pseudoreciprocity (at least one partner invests to cooperation), and biological market theory (partners exchange goods or commodities but differ in the degree of controlling theme); for details on above concepts see Grutter and Irving (2007).

Sea anemones - hermit crabs mutualistic symbiosis is characterized by increased complexity being affected by a great variety of factors (e.g. shell resource availability, predation, behavioural patterns), as thoroughly discussed in previous sections. The development of the symbiosis depends on both members, as aforementioned (see previous section), with both sea anemones and hermit crabs exhibiting behavioural patterns enhancing their symbiosis. Therefore, their interaction constitutes a model case to examine species co-evolution under symbiosis, and in particular under mutualism.

Evolutionary aspects of hermit crab symbiotic interactions have been thoroughly investigated by Williams and Dermott (2004). According to the latter authors, and despite the poor representation of hermit crab exoskeletons in the fossil records, hermit crabs seem to have provided a new niche for epibiotic organisms in marine ecosystem during the middle Jurassic (Walker, 1992). Shell resource utilization by hermit crabs has been hypothesized to develop initially for refuge and protection of their abdomen which became decalcified when posterior pereopods and uropods were modified to fit the animal in shells and pleopods were placed on one side to maximize utilization of gastropod lumen during reproduction (McLaughlin, 1983).

The knowledge on the shared evolutionary history of sea anemones and hermit crabs remains limited and it is mostly based on behavioural patterns followed by the symbionts. Ross (1974a, 1983) in his pioneer work of sea anemone - hermit crab symbiosis comprehensively studied evolutionary aspects, tried to elucidate possible drivers and hypothesized that the symbiosis evolved independently multiple times.

This latter hypothesis has been recently supported by molecular data presented by Gusmão and Daly (2010) who provided strong evidences of at least two independent origins of the sea anemones - hermit crab symbiosis, by constructing a phylogenetic tree of the sea anemone family Hormathiidae (a family that includes the vast majority of sea anemone genera having symbiotic interactions with hermit crabs).

Moreover, the widely accepted idea of close evolutionary relation among sea anemone genera symbiotic with hermit crabs, which has been assumed on the basis of common morphological and behavioral patterns, has been currently rejected on the basis of phylogenetic data; monophyly in the origin of the three symbiotic with hermit crabs sea anemone genera examined, i.e. *Calliactis*, *Adamsia* and *Paracalliactis*, has not been supported but evidences of paraphyly emerged (Gusmão and Daly, 2010). Accordingly, the reported similarities in morphology and behaviour of some sea anemone genera forming symbiotic interactions with hermit crabs is not due to shared evolutionary history but due to the necessary ways for the development and maintenance of symbiosis, as explicitly stated by Gusmão and Daly (2010).

Two main hypotheses have been suggested by Ross (1974a) to explain possible leading factors to the development of the sea anemones - hermit crabs symbiosis: (i) the "crab-driven" and (ii) the "shell-response" hypotheses which are driven by the behaviour of hermit crab and sea anemone, respectively, and have been subsequently adopted and analysed by other authors (Williams and Dermott, 2004; Gusmão and Daly, 2010).

According to the first hypothesis the initial establishment of the symbiosis is founded on hermit crabs behaviour of placing sea anemones on their residence shells to be protected, i.e. hidden from predators by camouflage, which, however, evolved afterwards to an actual mechanism of defence. Under this hypothesis a clear benefit emerges for the hermit crab increasing its fitness (Gusmão and Daly, 2010). According to the second hypothesis the development of the symbiosis is based on the sea anemone behaviour of shell mounting. In this case sea anemones firstly settled on living gastropod shells and later started also to utilize shells occupied by hermit crabs as the settlement of the anemone is stimulated by a shell factor stronger on alive than on discarded gastropod shells. Sea anemones, besides gaining novel habitat, benefit by transportation; thus settlement behaviour reinforced toward shells occupied by hermit crabs, since they are much more mobile than gastropods.

The most important evidence supporting the first hypothesis is that in most cases the symbiosis of sea anemones with a hermit crab is initiated under the activity of the crab, while sea anemones are more frequently found on shells occupied by crabs than on living gastropods, even in areas with dense gastropod populations. In favor of the second hypothesis is the exclusive presence of some anemones on living gastropods, such as the species *Allantactis parasitica* and *Hormanthia digitata*, the ability of some other anemones to actively move on gastropod shells, and the equal presence of some other species, e.g. *Calliactis conchiola*, on both living gastropods and shells occupied by hermit crabs (Hand, 1975). Whatever was the initial behavioural pattern stimulating the establishment of sea anemones - hermit crabs symbiosis, both patterns positively responded. Hermit crabs, after having their residence shells being occupied by sea anemones, started to benefit under their protection against predators, and evolved a specialized behaviour of actively enhancing anemone colonization of their shells. Sea anemones, after being picked up by the hermit crab, started to benefit from transportation, and evolved a positive respond to their stimulation by the crab, as Ross (1974b) showed with manipulative laboratory experiments (i.e. only those anemones that were previously symbiotic with hermit crabs responded to tactile stimulation by the latter). Overall, the limited number of sea anemones living on gastropods or inactive crabs (Gusmão and Daly, 2010) and the very strong pattern manifested by several hermit crabs of stealing sea anemones from other ones (Ross, 1979), argue against the "shell-response" hypothesis, which however, has been preferred to some extent by Ross (1974a).

Sea Anemones - Hermit Crabs Symbiosis: Summarized Conclusive Remarks

The sea anemones - hermit crabs symbiosis represents a clear example of mutualism, as it has reciprocal advantages for both symbionts. The partners' interaction is characterized by increased complexity as the establishment of the symbiosis depends on a large variety of factors such as shell resource availability, predation pressure and environmental constraints, and involves the cooperation of both participants in most cases. Well-developed behavioural patterns exhibited by both symbionts, including from the sea anemones' active transfer on shells inhabited by hermit crabs to the behavioural plasticity of crabs in view of shell utilization and gathering of sea anemones, determine the development of the symbiosis and confirm its importance for both participants, making them excellent models to examine species co-evolution under a mutual symbiotic context. Several species of sea anemones and hermit crabs frequently form symbiotic interactions in temperate marine environments providing benefits, not only to the directly involved partners, but also to other organisms, which colonize this complex biotic formation. Thus, through the intermingle processes of epibiosis and ecosystem engineering, sea anemones - hermit crabs symbiosis contribute to the diversity of marine benthic ecosystems by supporting diverse micro-communities.

References

Abrams PA, 1982. Frequencies of interspecific shell exchanges between hermit crabs. *Journal of Experimental Marine Biology and Ecology* 61: 99-109.

Ates RML, 1995. Pagurus prideaux and Adamsia palliata are not obligate commensals. *Crustaceana* 68: 522-524.

Ates AS, Katagan T, Kocatas A, 2007. Gastropod shell species occupied by hermit crabs (Anomura: Decapoda) along the Turkish coast of the Aegean Sea. *Turkish Journal of Zoology* 31: 13-18.

Bach CE, Herrnkind WF, 1980. Effects of predation pressure on the mutualistic interactions between the hermit crab *Pagurus pollicaris* Say, 1817 and the sea anemone *Calliactis tricolor* (Lesueur, 1817). Crustaceana 38: 104-108.

Balasch J, Mengual V, 1974. The behavior of Dardanus arrosor in association with Calliactis parasitica in artificial habitat. *Marine Behaviour and Physiology* 2: 251-260.

Balasch J, Cuadras J, 1976. Role of association with *Calliactis parasitica* (Couch) in social behaviour of *Dardanus arrosor* (Herbst). Vie et Milieu 26: 281-291.

Balasch J, Cuadras J, Alonso G, 1977. Distribution of Calliactis parasitica on gastropod shells inhabited by Dardanus arrosor. *Marine Behaviour and Physiology* 5: 37-44.

Balss H, 1924. Uber anpassungen und symbiose der Paguriden eine zusammenfassende uber-sicht. Zeitschriften Okologie Morphologie Tiere 1: 752-792.

Barnes DKA, 1999. Ecology of tropical hermit crabs at Quirimba Island, Mozambique: shell characteristics and utilisation. *Marine Ecology Progress Series* 183: 241–251.

Barnes DKA, 2003. Local, regional and global patterns of resource use in ecology: hermit crabs and gastropod shells as an example. *Marine Ecology Progress Series* 246: 211–223.

Bertness MD, 1981. The influence of shell-type on hermit crab growth rate and clutch size (Decapoda. Anomura). *Crustaceana* 40: 197-205.

Borjesson DL, Szelistowski WA, 1989. Shell selection, utilization and predation in the hermit crab Clibanarius panamensis Stimpson in a tropical mangrove estuary. *Journal of Experimental Marine Biology and Ecology* 133: 213-228.

Boucher DH, James S, Keeler KH, 1982. The ecology of mutualism. *Annual Review of Ecology and Systematics* 13: 315-347.

Brooks WR, 1988. The influence of the location and abundance of the sea anemone Calliactis tricolor (Lesueur) in protecting hermit crabs from octopus predators. *Journal of Experimental Marine Biology and Ecology* 116: 15-21.

Brooks WR, 1989. Hermit crabs alter sea anemone placement patterns for shell balance and reduced predation. *Journal of Experimental Marine Biology and Ecology* 132: 109–121.

Brooks WR, 1991. Chemical recognition by hermit crabs of their symbiotic sea anemones and a predatory octopus. *Hydrobiologia.* 216-217: 291-295.

Brooks WR, Mariscal RN, 1986. Population variation and behavioral changes in two pagurids in association with the sea anemone *Calliactis tricolor* (Lesueur). *Journal of Experimental Marine Biology and Ecology* 103: 275-289.

Brooks WR, Gwaltney CL, 1993. Protection of symbiotic cnidarians by their hermit crab hosts: evidence for mutualism. *Symbiosis* 15: 1-13.

Brunelli G, 1910. Osservasioni ed esperienze sulla simbiosi dei Paguridi e delle Attinie. Atti della Academia Nazionale dei Lincei, Classe si Scienze Fisiche, Mathematiche e Naturali, Rendiconti 19: 77-82.

Bruno JF, Stachowicz JJ, Bertness MD, 2003. Inclusion of facilitation into ecological theory. *Trends in Ecology and Evolution* 18: 119-125.

Buckley WJ, Ebersole JP, 1994. Symbiotic organisms increase the vulnerability of a hermit crab to predation. *Journal of Experimental Marine Biology and Ecology* 182: 49-64.

Caruso T, Chemello R, 2009. The size and shape of shells used by hermit crabs: a multivariate analysis of Clibanarius erythropus. *Acta Oecologica* 35: 349-354.

Caruso T, Falciai L, Zupo V, 2003. Decapoda Anomura Paguridea: morpho-functional relationships and influence of epibiotic anemones on shell use along a bathymetric cline. *Crustaceana* 76: 149–166.

Childress JR, 1972. Behavioral ecology and fitness theory in a tropical hermit crab. *Ecology* 53: 960-964.

Chintiroglou C, Koukouras A, 1991. Observations on the feeding habits of *Calliactis parasitica* (Couch, 1842) (Anthozoa, Cnidaria). *Oceanologica Acta* 14: 389-396.

Chintiroglou C, Doumenc D, Koutsoubas D, 1992. Allométrie d'une nouvelle association entre le Décapode Anomure Pagurus excavatus (Herbst, 1791) et l'Actinie Acontiaire Sagartiogeton undatus (Müller, 1788). *Crustaceana* 62: 1-12.

Christidis J, Chintiroglou C, Culley MB, 1997. A study of the populations of *Calliactis parasitica* (Couch, 1842) in symbiosis with anomuran decapods in Thermaikos Gulf (N. Aegean Sea). *Crustaceana* 70: 227-238.

Conover MR, 1975. Prevention of shell burial as a benefit hermit crabs provide to their symbionts (Decapoda, Paguridae). *Crustaceana* 29: 311-313.

Conover MR, 1978. The importance of various shell characteristics to the shell selection behavior of hermit crabs. *Journal of Experimental Marine Biology and Ecology* 32: 131-142.

Conover MR (1979) Effect of gastropod shell characteristics and hermit crabs on shell epifauna. *Journal of Experimental Marine Biology and Ecology* 40: 81-94.

Côté IM, Reverdy B, Cooke PK, 1998. Less choosy or different preference? Impacts of hypoxia on hermit crab shell assessment and selection. *Animal Behaviour* 56: 867-873.

Cowles RP, 1919. Habits of tropical crustacean: III Habits and reactions of hermit crabs associated with sea anemones. *Philippine Journal of Science* 15: 81-90.

Cutress CE, Ross DM, 1969. The sea anemone Calliactis tricolor and its association with the hermit crab Dardanus venosus. *Journal of Zoology* 158: 225-241.

Dauvin J.C., Bellan G., Bellan-Santini D., 2008. The need for a clear and comparable terminology in benthic ecology. Part I. Ecological concepts. Aquatic Conservation: *Marine and Freshwater Ecosystems* 18: 432-445.

Davenport D, Ross DM, Sutton L, 1961. The remote control of nematocyst discharge in the attachment of *Calliactis parasitica* to shells of hermit crabs. Vie et Milieux 12: 197-209.

Douglas A.E., 1994. *Symbiotic interactions*. Oxford University Press, Oxford, New York.

Douglas A.E., 2010. *The Symbiotic Habit*. Princeton University Press, Princeton, New Jersey.

Doumenc D, 1975. Actinies bathyales et abyssales de l'océan Atlantique nord. Families des Hormathiidae (genres *Paracalliactis* et *Phelliactis*) et des Actinostolidae (genres *Actinoscyphia* et *Sicyonis*). Bulletin de Musée Nationale de l'Histoire Naturelle 287: 1-201.

Dowds BM, Elwood RW, 1983. Shell wars: assessment strategies and the timing of decisions in hermit crab shell fights. *Behaviour* 85: 1-24.

Dowds BM, Elwood RW, 1985. Shell wars II: the influence of relative size on decisions made during hermit crab shell fights. *Animal Behaviour* 33: 649-656.

Dunn DF, Liberman MH, 1983. Chitin in sea anemone shells. *Nature* 221: 157-159.

Dunn DF, Devaney DM, Roth B, 1980. Stylobates: a shell-forming sea anemone (Coelenterata, Anthozoa, Actiniidae). *Pacific Science* 34: 379-388.

Elwood RW, McClean A, Webb L, 1979. The development of shell preference by the hermit crab *Pagurus bernhardus*. *Animal Behaviour* 27: 940-946.

Elwood RW, Mark N, Dick JTA, 1995. Consequences of shell-species preferences for female reproduction success in the hermit crab Pagurus bernhardus. *Marine Biology* 123: 431-434.

Faurot L, 1910. Etude sur les associations entre les Pagures et les Actinies; *Eupagurus prideauxi* Heller et *Adamsia palliata* Forbes, *Pagurus striatus* Latreille et *Sagartia parasitica* Cosse. Archives de Zoologie Experimentale et Generale 5: 421-486.

Fautin DG, 1987. *Stylobates loisetteae*, a new species of shell-forming sea anemone (Coelenterata: Actiniidae) from Western Australia. *Proceedings of the California Academy of Sciences* 45, 1-7.

Fautin DG, 1992. A shell with a new twist. Natural History 4: 50-57.

Fotheringham N, 1976. Population consequences of shell utilization by hermit crabs. *Ecology* 57: 570-578.

Fox HM, 1965. Confirmation of old observations on the behaviour of a hermit crab and its commensal sea anemone. *Annals and Magazine of Natural History* 13: 173-175.

Gherardi F, 2006. Fighting behavior in hermit crabs: the combined effect of resource-holding potential and resource value in Pagurus longicarpus. *Behavioral Ecology and Sociobiology* 59: 500–510.

Gilchrist SL, 1985. Ecology of juvenile hermit crabs shell use: Field and laboratory comparisons. *American Zoologist* 25: 60.

Giraud C, 2011. Intraspecific competition, stealing and placement of the symbiotic sea anemone *Calliactis tricolor* by the hermit crab *Dardanus pedunculatus*. Student Research Papers, Fall 2011, UCB Moorea Class: Biology and Geomorphology of Tropical Islands, Berkeley Natural History Museum, UC Berkeley.

Grutter AS, Irving AD, 2007. Positive interactions in marine communities. In: Conell SD, Gillanders BM (eds) Marine Ecology, Oxford, pp 110-137.

Gusmão LC, 2010. Systematics and evolution of sea anemones (Cnidaria: Actiniaria: Hormathiidae) symbiotic with hermit crabs. *Doctorate Dissertation*, Ohio State University.

Gusmão LC, Daly M, 2010. Evolution of sea anemones (Cnidaria: Actiniaria: Hormathiidae) symbiotic with hermit crabs. *Molecular Phylogenetics and Evolution* 56: 868-877.

Gutiérrez JL, Jones CG, Strayer DL, Iribarne OO, 2003. Mollusks as ecosystem engineers: the role of shell production in aquatic habitats. *Oikos* 101: 79-90.

Hahn DR, 1998. Hermit crab shell use patterns: response to previous shell experience and to water flow. *Journal of Experimental Marine Biology and Ecology* 228: 35-51.

Hand C, 1975. Behaviour of some New Zealand sea anemones and their molluscan and crustacean hosts. *New Zealand Journal of Marine and Freshwater Research* 9: 529-538.

Hazlett BA, 1978. Shell exchanges in hermit crabs: aggression, negotiation or both? *Animal Behavior* 26: 1278-1279.

Hazlett BA, 1981. The behavorial ecology of hermit crabs. *Annual Revue of Ecology and Systematics UK* 12: 1-22.

Hazlett BA, 1984. Epibionts and shell utilization in two sympatric hermit crabs. *Marine Behaviour and Physiology* 11: 131-138.

Hazlett BA, 1992. The effect of past experience on the size of shells selected by hermit crabs. *Animal Behaviour* 44: 203-205.

Henry SM, 1966. *Symbiosis*. I Associations of microorganisms, plants and marine organisms. New York and London, Academic Press.

Imafuku M, Yamamoto T, Ohta M, 2000. Predation on symbiont sea anemones by their host hermit crab Dardanus pedunculatus. *Marine Freshwater Behaviour and Physiology* 33: 221-232.

Jones CG, Gutiérrez JJ, 2007. On the purpose, meaning, and usage of the physical ecosystem engineering concept. In: Cuddington K, Byers JE, Wilson WC, Hastings A (eds) Ecosystem Engineers. Plant to protists, *Theoretical Ecology Series*, Academic Press, Elsevier, pp 3-24.

Jones CG, Lawton JH, Shachak M, 1997. Positive and negative effects of organisms as physical ecosystem engineers. *Ecology* 78: 1946-1957.

Kareiva P, 1982. Insects and Adaptions. *Science* 215: 658-659.

Kellogg CW, 1976. Gastropod Shells: a potentially limiting resource for hermit crabs. *Journal of Experimental Marine Biology and Ecology* 22: 101-111.

Lawn ID, 1976. The Marginal Sphincter of the Sea Anemone Calliactis parasitica. I. Responses of Intact Animals and Preparations. *Journal of Comparative Physiology* 105: 287-300.

Liszka D, Underwood AJ, 1990. An experimental design to determine preferences for gastropod shells by a hermit crab. *Journal of Experimental Marine Biology and Ecology* 137: 47-62.

Lom, J. 2001. *Protozoan Symbioses*. eLS.

Mainardi D, Rossi AC, 1969. Relations between social status and activity toward the sea anemone *Calliactis parasitica* in the hermit crab *Dardanus arrosor*. Atti della Academia Nazionale dei Lincei, Classe si Scienze Fisiche, Mathematiche e Naturali, Rendiconti 47: 116-121.

Martin D., Britayev T.A., 1998. Symbiotic polychaetes: review of known species. *Oceanography and Marine Biology an Annual Review* 36: 217-340.

McClintock TS, 1985. Effects of shell condition and size upon the shell choice behavior of a hermit crab. *Journal of Experimental Marine Biology and Ecology* 88: 271-285.

McFarlane ID, 1976. Two slow conduction systems co-ordinate shell-climbing behaviour in the sea anemone Calliactis parasitica. *Journal of Experimental Biology* 64: 431-445.

McLaughlin PA, 1983. Hermit crabs—are they really polyphyletic? *Journal of Crustacean Biology* 3: 608-621.

McLean R, 1983. Gastropod shells: a dynamic resource that helps shape benthic community structure. *Journal of Experimental Marine Biology and Ecology* 69: 151-174.

Nyblade CF, 1966. The association between Pagurus floridanus (Benedict) and Calliactis polypus (Forskal). *Journal of the Mississippi Academy of Sciences* 7: 232-241.

Osomo LL, Fernández_Casillas L, Rodríguez-Juárez C, 1998. Are hermit crabs looking for light and large shells?: Evidence from natural and field induced shell exchanges. *Journal of Experimental Marine Biology and Ecology* 222: 163-173.

Patzner RA, 2004. Associations with sea anemones in the Mediterranean Sea: a review. *Ophelia* 58: 1-11.

Reese ES, 1962. Shell selection behaviour of hermit crabs. *Animal behaviour* 10: 347-360.

Rhode K., 1981. The nature of parasitism. In: Australian Ecology Series: Ecology of marine parasites, H. Heatwole (ed), *University of Queensland Press*, 4-5.

Riemann-Zórneck K, 1994. Taxonomy and ecological aspects of the Subarctic sea anemones Hormathia digitata, Hormathia nodosa and Allantactis parasitica (Coelenterata, Actiniaria). *Ophelia* 39: 197-224.

Ross DM, 1959. The sea anemone (Calliactis parasitica) and the hermit crab (Eupagurus bernhardus). *Nature* 184: 1161–1162.

Ross DM, 1960. The association between the hermit crab Eupagurus bernhardus (L.) and the sea anemone Calliactis parasitica (Couch). *Proceedings of the Zoological Society of London* 134: 43-57.

Ross DM, 1965. Complex and modifiable behavior patterns in Calliactis and Stomphia. *American Zoologist* 5: 573-580.

Ross DM, 1967. Behavioral and ecological relationships between sea anemones and other invertebrates. *Oceanography and Marine Biology: an Annual Review* 5: 291-316.

Ross DM, 1970. The commensal association of Calliactis polypus and the hermit crab Dardanus gemmatus in Hawaii. *Canadian Journal of Zoology* 48: 351-357.

Ross DM, 1971. Protection of hermit crabs (Dardanus spp.) from octopus by commensal sea anemones (Calliactis spp.). *Nature* 230: 401-402.

Ross DM, 1974a. Evolutionary aspects of associations between crabs and sea anemones. In: Vernberg WB (ed) Symbiosis in the sea. University of South Carolina Press, Columbia, pp.111-125.

Ross DM, 1974b. Behavior patterns in associations and interactions with other animals. In: Muscatine L, Lenhoff HM (eds) Coelenterate Biology: *Reviews and New Perspectives.* Academic Press, New York, pp. 281-312.

Ross DM, 1979. "Stealing" of the symbiotic anemone, Calliactis parasitica, in interspecific and intraspecific encounters of three species of Mediterranean pagurids. *Canadian Journal of Zoology* 57: 1181-1189.

Ross DM, 1983. Symbiotic relations. In: Bliss D (ed) The Biology of the Crustacea, vol. 7. Academic Press, New York, pp. 163-212.

Ross DM, 1984. The symbiosis between the "cloak anemone" Adamsia carciniopados (Otto) (Anthozoa-Actinaria) and Pagurus prideauxi Leach (Decapoda-Anomura). *Bolletino di Zoologia* 51: 413-421.

Ross DM, Sutton L, 1961. The association between the hermit crab Dardanus arrosor (Herbst) and the sea anemone Calliactis parasitica (Couch). *Proceedings of the Royal Society B: Biological Sciences* 155: 282-291.

Ross DM, Sutton L, 1963. A sea anemone, a hermit crab and a shell. – An ecological triangle. *Proceedings of the International Congress of Zoology* 1,62 pp.

Ross DM, Boletzky S, 1979. The association between the pagurid Dardanus arrosor and the actinian Calliactis parasitica. Recovery of activity in "inactive" D. arrosor in the presence of cephalopods. *Marine Behaviour and Physiology* 6: 175-184.

Ross DM, Zamponi MO, 1982. A symbiosis between *Paracalliactis mediterranea* n. sp. (Anthozoa-Actiniaria) and *Pagarus variabilis* A. Milne-Edwards and Bouvier. Vie et Milieu 32: 175-181.

Roughgarden J, 1975. Evolution of Marine Symbiosis - A Simple Cost-Benefit Model. *Ecology* 56: 1201-1208.

Smith D.C., 2001. Symbiosis research at the end of the millennium. *Hydrobiologia* 461: 49-54.

Stachowicz JJ, 2001. Mutualism, facilitation, and the structure of ecological communities. *Bio. Science* 51: 235-246.

Stachowitsch M, 1979. Movement, activity pattern, and role of a hermit crab population in a sublittoral epifaunal community. *Journal of Experimental Marine Biology and Ecology* 39: 135-150.

Stachowitsch M, 1980. The epibiotic and endolithic species associated with the gastropod shells inhabited by the hermit crabs Paguristes oculatus and Pagurus cuanensis. *Marine Ecology* 1: 73-101.

Styron CE, 1977. An Ecological Triangle. *The American Biology Teacher* 39: 102-104.

Trivers RL, 1971. The evolution of reciprocal altruism. *The Quarterly Review of Biology* 46: 35-57.

Turra A, 2003. Shell condition and adequacy of three sympatric intertidal hermit crab populations. *Journal of Natural History* 37: 1781-1795.

Vafeiadou AM, Antoniadou C, Chintiroglou C, 2011. Symbiosis of sea anemones and hermit crabs: different resource utilization patterns in the Aegean Sea. *Helgoland Marine Research*, in press.

Vance RR, 1972. Competition and mechanism of co-existence in three sympatric species of intertidal hermit crabs. *Ecology* 53: 1062-1074.

Wada S, Ohmori H, Goshima S, Nakao S, 1997. Shell-size preference of hermit crabs depends on their growth rate. *Animal Behaviour* 54: 1-8.

Walker SE, 1992. Criteria for recognizing marine hermit crabs in the fossil record using gastropod shells. *Journal of Paleontology* 66: 535-558.

Williams JD, McDermott JJ, 2004. Hermit crab biocoenoses: a worldwide review of the diversity and natural history of hermit crab associates. *Journal of Experimental Marine Biology and Ecology* 305: 1-128.

Wortley S, 1863. On the habits of Pagurus prideauxii and Adamsia palliata. *Annals and Magazine of Natural History* 12: 388-390.

Xu FL, Zhao SS, Dawson RW, Hao JY, Zhang Y, Tao S, 2006. A triangle model for evaluating the sustainability status and trends of economic development. *Ecological Model* 195: 327-337.

Yasuda C, Suzuki Y, Wada S, 2011. Function of the major cheliped in male-male competition in the hermit crab Pagurus nigrofascia. *Marine Biology* 158: 2327-2334.

Yoshino K, Koga T, Oki S, 2011. Chelipeds are the real weapon: Cheliped size is a more effective determinant than body size in male-male competition for mates in a hermit crab. *Behavioral Ecology and Sociobiology* 65: 1825-1832.

Reviewed by: Dimitris Vafidis, Department of Ichthyology and Aquatic Environment, School of Agricultural Sciences, University of Thessaly, Nea Ionia, Magnesia, Greece.

In: Symbiosis: Evolution, Biology and Ecological Effects   ISBN: 978-1-62257-211-3
Editors: A. F. Camisão and C. C. Pedroso   © 2013 Nova Science Publishers, Inc

Chapter 5

MEMBRANE COMPONENTS ARE DETERMINANTS IN THE RESPONSE OF LEGUME-RHIZOBIA SYMBIOSIS AT THE ENVIRONMENTAL STRESSES

Natalia S. Paulucci, Daniela B. Medeot, Yanina B. Reguera,
Marta S. Dardanelli and Mirta B. García de Lema[*]

Departamento de Biología Molecular, Facultad de Ciencias Exactas,Físico-Químicas y
Naturales, Universidad Nacional de Río Cuarto, Río Cuarto, Córdoba, Argentina

Abstract

Legumes are able to fix nitrogen because of the bacterial symbionts (rhizobia) that inhabit nodules on their roots. The amount of ammonia produced by rhizobial fixation of nitrogen rivals that of the world's entire fertilizer industry. Consequently, this symbiotic relationship between legumes and rhizobia is of great agronomic and ecological importance.

Typical environmental stresses faced by the legume and their symbiotic partner may include, water stress, salinity and temperature and influence the survival in the soil.

In the Rhizobia-legume symbiosis, the host plant also influence rhizobial survival. In Arachis hypogaea rhizobia symbiosis is known that different abiotic stresses affect the viability, trehalose and membrane components content of rhizobia. Also, the attachment ability of peanut rhizobia is affected under abiotic stresses.

This chapter addresses the idea that the rhizobia and the plants must be able to adapt to survive to the environmental conditions. Our hypothesis is that rhizobia survival in the soil environmental because they are able to modify fatty acid and phospholipid components of their membranes, as well as other molecules with important roles in stress tolerance.

---

[*] E-mail address: mgarcia@exa.unrc.edu.ar, Tel. +54-358-4676114, Fax.+54-358-4676232, Address: Departamento de Biología Molecular, Facultad de Ciencias Exactas, Físico-Químicas y Naturales, Universidad Nacional de Río Cuarto, Ruta Nacional Nº36 Km 601, CP X5804BYA, Río Cuarto, Córdoba, Argentina., (Corresponding Author)

Introduction

Nitrogen (N) and water are the two major root-acquired resources that limit crop growth worldwide, and the availability of one can affect the utilization of the other (Medeot et al., 2010). From the 1960s until recently, the main aim of the most of the agricultural industry in developed countries was to optimize output per unit of land area and to achieve this, N fertilizer has been applied at, or close to, "economic optimum levels" on most crops (Firbank, 2005).

Legumes are widely used for food, fodder, fuel, timber, green manure, and as cover crops in different agricultural systems. In developing countries, legumes are often an integral part of forest, pastures and agricultural ecosystems. The availability of reduced nitrogenous compounds is a major limiting factor in plant growth and agricultural productivity. The microbiological process that converts atmospheric dinitrogen (N_2) into a plant-accessible species is known as biological nitrogen fixation (BNF). BNF reduces the degree of the requirement for external input of chemical N fertilizers to replenish soil N and improve internal resources (Peoples et al., 2002). Total global N_2 fixation from BNF has been estimated to 100–290 million tones N/year, with approximately 50–70 million tones N/year in agricultural systems, compared with 83 million tones N fixed industrially in fertilizer production (Graham and Vance, 2003).

Plants live in intimate association with a variety of microorganisms that can have profound repercussions on plant health by affecting nutrition and disease. Soil bacteria comprising members of the genera *Rhizobium*, *Bradyrhizobium*, *Mesorhizobium*, *Sinorhizobium*, and *Azorhizobium*, commonly referred to as rhizobia, are taxonomically diverse members of the α and β subclasses of the Proteobacteria. They possess the ability to induce root nodules on legume plants and provide these plants with fixed nitrogen, enabling them to grow in nitrogen limited soils. The interaction between legumes and rhizobia is perhaps the most intensively studied plant-microbe system and is characterized by a multistep signal exchange process. The host range can be determined at different levels during the interaction. Phenolic plant exudates induce the bacterial production of lipochitooligosaccharides, the Nod factors (NFs). These signal molecules trigger the cortical cell division that eventually leads to the formation of newly formed plant organs, the nodules. While changing their cell envelope structure, compatible rhizobia are released, through passage of the infection thread, into the emerging nodules; differentiate into bacteroids which eventually convert atmospheric nitrogen into ammonia. The microsymbiont receives carbon sources from the plant and provides reduced nitrogen, as ammonia, to the macrosymbiont (Laeremans andVanderleyden, 1998).

Improvement in an agricultural sustainability requires optimal use and management of soil fertility and soil physical properties, and relies on soil biological processes and soil biodiversity (Choudhary et al., 2011). The continued use of chemical fertilizers and manures for enhanced soil fertility and crop productivity often results in unexpected harmful environmental effects, including leaching of nitrate into ground water, surface run-off of phosphorus and nitrogen run-off, and eutrophication of aquatic ecosystems.

Soil microorganisms may comprise of mixed populations of naturally occurring microbes that can be applied as inoculants to increase soil microbial diversity. Investigations have shown that the inoculation of efficient microbial community to the soil ecosystem improves soil quality, soil health, growth, yield and quality of crops. These microbial populations may consist of selected species including plant growth promoting rhizobacteria, N2-fixing microorganisms, plant disease suppressive bacteria and fungi, soil toxicant degrading microbes, actinomycetes and other useful microbes (Singh et al., 2011). Microbial inoculants are promising components for integrated solutions to agro-environmental problems because inoculants possess the capacity to promote plant growth, enhance nutrient availability and uptake, and support the health of plants (Adesemoye and Kloepper, 2009).

Bacteria belonging to different genera including *Rhizobium*, *Bacillus*, *Pseudomonas*, *Pantoea*, *Paenibacillus*, *Burkholderia*, *Achromobacter*, *Azospirillum*, *Microbacterium*, *Methylobacterium*, *Enterobacte*r, among others, provide tolerance to host plants under different abiotic stress environments (Egamberdieva and Kucharova, 2009). Use of these microorganisms (rhizobacteria) per se can alleviate stresses in an agriculture thus opening a new and emerging application of microorganisms (Choudhary et al., 2011). Rhizobacteria are constantly faced with environmental stimuli stresses and should be responding to a wide range of factors through signal transduction pathways that convert extracellular information into intracellular forms. The rhizosphere, the soil zone influenced by plant roots, is dynamic. Its extent and properties are influenced by soil physical and chemical properties, weatherand plant-induced changes in soil water content, the composition and density of soil microbial populations, and the metabolic activities of plants and microbes (Miller and Wood, 1996). Legumes play a critical role in natural ecosystems, agriculture, and agroforestry, where their ability to fix N in symbiosis makes them excellent colonizers of low-N environments, and economic and environmentally friendly crop, pasture, and tree species. In addition to traditional food and forage uses, legumes can be used to make bread, doughnuts, tortillas, chips, spreads, and extruded snacks or used in liquid form to produce milks, yogurt, and infant formula (Garcia et al., 1998, Graham and Vance, 2003). Legumes (predominantly soybean and peanut (*Arachis hypogaea*) provide more than 35% of the world's processed vegetable oil, and soybean and peanut are also rich sources of dietary protein for the chicken and pork industries (Graham and Vance, 2003).

Peanut is cultivated around the globe in different agronomical systems and the worldwide production was estimated in 33.1 million tons. According to USDA, China leads the production of peanuts having a share of about 37.5% of overall world production, followed by India, United States of America, Argentina and Vietnam (Fabra et al., 2010). In Argentina, about 87% of peanut production takes place in the province of Córdoba. In this country, peanut obtained is of very high quality and almost all the production is exported to European Union, Indonesia, and Canada (Fabra et al., 2010). In spite of its agronomic importance, studies about peanut-rhizobacteria diversity are relatively scarce. The analysis through morphophysiological and molecular methods of peanut symbionts obtained from different geographical regions revealed high level of diversity and heterogeneity (Fabra et al., 2010). Besides nitrogen fixing peanut rhizobia, several authors informed that other peanut associated beneficial bacteria have multiple plant promoting activities (Dey et al., 2004).

To survive in different conditions, rhizobacteria need to adapt by accumulation or releasing specific solutes and by change in their membranes.

Bacterial cytoplasmic membranes are both functionally and structurally diverse. Cytoplasmic membranes define cells from the external environment; contain the cytoplasm and other cellular constituents. They regulate the movement of substances entering or exiting cells and catalyze exchange reactions. Membranes also play a role in energy transduction and in the maintenance of ion and solute gradients to maintain a constant intracellular environment. They also provide a milieu where biological reactions can occur and act to regulate cellular growth and metabolism. Cytoplasmic membranes stabilize protein structure, which is important with regard to the function of membrane embedded enzymes. In addition, membrane embedded molecules and receptors add to the role of the membrane by allowing it to provide for intercellular communication and detection of cellular signals (Dowhan, 1997; Denich et al., 2003). These cellular functions require a membrane that maintains optimal or near optimal structure and function in spite of an often variable and potentially damaging external environment. Collectively, optimal membrane state is most frequently defined in terms of fluidity, a broad term that encompasses lipid order which includes structure and microviscosity as well as membrane phase, which includes lipid shape, packing and curvature (Rilfors et al., 1984; van de Meer, 1984; Bloom et al., 1991).

In a bacterial cytoplasmic membrane, limits of fluidity are generally defined by the thresholds beyond which the cell can no longer function. Physiological factors such as a structurally unstable lipid bilayer or the inability to maintain membrane protein functions (McElhaney and Souza, 1976) have been implicated as determinants of finite levels of membrane fluidity. However, the range of cytoplasmic membrane fluidity that each bacterial species can have and still grow and divide is often not known. Changes in cytoplasmic membrane fluidity can occur as a result of the physical and chemical interaction of membrane lipids and environmental factors such as temperature, pH, pressure, ions, water availability, nutrients, and chemicals. Membrane perturbations elicit an adaptive response from the bacteria that must compensate for the effects that create sub-optimal membrane conditions. Bacterial adaptation to stress is a multi-factorial celular process, whereby the cytoplasmic membrane and alterations therein are a primary response mechanism. Lipids which constitute a primary component of biological membranes are varied in their structure and physical properties. They exhibit polymorphism, or phase preference, aggregating into different structures and assuming structurally distinct phases (Cullis and de Kruijff, 1979; Gruner et al., 1985; Dowhan, 1997).

Bacteria use lipid diversity to alter their membranes in response to environmental stress ensuring that the cell membrane remains within optimal membrane state and continued functionality (Finean and Michell, 1981; Hazel and Williams, 1990; Dowhan, 1997). In general, perturbation of membrane fluidity by extrinsic chemical agents or other factors initiates an active response based on intrinsic chemical changes such as the modification of existing lipids and the de novo synthesis that tend to counteract the perturbation (Soltani, et al., 2005, Denich et al., 2003).

In biological systems there are competing theories on how organisms alter their cytoplasmic membrane to maintain optimal membrane function. The most widely employed theory when interpreting environmentally induced membrane restructuring is that of Sinensky, (1974) who initially described that membrane fluidity is maintained at a constant value under changing conditions, namely temperature, as homeoviscous adaptation. However, McElhaney, (1974) demonstrated that not all organisms change fatty acid (FA) composition in response to growth temperature and that cytoplasmic membrane fluidity does not always remain constant. Instead it was suggested that perhaps not an exact level of fluidity must be maintained but rather that the temperature range of the thermal transition be adjusted so that the majority of the lipids are in the correct liquid crystalline phase. Later, Silvius et al., (1980) attempted to define limits of fluidity and concluded that although an organism will not grow with its membrane in the gel-state, it could tolerate a wide range of FA compositions and different fluidities within the liquid crystalline phase.

Bacterial membranes consist of proteins that are embedded in a lipid matrix that closely approximates a phospholipid (PL) bilayer.

Although there is a considerable diversity of PL structures in the bacterial world, most membrane PL are glycerolipids that contain two FA chains. These PL acyl chains determine the viscosity of the membrane, which, in turn, influences many crucial membrane associated functions, such as the passive permeability of hydrophobic molecules, active solute transport and protein–protein interactions (Zhang and Rock, 2008).

The ability of bacteria to modify their membrane composition in response to environmental changes, such as in temperature, osmolarity, salinity and pH, was determined early in the study of bacterial lipid metabolism (Cronan Jr and Gelmann, 1975)

In bacteria, the membrane lipid composition has been relatively well defined with predominant polar lipids including phosphatidylethanolamine (PE) and phosphatidylglycerol (PG) with varying arrangements of chain length, degree of saturation, isomer conformation, branching, and cyclization of FA (Ratledge and Wilkinson, 1988; Sajbidor, 1997; Denich et al., 2003). The biologically active state of the membrane is a lamellar liquid crystalline (La) bilayer (Sinensky, 1974; McElhaney, 1982; Cullis et al., 1985). The PL is arranged with the polar heads oriented externally with the lipid acyl chains directed toward the interior of the bilayer (Singer and Nicolson, 1972; Rilfors et al., 1984).

(PE), (PG), and cardiolipin (CL) are characteristic PL of most Gram-negative bacteria. Phosphatidylcholine (PC) has also been found in bacteria, and it is more widespread than originally thought, although its role is unclear (López Lara and Geiger, 2001).

PL changes have been reported in response to environmental stress such as that produced by temperature (Russell, 1992) and by salt (Sutton et al., 1991). In addition, low-oxygen conditions have been shown to increase PE and PG amounts and also to decrease PC amount (Tang and Hollingsworth, 1998).

It was speculated that PC might serve a special function during host–pathogen/symbiont interactions. The importance of PC for the establishment of successful interactions with eukaryotic hosts is exemplified by the fact that *Agrobacterium tumefaciens* mutants lacking PC are unable to form tumours in susceptible plants (Wessel et al., 2006), that the human pathogen *Brucella abortus* requires PC for full virulence (Comerci et al., 2006) and that PC synthesis is required for optimal function of virulence determinants in *Legionella pneumophila* (Conover et al., 2008).

Additional mechanisms to stabilize membrane fluidity in bacteria involve changes in FA, the major component of PL. Such mechanisms, which may occur in combination, involve changing the ratio of saturation to unsaturation; cis to trans unsaturation; branched to unbranched structures, type of branching; acyl chain length and formation of cyclopropane (Ramos et al., 1997; Donatto et al., 2000).

Increased degree of unsaturation in response to reduced temperature has been described for many microorganisms (Russell and Fukunaga, 1990; Suutari et al., 1990), and can be regarded as a universally conserved adaptation response (Suutari and Laakso, 1994). In *Aeromonas*, alteration of growth temperature induced changes in unsaturation, branching, and chain length of the FA. At temperaturas below 15 °C or above 25 °C, three species of *Aeromonas, A. caviae, A. hydrophila* and *A. sobria*, showed significant decrease of cis-vaccenic acid (18:1n-7) content. In cells exposed to high NaCl concentration, maintenance of growth ability was related to a reduced ratio of unsaturated to saturated FA, reflecting membrane rigidification (Chihib et al., 2005).

How the FA composition of membrane lipids is altered in response to change of growth temperature depends on the mechanism of unsaturated FA (UFA) synthesis (Keweloh and Heipieper, 1996). In bacteria, UFA synthesis involves both anaerobic and aerobic mechanisms. UFA synthesis in response to low temperature was characterized in vivo for the Gram-positive bacteria *Bacillus subtilis*, which desaturates palmitate to delta 5-hexadecenoate (Aguilar et al., 1998). The molecular mechanism of UFA synthesis in response to temperature change has been well studied in the Gram-negative bacteria *Escherichia coli*.

The PL pattern of rhizobia still is not entirely clear, but there is general agreement that PE and PC are present in apreciable amounts and that PG is also consistently present. In addition, monomethylphosphatidylethanolamine (MMPE) and dimethylphosphatidylethanolamine (DMPE) may also be formed (Wilkinson, 1988). Among the PL mentioned, PC may play a particularly important role, because it has been found that in *Bradyrhizobium japonicum* PC participates in successful interaction with the eukaryotic host (Minder et al., 2001).

In *Sinorhizobium (Ensifer)meliloti*, which can form nitrogen-fixing nodules on its host plant alfalfa, PC can be synthesized by two entirely different biosynthetic pathways. In the methylation pathway, the enzyme phospholipid N-methyltransferase (PmtA) forms PC by three successive methylations of PE (de Rudder et al., 2000). The second pathway is dependent on the supply of choline and consists of the direct condensation of choline and CDP-diacylglycerol in a reaction catalysed by PC synthase (Pcs) (Sohlenkamp et al., 2000). *S. meliloti* mutants deficient in either pathway show wild-type-like PC levels when grown on complex medium while a mutant defective in both pathways does not form PC and shows a severe reduction of the growth rate with respect to the wild-type (de Rudder et al., 2000). Furthermore, the *S. meliloti* mutant lacking PC is unable to form nodules on alfalfa (Sohlenkamp et al., 2003). In contrast to *S. meliloti*, in a pmtA-deficient *B. japonicum* mutant, the PC content is reduced from 52% to 6%. This reduction in the PC content did not prevent nodule formation, but drastically reduced nodule occupancy and nitrogen-fixation ability (Minder et al., 2001).

Little is yet known about the biochemical and physiological basis of saline and temperature tolerance by rhizobia nodulating *Arachis hypogaea* (peanut) roots. Is also unknown the control of FA synthesis in legume-nodulating rhizobia under abiotic stress.

The FA composition profiles of *Bradyrhizobium* and *Rhizobium* are quite different, and have been used for chemotaxonomic purposes (Tighe et al., 2000). Effects of growth phase (Boumahdi et al., 1999; 2001) and low temperature (Drouin et al., 2000; Théberge, 1996) on FA synthesis and composition in these genera have been studied, but not the effects of high temperature or high salinity.

As well as the organisms mentioned previously, rhizobacteria are constantly under stress and must have diferent strategies to cope with it.

Our purpose was to clarify the role of cell membrane modifications in resistance and adaptation of these rhizobia, *Bradyrhizobium* SEMIA6144 (SEMIA6144) and *Bradyrhizobium* TAL1000 (TAL1000) by characterizing the physiological and metabolic response to environmental stresses. We also have identified the genes involved in PC biosynthesis and have mutated the homologous *pmtA* gene, in order to elucídate the role of

PC in a peanut-nodulating *Bradyrhizobium*. The results may identify new strategies for increasing symbiotic efficiency between rhizobia and peanut.

Effects of High Growth Temperature (37 °C) and Salinity on TAL1000 and SEMIA6144 Survival and Phospholipid Metabolism

Since soils are subjected to high temperature stress in summer, saline concentration also increase, which may have detrimental effects on the introduced rhizobia. Temperature can affect rhizobial persistence in inoculants during shipment or in storage. Also temperature and salt stress can influence survival in soil and can limit both nodulation and nitrogen fixation. An understanding of the growth of *Rhizobium* is likely when the physiology of these organisms has been carefully studies under these suboptimal conditions (Abdelmounmen et al., 1999; Kulkarni and Nautiyal, 2000).

In peanut rhizobia (*Bradyrhizobium* sp ATCC10317, SEMIA6144 and TAL1371 strains) high growth temperature (37 °C) provoked a slightly reduced biomass production, an increase in the cellular content of low molecular weight oligosaccharides and fully suppressed the synthesis of neutral glucans (Dardanelli et al., 1997).

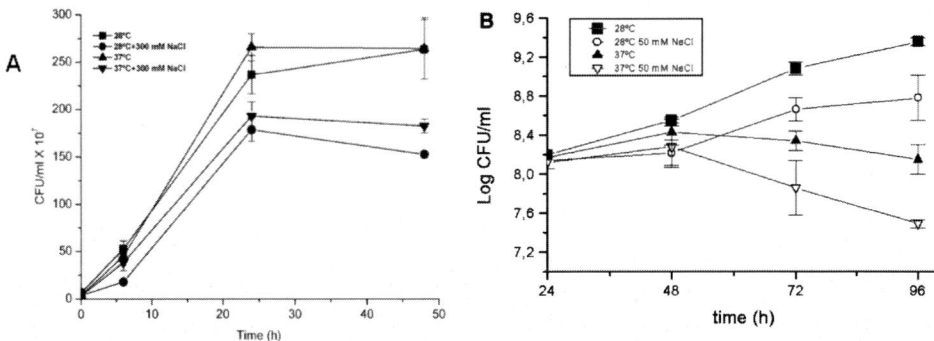

Figure 1. Effect of NaCl and temperature on viability of fast-growing *Bradyrhizobium* strain TAL1000 (a) and slow-growing *Bradyrhizobium* SEMIA6144 (b). Viability is expressed as CFU ml^{-1}. Values represent means ± SEM from three independent experiments.
Source: a) Paulucci et al., 2011: b) Medeot et al., 2007

Salt and combined conditions, affected negatively the vaibility of SEMIA6144 and the TAL1000 viability. Both rhizobia survived under stress conditions and they showed different saline tolerance (300 mM TAL1000 and 50 mM SEMIA6144, Fig 1a and 1b). It was observed that viability of salt tolerant TAL1000 was unaffected by high temperature, consistent with studies that indicate a relationship between salt tolerance and temperature tolerance in rhizobial strains. In certain genera of rhizobia, tolerance to salinity has been associated with tolerance to high temperatures, previous studies have shown that strains of *S. meliloti*, salt-tolerant can grow well at elevated temperatures (Zhang et al., 1991) while

Rhizobium leguminosarum, salt-sensitive is much less tolerant to higher growth temperatures (Lindstron and Lehtomaki, 1988).

Previous studies relating to the ability of rhizobial strains to grow in presence of salts have demonstrated a marked variation in tolerance. Some strains such as *Bradyrhizobium japonicum* are inhibited by concentrations below 100 mM (Elsheikh and Wood, 1990). By contrast, the growth of various strains of *S. meliloti* can occur at concentrations above 300 mM NaCl (Bernard et al., 1986), *Rhizobium* spp isolated from plants *Hedysarum* nodules, *Acaciacon, Prosopis*, and *Leucaena* can tolerate up to 500 mM NaCl (Zhang et al., 1991).

In relation to lipid metabolism of nodulating peanut, radioactive acetate sodium salt was incorporated mostly (88-92%) into PL and the rest into neutral lipid fraction of both rhizobia strains (Table 1). The predominat PL was the zwitterionic PC and represent the same pocentage in both strains. PE, the other zwitterionic PL was the second in SEMIA6144 but not in TAL1000 where the second was the anionic PL, PG. TAL1000 showed greater proportion of anionic PL than SEMIA6144, so that the ratio Z/A was less. It has been reported that, only cell growth and not cell viability is dependent on the anionic phospholipid content. This indicates that lack of anionic PL does not compromise membrane integrity but does limit other functions. One such function is a source of components for modification of other membrane components. Anionic phospholipid requirement is not the same for all organisms. *E. coli* has an absolute requirement for PC, but there are several processes that require anionic PL at different levels.

Table 1. Effect of salinity and temperature stress on the incorporation of [^{14}C]acetate into phospholipids of two peanut-nodulating *Bradyrhizobium*.

PL (%)	Growth condition			
	28°C SEMIA-TAL	28°C+NaCl SEMIA-TAL	37°C SEMIA-TAL	37°C+NaCl SEMIA-TAL
PC	43.7-44.5	50.2*-45.7	54.6*-49.4*	57.1*-54.0*
DMPE	2.9-9.9	1.9-9.70	2.3-10.9	2.0-10.8
LPE	ND-0.88	ND-2.50	ND-3.60*	ND-2.50*
PE	24.4-13.7	21.4*-13.5	19.0*-7.60*	15.5*-8.10*
CL	6.6-4.20	6.1-3.90	3.1*-3.00*	7.0-4.10
PG	10.1-17.5	8.9-15.4	6.4*-18.2	6.8*-12.4*
PA	1.2-ND	0.9-ND	1.8-ND	1.0-ND
NL	11.1-9.30	10.6-9.60	12.7-7.80	10.5-9.00

Bradyrhizobium SEMIA6144 (SEMIA) and *Bradyrhizobium*TAL1000 (TAL) were harvested at late exponential phase. Values represent means of three independent experiments.
* Difference form control (28 °C) value statistically significant at $P<0.05$ level.
NL: neutral lipids.
(Data are from Medeot et al., 2007 and Paulucci et al., 2011)

Saline stress provoked changes in the radioactive label of the PL of SEMIA6144 but was unable to alter the label in the PL of TAL1000. This is consistent with the tolerance to saline stress of both strains such as detailed above. However, high growth temperature (37 °C) and combined conditions modified PL patterns of SEMIA and TAL. So, with high growth temperature, PC label was increased significantly (25% and 11%), also LPE label increased (ND and 316%), while PE label was decreased (22% and 45%), CL label was decreased (53% and 29%) and PG label decreased (37% and unchanged), for SEMIA6144 and TAL1000, respectively. The combined conditions produced a similar effect as the high growth temperature. Thus, it was observed an increase in the label of PC (30% and 21%) and in the label of LPE (ND and 184%), while the label of PE (37% and 41%) and of PG (33% and 30%) decreased for SEMIA6144 and TAL1000, respectively. The indicated changes caused modifications in the zwitterionic/anionic (Z/A) ratio. All the stresses applicated were able to increase Z/A ratio: SEMIA was more sensitive to temperature and responded with an increase of 72% in the ratio Z/A, while TAL1000 was more affected by the combination of both stresses which caused an increase of 42% in this ratio.

The modifications may be a consequence of the decreased amount of the anionic PL, suggesting a decreased of negative charges on the membrane provoked by temperature stress for SEMIA6144 and by combined stresses for TAL1000 (Enomoto and Koyama, 1999). In relation to that, Bakholdina et al., (2004) observed changes in the relative percentages of each PL and in the ratio between Z/A PL with growth temperature.

Because in all conditions tested the PC amount was always modified and PC was the major PL, we suggest that this PL may be involved in the bacterial response to environmental conditions. The importance of PC in other responses like the successful interaction with the eukaryotic host has been noted by Minder et al., (2001). The other PL, i.e., PE, which would be the substrate for PC synthesis in this organism, was modified in every condition. The PE variation in the proportion labeling would implicate the PE participation in the response to stress, as noted for other factors (Correa et al., 1999; Enomoto and Koyama, 1999).

Latter we designed a work to identify the genes involved in PC biosynthesis as well as to isolate and mutate the homologous *pmtA* gene in SEMIA 6144 in order to elucidate the role of PC in free-living bacteria and during symbiosis with peanut plants.

Phosphatidylcholine Levels of Peanut-Nodulating Bradyrhizobium Sp. SEMIA6144 and the Relation with the Cell Size and Motility

Pmt and Pcs activities were determined in vitro using cell-free protein crude extracts and radiolabelled substrates. A major Pmt activity was found in SEMIA6144 (Fig. 2a), while only a minor Pcs activity was detected (Fig. 2b). This is similar to the situation found for cell extracts of *B. japonicum* USDA 110 (Martínez-Morales et al., 2003). SEMIA6144 was incapable of incorporating [^{14}C] choline from the medium consistent with a previous study that had shown that all the rhizobial strains tested, except *B. japonicum*, possess a choline uptake activity and can use choline as a carbon, nitrogen and energy source for growth (Boncompagni et al., 1999).

Genes involved in PC biosynthesis from *S. meliloti* (*pmtA* and *pcs*) and from *B. japonicum* (*pmtA*, *pmtX1*, *pmtX2*, *pmtX3* and *pcs*) were used as probes against SEMIA6144 genomic DNA. Only *pcs, pmtA, pmtX1 and pmtX2* probes from *B. japonicum* hybridized. No hybridization was observed when *S. meliloti* probes were used (data not shown). This is in agreement with the genetic and physiological similarities between *B. japonicum* USDA 110 and SEMIA6144 (Gomes-Germano et al., 2006).

A *pmtA* gene was cloned from SEMIA6144 and created a *pmtA*-deficient mutant.

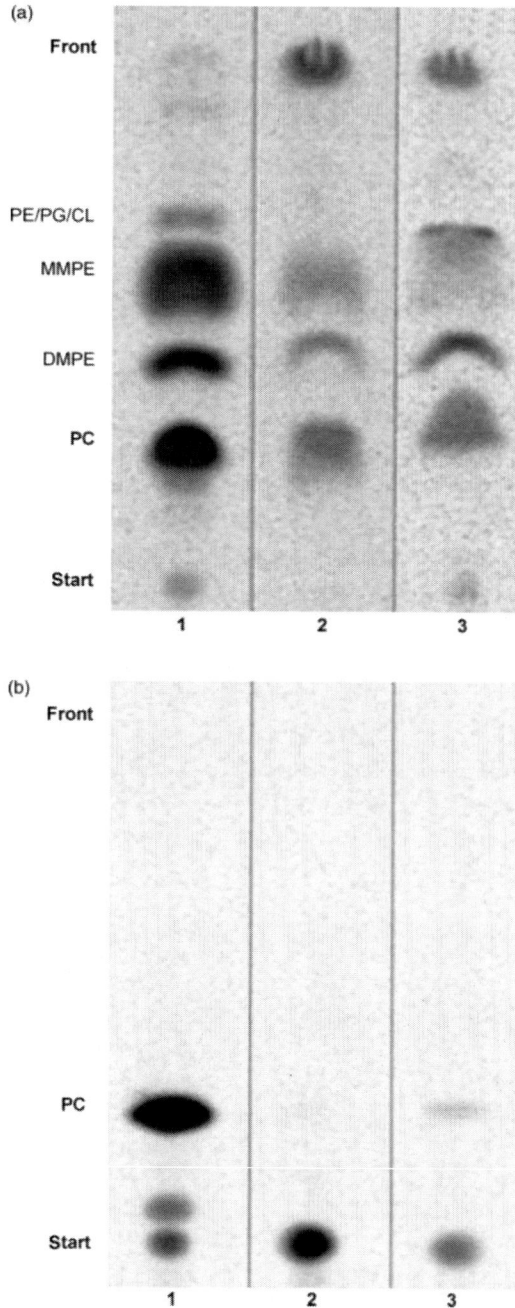

Figure 2. Determination of Pmt and Pcs activities. The Pmt activity (a) was performed according to de Rudder et al. (1997) using S-adenosyl-L- [methyl-^{14}C]methionine (SAM) as a radiolabelled substrate. The Pcs activity (b) was determined according to Martínez-Morales et al. (2003) using the substrates [methyl-^{14}C]choline and CDP-diacylglycerol. Bacterial lipids formed in vitro were extracted according to Bligh & Dyer (1959) and analysed by TLC. Each lane in the figure is derived from different TLCs developed in parallel. (1) *Sinorhizobium meliloti* 1021, (2) *Bradyrhizobium japonicum* USDA 110spc4, (3) *Bradyrhizobium* sp. SEMIA 6144. PE, phosphatidylethanolamine; PG, phosphatidylglycerol; CL, cardiolipin; MMPE, monomethylphosphatidylethanolamine; DMPE, dimethylphosphatidylethan-olamine; PC, phosphatidylcholine. Source: Medeot et al., 2010.

A 2.5-kb HindIII fragment hybridizing with *pmtA~Bj~* (data not shown) was cloned into pUC18, resulting in plasmid pDBM01.

The DNA sequence of SEMIA6144 *pmtA* showed high identity (92%) with the *pmtA* sequence of *B. japonicum* USDA 110 (Y09633). The SEMIA6144 *pmtA* gene is located downstream of the heat shock-controlled *dnaKJ* chaperone operon (data not shown), which is the same gene organization as in *B. japonicum* (Minder et al., 2001). Comparison of the predicted amino acid sequence of SEMIA6144 PmtA with other rhizobial PmtA sequences (data not shown) revealed the presence of the motif VVEXGXGXG, which is the same consensus motif found in PmtA of *B. japonicum* for the S-adenosylmethionine (SAM)-binding site present in SAM-dependent methyltransferases (Minder et al., 2001; Sohlenkamp et al., 2003).

To understand the biological role of PC in the peanut-nodulating strain SEMIA6144, a *pmtA*-deficient mutant (DBM13) was constructed. Table 2 shows that the wild-type strain possesses PC (47.6% of total PL) and PE (27.5%) as major PL. In contrast, DBM13 showed a marked decrease of PC and a concomitant increase of PE (24.8% and 57.6%, respectively), indicating that *pmtA* plays a major role in PC biosynthesis in SEMIA6144. Probably, the significant amounts of PC still remaining in DBM13 are due to activities encoded by other functional *pmt* genes. In a similar way, the biosynthesis of PC in *B. japonicum* USDA 110 is achieved through the action of different Pmt activities (Hacker et al., 2008).

Table 2. Membrane lipid composition of *Bradyrhizobium* sp. SEMIA 6144 wild type, pmtA-deficient mutant (DBM13), DBM13 complemented with plasmid pDBM07 and DBM13 harbouring the vector pBBR1MCS-5

Lipid labelled (% of total ^{14}C $\pm$ SD)				
	SEMIA 6144 wild type	*pmtA*-deficient mutant	*pmtA*-deficient mutant × pDBM07	*pmtA*-deficient mutant × pBBR1 MCS-5
PA	2.4 ± 1.0	1.3 ± 0.4	3.3 ± 1.3	2.0 ± 0.6
PC	47.6 ± 3.9	24.8 ± 3.8	56.6 ± 2.5	25.0 ± 3.9
PG	7.0 ± 2.3	7.2 ± 1.1	6.4 ± 0.1	7.2 ± 2.2
PE	27.5 ± 6.5	57.6 ± 5.2	20.1 ± 1.4	56.4 ± 8.0
CL	13.4 ± 3.3	5.3 ± 0.7	9.6 ± 2.2	5.8 ± 0.7
UL	4.1 ± 0.7	3.0 ± 0.6	3.9 ± 0.5	3.6 ± 0.5

Bradyrhizobium sp. SEMIA 6144 strains were grown in B⁻ medium. [1-^{14}C]acetate, sodium salt (37 kBq mL^{-1}) was added. Aliquots of the total lipid extracts were analysed by TLC and the fractions were quantified by radioactivity measured in a liquid scintillation counter (Beckman LS 60001 C). Values represent means ± SD of three independent experiments. PA, phosphatidic acid; PC, phosphatidylcholine; PG, phosphatidylglycerol; PE, phosphatidylethanolamine; CL, cardiolipin; UL, unidentified lipid(s) migrating with the solvent front.

Source: Medeot et al., 2010

Also CL level was reduced in the mutant DBM13 (Table2). A slight reduction in CL had also been observed in the *pmtA* mutant of *B. japonicum* (Minder et al., 2001).

We noticed that wild-type colonies were larger than colonies of the mutant strain. Determination of cell size under the light microscope showed that wild-type cells were longer than DBM13 cells (Table 3). Both phenotypes, colony and cell size were recovered when plasmid pDBM07 was introduced into DBM13. The recovery in cell and colony size of the complemented mutant correlates with the recovery of its PC levels (Table 2). The formation of CL domains at the cell pole and the division site plays an important role in selection and recognition of the division site by cell cycle and cell divisíon proteins in *E. coli* (Mileykovskaya et al., 2009). Because the level of CL was reduced to more than half in DBM13 with respect to wild-type cells (Table 2), it is possible that the decrease in cell size is due to the reduction of CL. Bernal et al., (2007) found a similar decrease in the cell length in a CL synthase-deficient mutant of *Pseudomonas putida* and concluded that changing the amount of anionic PL led to cell division before the cell had reached the average size of wild-type cells.

Table 3. Average size of *Bradyrhizobium* sp. SEMIA 6144 wild type and DBM13 cells.

Strain	Size (μm)
SEMIA 6144 wild type	2.31 ± 0.38
SEMIA 6144 *pmtA*-deficient mutant (DBM13)	1.86 ± 0.25
DBM13 × pDBM07	2.25 ± 0.37
DBM13 × pBBR1MCS-5	1.48 ± 0.16

Cells were examined under the Axiophot (Carl Zeiss) light microscope. The images were acquired with a Canon PC1089 Powershot G6 7.1-megapixel digital camera (Canon Inc., Japan) and processed with the AXIOVISION 4.1 software (Carl Zeiss).

Source: Medeot et al., 2010

The zones of swimming of the mutant were smaller than that of the wild type (Fig. 3). Complementation experiments confirmed the correlation between defective motility and the mutation in the *pmtA* gene (Fig. 3c). The reduced diameter of *pmtA*-deficient mutant colonies suggests that they were impaired in motility and/or chemotaxis. The defects in motility observed in our work are in agreement with data reported in other bacteria. Mutants of *L. pneumophila* lacking PC are unable to transit to a motile state and have low levels of flagellin protein (Conover et al., 2008). Also in *A. tumefaciens*, the loss of PC resulted in reduced motility (Klüsener et al., 2009). *pmtA*-deficient mutant showed increased biofilm formation and higher aggregation capacity than the wild strain (unpublished results)

The PC level encountered in DBM13 (Table 2) was sufficient to develop functional nitrogen-fixing nodules. Hacker et al., (2008) reported wild-type-like symbiotic characteristics for soybean plants infected with *B. japonicum pmtX2, pmtX3, pmtX4* or *pcs* mutants, but all of which showed wild-type levels of PC. On the other hand, soybean plants inoculated with *pmtA* mutants of *B. japonicum*, which were severely affected in PC biosynthesis, showed drastic nitrogen-fixation defects (Minder et al., 2001).When peanut roots were coinoculated with the wild-type and DBM13 strains in a 1:1 inoculum ratio, DBM13 was detected in only 27.8% of the total nodules, indicating a defect in their nodulation competitiveness. We related this defect of DBM13 to its lack of motility and/or chemotaxis because many earlier reports indicate their importance for competitive nodulation (Alexandre et al., 2004; Miller et al., 2007). Therefore, wild-type levels of PC could be important for the competitive abilities of SEMIA6144 in the rhizosphere.

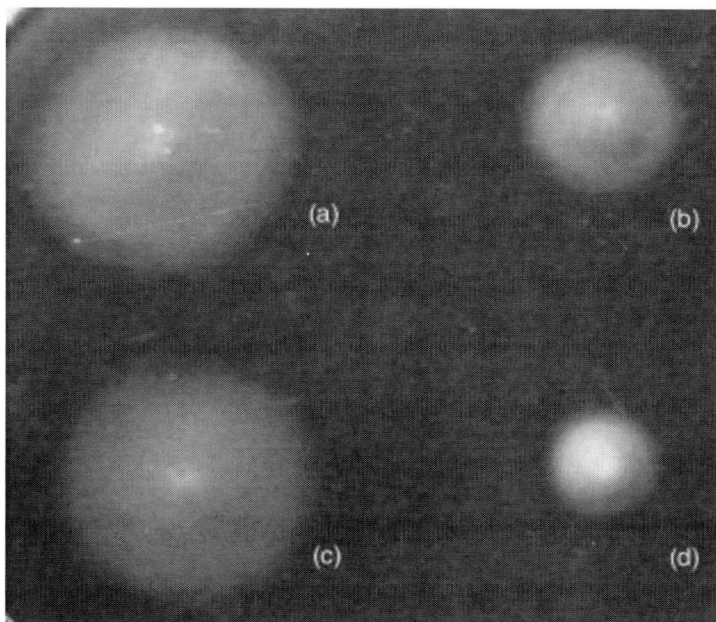

Figure 3. Swimming motility of *Bradyrhizobium* sp. SEMIA 6144 strains. The assay was performed at 28 °C on YEM swim plates (0.3% agar). (a) SEMIA 6144 wild type, (b) DBM13, (c) DBM13-pDBM07 and (d) DBM13-pBBR1MCS-5. Similar results were obtained in three indeendent experiments.

Source: Medeot et al., 2010

In the *pmtA*-deficient mutant PE is the most abundant PL instead of PC. It should cause major changes in the membrane properties. PE has a smaller headgroup, can hydrogen bond through its ionizable amine and has the unique property of undergoing a bilayer to-nonbilayer physical transition influenced by its FA content and the temperature (Dowhan, 1997). These changes could lead to modifications in the structure of transmembrane a-helices of membrane proteins, altering the packing of these helices (Dowhan, 1997). As a consequence, membrane-associated functions of DBM13, such as motility, might be affected. In addition, the amount of CL is strongly reduced in the *pmtA*-deficient mutant.

This reduction might be a direct effect of the decrease in PC and the increase in PE. Possibly, by decrease of CL, the cell size might be affected. Finally, the change in the proportion between anionic and zwitterionic lipids could be important in seemingly diverse membrane-associated processes.

Since FA are the major components of PL and in order to clarify the membrane role, we also described the FA composition and the participation of FA in resistence and adaptation of SEMIA6144 and TAL1000 to environmental stresses.

Fatty Acid Composition of TAL1000 and SEMIA6144 and Effect of Growth Conditions on Fatty Acid Composition and Metabolism of TAL1000 and SEMIA6144

Major FA detected in TAL1000 and SEMIA6144 were cisvaccenic (18:1n-7) + oleic (18:1n-9), stearic acid (18:0) and palmitic acid (16:0). Eicosatrienoic acid (20:3n-6) and cyclopropane FA (19:0cyclo) were only detected in TAL1000, while palmitoleic acid (16:1n-7) was only detected in SEMIA6144 (Table 4)

Under high growth temperature 18:1n-7 + 18:1n-9 percentage declined (63.3 to 8.2%), while the saturated FA (SFA), 16:0 (8.4 to 20%), 18:0 (12.6 to 24%) and 19:0cyclo (3.4 to 10%) increased. On the other hand, combined conditions provoked decline of 18:1n-7 + 18:1n-9 percentage (63.3 to 4.3%), increase of 16:0 (8.4 to 16.1%), of 18:0 (12.6 to 29%) and increase of 19:0cyclo (3.4 to 14.5%).

Of all the tested conditions, the growth temperature increase was the one causing the most significant changes in the level of FA in TAL1000.

Table 4. Effects of temperature and salinity stress on fatty acid composition of two peanut-nodulating rhizobia.

Fatty acid type (%)	Strain	Growth condition			
		28 °C	28 °C + NaCl	37 °C	37 °C + NaCl
Saturated					
Stearic acid (18:0)	TAL1000	12.6 ± 1.5	14.6 ± 1.6*	24.0 ± 2.7*	29.0 ± 0.9*
	SEMIA6144	1.40 ± 0.3	1.66 ± 0.3	2.00 ± 0.6	2.16 ± 0.9
Palmitic acid (16:0)	TAL1000	8.40 ± 1.7	9.30 ± 1.5	20.0 ± 2.3*	16.1 ± 1.8*
	SEMIA6144	11.0 ± 1.2	12.6 ± 2.7	18.6 ± 3.9*	19.7 ± 3.5*
Unsaturated					
Palmitoleic acid (16:1n-7)	TAL1000	ND	ND	ND	ND
	SEMIA6144	0.42 ± 0.0	0.90 ± 0.4	0.65 ± 0.2	0.56 ± 0.2
cis-vaccenic acid + oleic acid (18:1)	TAL1000	63.3 ± 5.4	55.8 ± 2.7	8.20 ± 0.4*	4.30 ± 0.0*
	SEMIA6144	84.0 ± 2.2	82.4 ± 2.9	73.5 ± 3.4*	68.5 ± 6.8*
Eicosatrienoic acid (20:3)	TAL1000	6.30 ± 1.9	9.20 ± 1.2	15.8 ± 1.6*	19.0 ± 4.7*
	SEMIA6144	ND	ND	ND	ND
Cyclopropane					
19:0cyclo	TAL1000	3.40 ± 0.7	3.85 ± 0.3	10.0 ± 1.5*	14.5 ± 1.7*
	SEMIA6144	ND	ND	ND	ND
Others	TAL1000[a]	6.70 ± 1.6	7.10 ± 0.8	23.2 ± 2.6*	17.1 ± 1.9*
	SEMIA6144	1.10 ± 0.0	2.55 ± 0.7	5.00 ± 0.2	5.30 ± 1.3
U/S[b]	TAL1000	3.3	2.7	0.5	0.5
	SEMIA6144	7.0	6.1	3.6	3.1

Lipids were extracted, and fatty acids of total lipid were converted to methyl esters and analyzed by GC, as described in the text

Percentage of each fatty acid is relative to total fatty acids defined as 100%. Values represent means ± SEM of three independent experiments

ND not detected

[a] Correspond to two peaks of retention times of 36 min and 37.4 min. Such peaks could correspond to FA of more than 18 carbon atoms

[b] Ratio between sum of unsaturated and sum of saturated fatty acids

* Difference from control (28 °C) value statistically significant at $P < 0.05$ level

Source: Paulucci et al., 2011

In SEMIA6144, 18:1n-7 + 18:1n-9 decreased (84 to 73.5%) and 16:0 increased (11 to 18.6%) under temperature stress. Under combined stresses, 18:1n-7 + 18:1n-9 decreased (84 to 68.5%) and 16:0 increased (11 to 19.7%). FA values under NaCl stress alone were not significantly different from control values. The changes in FA percentages led to alteration of the ratio between unsaturated to saturated FA (U/S), (Table 4), which decreased in all experimental conditions for TAL1000 and SEMIA6144, but in SEMIA6144 it was not as remarkable as in TAL1000.

The FA composition of total lipids in the two strains was different, since TAL1000 contained 20:3n-6 and 19:0cyclo FA, which were not present in SEMIA6144. The FA profile of TAL1000 is similar to that reported for *Rhizobium* (Tighe et al., 2000).

High growth temperature and combined conditions caused significant reduction of 18:1 and increase of 18:0 and 16:0 (in TAL1000), or 16:0 (in SEMIA6144). These changes, more pronounced in TAL1000, provoked modifications at the level of FA unsaturation degree coincident with results obtained for other rhizobia in which different environmental changes caused modifications in the U/S ratio (Drouin et al., 2000; Boumahdi et al., 2001). TAL1000 at 37 °C showed a decrease in the U/S ratio and enhanced formation of 19:0cyclo (Table 4). Cyclic FA in the membranes of rhizobia could represent a mechanism to reduce membrane fluidity, similar to lactobacillus (Guerzoni et al., 2001).

SEMIA6144, showed change in degree of FA unsaturation and increased 16:0/18:0 ratio under temperature stress, which may reflect decreased chain length and it may alter transition temperature for change from gel to liquid crystalline phase (Suutari and Laakso, 1994).

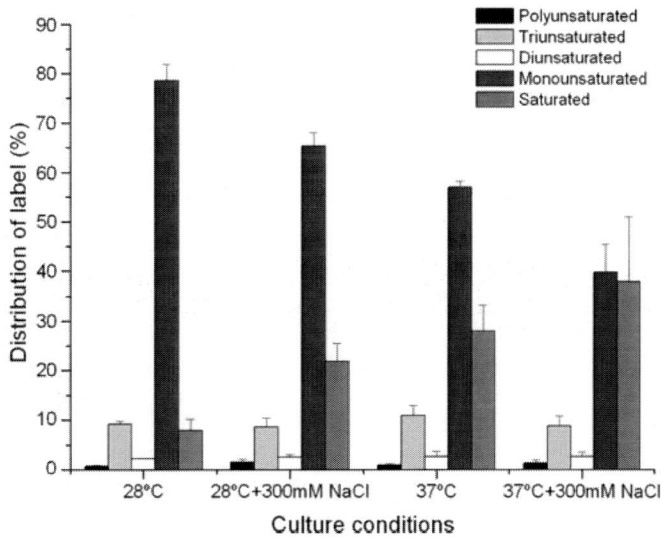

Figure 4. Effect of NaCl and temperature on incorporation of [1-^{14}C]acetate in fatty acids of *Bradyrhizobium* TAL1000. FAME were prepared from total lipids, and separated according to unsaturation degree using TLC plates impregnated with 10% AgNO$_3$. Results are expressed as the percentage of total radioactivity incorporated in each FA fraction. Values represent means ± SEM from three independent experiments.

Source: Paulucci et al., 2011

Under control conditions (28 °C), labeling in TAL1000 was found predominantly in monounsaturated FA (MUFA), followed by triunsaturated (TriUFA), saturated (SFA), and diunsaturated (DiUFA) fractions. Consistent with findings for FA composition (Table 4), the [1-^{14}C]acetate incorporation in MUFA decreased (27.4% and 49.4%) while [1-^{14}C]acetate incorporation in SFA increased (3.5-fold and 4.7-fold) by high growth temperature and by combined conditions, respectively. [1-^{14}C]acetate incorporation in MUFA decreased (78 to 65%) and that of SFA increased (8 to 22%) by NaCl stress (Fig. 3). The U/S ratio decreased from 11.5 to 3.5 for NaCl stress, from 11.5 to 2.5 for high growth temperature, and from 11.5 to 1.6 for combined conditions.

In SEMIA6144 the labeling was observed primarily in MUFA, followed by SFA, DiUFA and TriUFA. [1-^{14}C]acetate incorporation in SFA increased (28%, 39% and 45%) while radioactive incorporation in DiUFA decreased (73%, 66% and 82%) under NaCl stress, at 37 °C, and under combined conditions, respectively. So, the U/S ratio decreased from 14.7 to 11.5 for NaCl stress, from 14.7 to 10.7 for high growth temperature, and from 14.7 to 10 for combined conditions.

Results using [1-^{14}C]acetate labeling are consistent with FA composition studies and similar with studies in other Gram-negative bacteria (Wada et al., 1989; Ghaneker and Nair, 1973) since showed synthesis of both SFA and MUFA. Tested experimental conditions altered incorporation of labeled acetate in FA of both rhizobial strains. We found no studies on effects of abiotic stress on FA synthesis in rhizobia, but a study showing increase of SFA synthesis at high temperature in *B. subtilis* was consistent with our results (Aguilar et al., 1998).

Conclusion

The variation in the relative amount of zwitterionic PL and anionic PL observed in the two nodulant-peanut strains might be considered as a factor of stabilization for membrane lipid bilayer. The PC and PE variation would implicate the participation of both PL in the response to stress.

We suggest that the two strains, although possessing similar PL composition, have different mechanisms for stabilizing membrane fluidity.

Our results suggest that PC formation in *Bradyrhizobium* sp. SEMIA6144 is mainly due to the PL methylation pathway. Southern blot analysis using *pmt* and *pcs* probes of *B. japonicum* USDA 110 revealed a *pcs* and multiple *pmt* homologues in *Bradyrhizobium* sp. SEMIA 6144. A *pmtA* knockout mutant was constructed in *Bradyrhizobium* sp. SEMIA6144 that showed a 50% decrease in the PC content in comparison with the wild-type strain. The mutant was severely affected in motility and cell size, but formed wild-type-like nodules on its host plant. However, in coinoculation experiments, the *pmtA*-deficient mutant was less competitive than the wild type, suggesting that wild-type levels of phosphatidylcholine are required for full competitivity of *Bradyrhizobium* in symbiosis with peanut plants.

Peanut-nodulating rhizobia strains are able to adapt to tested environmental conditions through FA modification, and that fast-growing TAL1000 is more efficient in this respect than slow-growing SEMIA6144. The most important mechanism for maintaining physical properties of the membrane is modification of the FA unsaturation degree. The ability of TAL1000 to alter its content of FA 19:0cyclo may account for its tolerance of high temperature, while adaptation to environmental stresses in SEMIA6144 may involve shortening of the FA chain length.

This chapter contributes to the understanding of the behavior of strains nodulating peanut against environmental stresses and supports the idea that, the FA composition of strains of rhizobia used in commercial formulations should be considered as an indicator of whether or not an organism can adapt to changing environmental conditions, since differences in the capacity of rhizobia to adapt to environmental conditions may be related to differences in the FA composition.

Outlook

Another stress that affects much of the land surface and has an adverse effect on plants and microorganisms is drying. Lack of water results in lack of plants, soil and microbial life, degradation, negatively influencing for any attempt to revegetation. Peanut plants and rhizobacteria associated with them are not exempt from suffering the effects of desiccation. Hence, it becomes important to find and characterize new microorganisms that are related to crop plants of interest and that are resistant to various stresses, including desiccation. Moreover, in recent years, the study of plant and microbial components hs been powered in order to achieve a technical-industrial application in agricultural production and in other respects. On the other hand it is considered important to undertake studies related to peanut because it is necessary to relate productivity, the demands of peanut in terms of value added and production of various commercial products.

Members of the genus *Paenibacillus* inhabit different niches, such as soil, roots, rhizosphere of various crops, forest trees and marine sediments (Guemouri et al., 2000). In the rhizosphere, some *Paenibacillus* are involved in fixing nitrogen, in the solubilization of phosphate, in the production of antibiotics and in the production of hydrolytic enzymes, among others. These spore-forming bacteria have advantages over those that do not form spores because are more stable both in formulations and in soil. Within this genus, *Paenibacillus polymyxa* has attracted much attention based on its great biotechnological potential in various industrial processes and sustainable agriculture. Like many other microorganisms *P. polymyxa* is a plant growth promoting activity (PGPR) suggested by the stimulation of wheat growth (Holl et al., 1988) showing a wide range of host plants. We have isolated a strain from *Arachis hypogaea* nodules that according to 16S rDNA would be a *Paenibacillus* spp. With this strain we propose:

- Conduct studies of physiological, biochemical and genetics.
- Check if this strain is able to synthesize AG and lipids from acetate $[1-^{14}C]$.
- Test whether the drying modifies some of the parameters (nitrogenase activity and biological control), determined to characterize the strain.
- Study the effect of drying on membrane components at FA and lipid biosynthesis level.

References

Abdelmounmen, H., Filali-Maltouf, A., Neyra, M., Belabed, A. & Missbah El Idrissi, M. (1999). Effect of high salts concentrations on the growth of rhizobia and responses to added osmotica. *Journal of Applied Microbiology*, 86, 889-898.

Adesemoye, A. O. & Kloepper, J. W. (2009). Plant-microbes interactions in enhanced fertilizer-use efficiency. *Applied Microbiology and Biotechnology*, 85, 1-12.

Aguilar, P., Cronan, J. & de Mendoza, D. (1998). A Bacillus subtilis gene induced by cold shock encodes a membrane phospholipid desaturase. *Journal of Bacteriology*, 180, 2194-2200.

Alexandre, G., Greer-Phillips, S. & Zhulin, I. B. (2004). Ecological role of energy taxis in microorganisms. *FEMS Microbiology Reviews*, 28, 113-126.

Bakholdina, S. I., Sanina, N. M., Krasikova, I. N., Popova, O. B., Solovéva, T. F. (2004). The impact of abiotic factors (temperature and glucose) on physicochemical properties of lipids from *Yersinia pseudotuberculosi. Biochimie*, 86, 875-881.

Beney, L. & Gervais, P. (2001). Influence of the fluidity of the membrane on the response of microorganisms to environmental stresses. *Applied Microbiology and Biotechnology*, 57, 34-42.

Bernard, T., Pocard, J. A., Perroud, B. & Le Rudulier, D. (1986). Variations in the response of salt-stressed Rhizobium strains to betaines. *Archieves of Microbiology*, 143, 359-364.

Bloom, M., Evans, E. & Mouritsen, O. G. (1991). Physical properties of the fluid lipid-bilayer component of cell membrane: a perspective. *Quarterly Reviews of Biophysics*, 24, 293-397.

Boncompagni, E., Østeras, M., Poggi, M. & Le Rudulier, D. (1999). Occurrence of choline and glycine betaine uptake and metabolism in the family Rhizobiaceae and their roles in osmoprotection. *Applied and Environmental Microbiology*, 65, 2072-2077.

Boumahdi, M., Mary, P. & Hornez, J. (1999). Influence of growth phases and desiccation on the degrees of unsaturation of fatty acids and the survival rates of rhizobia. *Journal of Applied Microbiology*, 87, 611-619.

Boumahdi, M., Mary, P. & Hornez, J. (2001). Changes in fatty acid composition and degree of unsaturation of (brady)rhizobia as a response to phases of growth, reduced water activities and mild desiccation. *Antonie van Leeuwenhoek*, 79, 73-79.

Chihib, N., Tierny, Y., Mary, P. & Hornez, J. (2005). Adaptational changes in cellular fatty acid branching and unsaturation of Aeromonas species as a response to growth temperature and salinity. *International Journal of Food Microbiology*, 102, 113-119.

Choudhary, D. K., Sharma, K. P. & Gaur, R. K. (2011). Biotechnological perspectives of microbes in agro-ecosystems. *Biotechnology Letters*, 33, 1905-1910.

Comerci, D. J., Altabe, S., de Mendoza, D. & Ugalde, R. A. (2006). Brucella abortus synthesizes phosphatidylcholine from choline provided by the host. *Journal of Bacteriology*, 188, 1929-1934.

Conover, G. M., Martínez-Morales, F., Heidtman, M. I., Luo, Z. Q., Tang, M., Chen, C., Geiger, O. & Isberg, R. R. (2008). Phosphatidylcholine synthesis is required for optimal function of Legionella pneumophila virulence determinants. *Cell Microbiology*, 10, 514-528.

Correa, O., Rivas, E. & Barneix, A. (1999). Cellular envelopes and tolerance to acid pH in *Mesorhizobium loti. Current Microbiology*, 38, 329-334.

Cronan, J. E. Jr. & Gelmann, E. P. (1975). Physical properties of membrane lipids: biological relevance and regulation. *Bacteriology Review* 39, 232-256.

Cullis, P. R. & de Kruijff, B. (1979). Lipid polymorphism and the functional roles of lipids in biological membranes. *Biochimica et Biophysica Acta* 55, 399-420.

Cullis, P. R., Hope, M. J., de Kruijff, B., Verkleij, A. J. & Tilcock, C. P. S. (1985). Structural properties in biological membranes. In: Kuo, J. F. (Ed.), *Phospholipids and cellular regulations*, vol. 1. CRC Press, Boca Raton, FL, pp. 1-60.

de Rudder, K. E. E., López-Lara, I. M. & Geiger, O. (2000). Inactivation of the gene for phospholipid N-methyltransferase in Sinorhizobium meliloti: phosphatidylcholine is required for normal growth. *Molecular Microbiology*, 37, 763-772.

Dardanelli, M. S., Woelke, M. R., González, P. S., Bueno, M. A. & Ghittoni, N. E. (1997). The effects of nonionic hyperosmolarity and of hight temperature on cell-associated low molecular weight saccharides from two peanut rhizobia strains. *Symbiosis* ISSN 0334-5114, 23: 73-84.

Denich, T. J., Beaudette, L. A., Lee, H. & Trevors, J. T. (2003). Effect of selected environmental and physico-chemical factors on bacterial cytoplasmic membranes. *Journal of Microbiological Methods*, 52, 149-182.

Dey, R., Pal, K. K., Bhatt, D. M. & Chauhan, S. M. (2004). Growth promotion and yield enhancement of peanut (Arachis hypogaea L). plant growth-promoting rhizobacteria. *Microbiology Research*, 159, 371-394.

Donato, M., Jurado, A., Antunes-Madeira, M. & Madeira, V. (2000). Membrane lipid composition of Bacillus stearo-thermophilus as affected by lipophilic environmental pollutants: an approach to membrane toxicity assessment. *Archieves Environmental Contamination Toxicology*, 39, 145-153.

Drouin, P., Prevost, D. & Antoun, H. (2000). Physiological adaptation to low temperatures of strains of Rhizobium leguminosarum bv. viciae associated with Lathyrus spp. *FEMS Microbiology Ecology*, 32, 111-120.

Egamberdieva, D. & Kucharova, Z. (2009). Selection for root colonizing bacteria stimulating wheat growth in saline soils. *Biology and Fertility of Soil*, 45, 563-571.

Enomoto, K. & Koyama, N. (1999). Effect of growth pH on the phospholipid contents of the membranes from alkaliphilic bacteria. *Current Microbiology*, 39, 270-273.

Fabra, A., Castro, S., Taurián, T., Angelini, J., Ibañez, F., Dardanelli, M. S., Tonelli, J., Bianucci, E. & Valetti, L. (2010). Interaction among Arachis hypogaea L. (peanut) and beneficial soil microorganisms: how much is it known? *Critical Reviews in Microbiology*, 36, 179-194.

Finean, J. B. & Michell, R. H. (1981). Isolation, composition, and general structure of membranes. In: Finean, J. B., Michell, R. H. (Eds.), *Membrane Structure.* Elsevier, New York, NY, pp. 19-25.

Firbank, L. G. (2005). Striking a new balance between agricultural production and biodiversity. *Annals of Applied Biology*, 146, 163-175.

Elsheikh, E. & Wood, M. (1990). Rhizobia and Bradyrhizobia under salt stress: possible role of trehalose in osmoregulation. *Letters Applied Microbiology*, 10, 127-129.

Garcia, M. C., Marina, M. L., Laborda, F. & Torre, M. (1998). Chemical characterization of commercial soybean products. *Food Chemistry*, 62, 325-331.

Ghaneker, A. & Nair, P. (1973). Evidence for the existence of an aerobic pathway for synthesis of monounsaturated fatty acids by *Alcaligenes faecalis*. *Journal of Bacteriology*, 114, 618-624.

Gomes-Germano, M., Menna, P., Mostasso, F. & Hungria, M. (2006). RFLP analysis of the rRNA operon of a Brazilian collection of bradyrhizobial strains from 33 legume species. *International Journal of Systematic and Evolution Microbiology*, 56, 217-229.

Gruner, S. M., Cullis, P. R., Hope, M. J. & Tilcock, C. P. S. (1985). Lipid polymorphism, the molecular basis of nonbilayer phases. *Annual Review of Biophysics and Biophysical Chemistry*, 14, 211-238.

Guerzoni, E., Lanciotti, R. & Cocconcelli, S. (2001). Alteration in cellular fatty acid composition as a response to salt, acid, oxidative and thermal stresses in *Lactobacillus helveticus*. *Microbiology*, 147, 2255-2264.

Hacker, S., Sohlenkamp, C., Aktas, M., Geiger, O. & Narberhaus, F. (2008). Multiple phospholipid N-methyltransferases with distinct substrate specificities are encoded in *Bradyrhizobium japonicum*. *Journal of Bacteriology*, 190, 571-580.

Hazel, J. R. & Williams, E. E. (1990). The role of alterations in membrane lipid composition in enabling physiological adaptation of organisms to their physical environment. *Progress in Lipid Research*, 29, 167-227.

Keweloh, H. & Heipieper, H. (1996). Trans unsaturated fatty acids in bacteria. *Lipids* 31, 129-136.

Klüsener, S., Aktas, M., Thormann, K. M., Wessel, M. & Narberhaus, F. (2009). Expression and physiological relevance of Agrobacterium tumefaciens phosphatidylcholine biosynthesis genes. *Journal of Bacteriology*, 191, 365-374.

Kulkarni, S. & Nautiyal, C. S. (2000). Effects of salt and pH stress on temperature-tolerant Rhizobium sp. NBRI330 nodulating *Prosopis juliflora*. *Current Microbiology*, 40, 221-226.

Kyriakidis, D. A. (2009). Bacterial signaling and adaptation. *Amino Acids*, 37, 441.

Laeremans, T. & Vanderleyden, J. (1998). Review: Infection and nodulation signalling in *Rhizobium-Phaseolus vulgaris* simbiosis. *World Journal of Microbiology and Biotechnology*, 14, 787-808.

Lindstrom, K. & Lehtomaki, S. (1988). Metabolic properties, maximum growth temperature and phage sensitivity of Rhizobium sp. (Galea) compared with other fast growing rhizobia. *FEMS Microbiology Letters*, 50, 277-287.

López Lara, I. M. & Geiger, O. (2001). Novel pathway for phosphatidylcholine biosynthesis in bacteria associated with eukaryotes. *Journal of Biotechnology*, 91, 211-221.

Martínez-Morales, F., Schobert, M., López-Lara, I. M. & Geiger, O. (2003). Pathways for phosphatidylcholine biosynthesis in bacteria. *Microbiology*, 149, 3461-3471.

McElhaney, R. N. (1982). Effects of membrane lipids on transport and enzymatic activities. In: Razin, S., Rottam, R. (Eds.), *Current Topics in Membranes and Transport*, vol. 17. Academic Press, London, pp. 317-380.

McElhaney, R. N. & Souza, K. A. (1976). The relationship between environmental temperature, cell growth, and the fluidity and physical state of the membrane lipids in *Bacillus stearothermophilus*. *Biochimica et Biophysica Acta*, 443, 348-359.

Medeot, D. B., Bueno, M. A., Dardanelli, M. S, & García de Lema M.B. (2007) Adaptational changes in lipids of Bradyrhizobium SEMIA 6144 nodulating peanut as a response to growth temperature and salinity. *Current Microbiology*, 54, 31-35.

Medeot, D. B., Sohlenkamp, C., Dardanelli, M. S., Geiger, O. García de Lema, M. B., & López-Lara, I.M. (2010) Phosphatidylcholine levels of peanut-nodulating Bradyrhizobium sp. SEMIA6144 affect cell size and motility. *FEMS Microbiology Letter,* 303, 123-131.

Medeot, D. B., Paulucci, N. S., Albornoz, A., Fumero, M. V., Bueno, M., Garcia, M. B., Woelke, M., Okon, Y. & Dardanelli, M. S. (2010). Plant growth-promoting rhizobacteria-rhizobia and legume improvement. In: *Microbes for Legume Improvement.* Saghir Khan, M. D., Zaidi, A. & Musarrat, (Eds.), J., Springer-Verlag (Germany), pp 473-494.

Mileykovskaya, E., Ryan, A., Mo, X., Lin, C., Khalaf, K. I., Dowhan, W. & Garrett, T. A. (2009). Phosphatidic acid and N-acyl phosphatidylethanolamine form membrane domains in Escherichia coli mutant lacking cardiolipin and phosphatidylglycerol. *The Journal of Biological Chemistry,* 284, 2990-3000.

Miller, L. D., Yost, C. K., Hynes, M. F. & Alexandre, G. (2007). The major chemotaxis gene cluster of Rhizobium leguminosarum bv. viciae is essential for competitive nodulation. *Molecular Microbiology, 63,* 348-362.

Miller, K. J. & Wood, J. M. (1996). Osmoadaptation by rhizosphere bacteria. *Annual Review of Microbiology, 50,* 101-36.

Minder, A. C., de Rudder, K. E., Narberhaus, F., Fischer, H. M., Hennecke, H. & Geiger, O. (2001). Phosphatidylcholine levels in Bradyrhizobium japonicum are critical for an efficient symbiosis with the soybean host plant. *Molecular Microbiology, 39,* 1186-1198.

Paulucci, N. S., Medeot, D. B., Dardanelli, M. S., & García de Lema, M. (2011) Growth Temperature and Salinity Impact Fatty Acid Composition and Degree of Unsaturation in Peanut-Nodulating Rhizobia. *Lipids, 46,* 435-441.

Peoples, M. B., Giller, K. E., Herridge, D. F. & Vessey, J. K. (2002). Limitations of biological nitrogen fixation as a renewable source of nitrogen for agriculture. In: Finan, T. M., O'Brian, M. R., Layzell, D. B., Vessey, J. K. & Newton, W. (Eds.), *Nitrogen fixation, global perspectives.* Wallingford, CABI International, pp 356-360.

Ramos, J., Duques, E., Rodriguez-Herva, J., Godoy, P., Haidour, A., Reyes, F. & Fernandez-Barrero, A. (1997). Mechanisms for solvent tolerance in bacteria. *The Journal of Biological Chemistry, 272,* 3887-3890.

Ratledge, C. & Wilkinson, S. G. (1988). Fatty acids, related and derived lipids. In: Ratledge, C., Wilkinson, S. G. (Eds.), *Microbial Lipids,* vol. 1. Academic Press, Toronto, Canada, pp. 23-52.

Rilfors, L., Lindblom, G., Wieslander, A. & Christiansson, A. (1984). Lipid bilayer stability in biological membranes. In: Kates, M., Morris, L. (Eds.), Biomembranes, vol. 12. *Membrane Fluidity.* Plenum Press, New York, pp. 206-245.

Russell, N. J. (1992). Physiology and molecular biology of psychrophilic microorganisms. In: Herbert, R. A., Sharp, R. J. (Eds.), *Molecular biology and biotechnology of extremophiles.* Glasgow and London: Blackie, pp 203-224.

Russell, N. & Fukunaga, N. (1990). A comparison of thermal adaptation of membrane lipids in psychrophilic and thermophilic bacteria. *FEMS Microbiology Reviews, 75,* 171-182.

Sajbidor, J. (1997). Effect of some environmental factors on the content and composition of microbial membrane lipids. *Critical Review in Biotechnology, 17,* 87-103.

Silvius, J. R., Mak, N. & McElhaney, R. N. (1980). Why do prokaryotes regulate membrane lipid fluidity? In: Kates, H., Kukis, A. (Eds.), Membrane fluidity, biophysical techniques and cellular regulation. *Humana Press,* Clifton, NJ, pp. 213-222.

Sinensky, M. (1974). Homeoviscous adaptation, a homeostatic process that regulates the viscosity of membrane lipids in Escherichia coli. *Proceedings of the National Academic of Sciences,* 71, 522-525.

Singer, S. J. & Nicolson, G. L. (1972). The fluid mosaic model of the structure of cell membranes. *Science,* 175, 720-730.

Singh, J. S., Pandey, V. C. & Singh, D. P. (2011). Efficient soil microorganisms: A new dimension for sustainable agriculture and environmental development. *Agriculture, Ecosystems Environment,* 140, 339-353.

Sohlenkamp, C., de Rudder, K. E. E., Röhrs, V., López-Lara, I. M. & Geiger, O. (2000). Cloning and characterization of the gene for phosphatidylcholine synthase. *Journal of Biological Chemistry,* 275, 18919-18925.

Sohlenkamp, C., López-Lara, I. M. & Geiger, O. (2003). Biosynthesis of phosphatidylcholine in bacteria. *Progress in Lipid Research,* 42, 115-162.

Soltani, M., Metzger, P. & Largeau, C. (2005). Fatty acid and hydroxy acid adaptation in three gram-negative hydrocarbon-degrading bacteria in relation to carbon source. *Lipids,* 40, 1263-1272.

Sutton, G. C., Russell, N. J. & Quinn, P. J. (1991). The effect of salinity on the phase behaviour of total lipid extracts and binary mixtures of the major phospholipids isolated from a moderately halophilic eubacterium. *Biochemica et Biophysica Acta,* 1061, 235-246.

Suutari, M., Liukkonen, K. & Laakso, S. (1990). Temperature adaptation in yeasts: the role of fatty acids. *Journal of General Microbiology,* 136, 1469-1474.

Suutari, M. & Laakso, S. (1994). Microbial fatty acid and termal adaptation. *Critical Review in Microbiology,* 20, 285-328.

Tang, Y. & Hollingsworth, R. (1998). Regulation of lipid synthesis in *Bradyrhizobium japonicum*: low oxygen concentrations trigger phosphatidylinositol biosynthesis. *Applied and Environmental Microbiology,* 64, 1963-1966.

Théberge, M., Prévost, D. & Chalifour, P. (1996). The effect of different temperatures on the fatty acids composition of *Rhizobium leguminosarum* bv. viciae in the faba bean symbiosis. *New Phytology,* 134, 657-664.

Tighe, S., de Lajudie, P., Dipietro, K., Lindström, K., Nick, G. & Jarvis, B. (2000). Analysis of cellular fatty acids and phenotypic relationships of Agrobacterium, Bradyrhizobium, Mesorhizobium, Rhizobium and Sinorhizobium species using the Sherlock Microbial Identification System. *International Journal of Systematic Evolution Microbiology,* 50, 787-801.

van de Meer, W. (1984). Physical aspects of membrane fluidity. In: Shinitzsky, M. (Ed.), *Physiology of membrane fluidity.* CRC Press, Boca Raton, FL, pp. 54-71.

Wada, M., Fukunaga, N. & Sasaki, S. (1989). Mechanism of biosynthesis of unsaturated fatty acids in Pseudomonas sp. strain E-3, a psychotropic bacterium. *Journal of Bacteriology,* 171, 4267-4271.

Wessel, M., Klüsener, S., Gödeke, J., Fritz, C., Hacker, S. & Narberhaus, F. (2006). Virulence of Agrobacterium tumefaciens requires phosphatidylcholine in the bacterial membrane. *Molecular Microbiology,* 62, 906-915.

Wilkinson, S. G. (1988). Gram-negative bacteria. In: Ratledge. C., Wilkinson, S. G. (Eds.), *Microbial lipids,* vol. 1. Academic Press, pp 355-356.

Zhang, Y. M. & Rock, C. O. (2008). Membrane lipid homeostasis in bacteria. *Nature Reviews Microbiology, 6,* 1-13.

Zhang, X., Harper, R., Karsisto, M. & Lindstrom, K. (1991). Diversity of *Rhizobium* bacteria isolated from the root nodules of leguminous trees. *International Journal of Systematic Bacteriology, 41,* 104-113.

In: Symbiosis: Evolution, Biology and Ecological Effects   ISBN: 978-1-62257-211-3
Editors: A. F. Camisão and C. C. Pedroso   © 2013 Nova Science Publishers, Inc

Chapter 6

BEHAVIORAL, PHYSIOLOGICAL AND ECOLOGICAL EFFECTS OF ORGANISMS IN SYMBIOTIC ASSOCIATIONS

W. Randy Brooks
Professor of Biology, Florida Atlantic University, FL, US

Abstract

The significance of coevolutionary adaptations by associated organisms at the cellular and molecular level has been the primary focus of much research in symbiology (e.g., endosymbiotic hypothesis for development of eukaryotes). Studies on the behavioral and physiological ecology of organisms involved in symbiotic associations have also demonstrated extraordinary examples of adaptation. These associations represent tremendous potential in demonstrating alternatives to competition as major evolutionary selective forces. Interactions vary tremendously within the commensalism, mutualism and parasitism subdivisions of symbiosis, and divergent examples will be discussed. For example, many organisms are limited by their anatomy to remove parasites and necrotic tissues, etc., and must rely on allogrooming by others. While terrestrial animals are usually better equipped for autogrooming, the "self-cleaning" problem is especially significant in marine environments where cohorts of animals switch from potential predators to symbiotic cleaning "clients." Organisms that possess significant, innate adaptations for protection are potentially attractive commensalistic or mutualistic hosts if a potential symbiont can inhibit or withstand the defensive mechanisms. Parasitism has also demonstrated bizarre behavioral outcomes, including examples where a larval form can "force" an intermediate host to alter its behavior to facilitate completion of the parasite's life history. Finally, the characteristics or phenoptypes of many organisms are actually based on composite genotypes of the host and any significant symbionts.

The question of why organisms associate with each other symbiotically can best be addressed by looking at the evolutionary and biological need for all organisms to exploit necessary resources. Ecologists typically partition the ecosystem resources into biotic and abiotic components, for which organisms typically compete either directly (interference) or indirectly (exploitative). Non-human species, however, probably have less need to make such formal distinctions between living and nonliving resources; unless the resource is recognized as a direct safety threat. Thus, from the tiniest microbes to the largest organisms, symbioses are a natural and expected outcome of this competition! As such, initially, a "pioneer" organism gets on or in a host. The symbiont, being typically much smaller than the host (which can technically be called a symbiont or symbiote, too), may go relatively unnoticed if it can gain resources (e.g., shelter, consumes castaway cells or tissues) without any significant effect or damage to the host. This would constitute commensalism. However, not all associations start off this innocuously; or if they do, may eventually or episodically move to a situation in which significant negative impacts to the host may occur. In such situations, we have a parasitic association.

At this point it is certainly in the host's best interest to minimize such deleterious effects by either ridding itself of the parasite, or neutralizing the effects. The former action is clearly not possible in many symbioses; therefore, the next best outcome is to alter the encounter to minimize harm. With many symbioses, the host – if given enough time, on an evolutionary scale – can successfully prevent significant damage. In cases where complete removal of the parasite, or prevention of infection, are impossible, an even more optimal outcome by the host would be to accrue some benefit concomitant to the neutralization of harmful effects. This, ultimately, could lead to a mutualistic encounter. Symbioses that result in benefits to both associates represent tremendous potential in demonstrating alternatives to competition as major evolutionary selective forces. For clarification, ecology also recognizes mutualisms, but does not have the requisite symbiosis. Thus, there are non-symbiotic interactions between species that are mutually beneficial, too.

In any case, long-term encounters between symbionts and hosts can certainly result in coevolutionary adaptations. Additionally, given that no organisms are devoid of symbionts (i.e., axenic or aposymbiotic)(Paracer & Ahmadijian, 2000), many evolutionarily fascinating and ecologically significant associations have been discovered. Coevolutionary adaptations of associated organisms at the cellular and molecular level have been the primary focus of much research in symbiology (e.g., endosymbiotic hypothesis for development of eukaryotes). In this chapter, the focus will be on behavioral, physiological and ecological adaptations of organisms involved in symbiotic associations.

As stated previously, organisms – especially relatively large ones – are under a constant barrage of attacks by opportunistic, smaller symbionts. Microscopically, these typically include prokaryotic, single-celled eukarotic (e.g., protists) and fungal agents. Tissue defense mechanisms (e.g., immune defenses) are certainly important in regulating these types of encounters, both inside and outside the host's body. Macroscopic ectosymbionts are more likely to trigger active, behavioral engagement by the potential host organism, as direct detection visually or tactilely can occur. Completely avoiding such initial infection would be the ideal situation, and there are documented behavioral mechanisms to attempt to do so, such as evasive movements, selecting habitats that minimize exposure to parasites, etc. (Combes, 2001; Moore, 2002). However, once the symbiont has successfully "connected," the typical reaction of most organisms would be to "clean" themselves of these potential parasites, especially if they cause physical "irritation" to the host. Some organisms are quite capable of removing ectosymbionts via autogrooming – or by themselves. Social species and interspecies groups also might supplement autogrooming with allogrooming, in which individuals groom others altruistically. In some cases, allogrooming plays a dominant role in maintaining the health of organisms and may fit the theoretical model called Biological Markets, which suggests that organisms can trade services and goods – such as cleaning or grooming – for mutual benefit. Like real markets, these interactions can be affected by current conditions of supply and demand (Noë & Hammerstein, 1995).

When different species are involved in allogrooming this is called cleaning symbioses, which Losey (1987) describes as "A three-party symbiotic relationship in which the cleaning organisms act as microcarnivores and use the body surfaces of their hosts as a feeding substratum." "Cleaning" adaptations in terrestrial situations, including flying organisms, will be discussed first, followed by adaptations found in aquatic environments.

To accomplish self-cleaning, innate, morphological structures that can be finely manipulated are typically required. Mammals are major targets by numerous arthropods, including insects such as lice, fleas; tics and spiders are also pests (cf., Hopla et al., 1994). Not surprisingly, all terrestrial mammals attempt to groom for removal of ectoparasites (Mooring et al., 2004); and those with extensive hair or fur are even more prone to successful parasitic attacks (Pagel & Bodmer, 2003). Thus, extensive capabilities for cleaning have evolved in mammals. Arguably, *Homo sapiens* – the Naked Ape – is one of most adept species at both autogrooming and allogrooming. Furthermore, humans may have evolved the "relatively" hairless condition to minimize ectoparasitic loads, including wearing clothing to further reduce parasite contact (Pagel & Bodmer, 2003). Perhaps axillary and pubic hair, which develop primarily at puberty, were more important for our naked (i.e., unclothed) early human ancestors to advertise both visually and chemically (with the active scent glands associated with hair follicles) in sexual selection situations. "Clothed" humans retain these vestigial characteristics of thick body hair in these areas. Cephalic hair (including eyebrows) is likely developed primarily for both thermoregulation and visual communication purposes. Still, having these areas of relatively thick hair make us more susceptible to certain parasites (e.g., lice).

When we see or feel symbionts on us, we can use our fingers and hands to pick off these parasites, including allogrooming others (especially young children). We can also supplement our manual abilities by applying topical medical treatments to rid ourselves of external (e.g., lice) and internal (e.g., worms) symbionts. Additionally, we can attempt to avoid areas or times which would make us more vulnerable to feeding symbionts, such as mosquitoes that commonly feed at crepuscular times of day. Psychologically, when we feel more vulnerable, we are more likely to autogroom and do less allogrooming (Thompson, 2010).

Closely related primates can also engage in cleaning strategies similar to humans; and the occurrence of thicker and more body hair presumably make detection and cleaning significantly more challenging. Although non-human primates are technically not bipedal, they have excellent dexterity with their forelimbs and hands (and sometimes also with their hindlimbs and toes). As such, many primate species engage in both autogrooming and allogrooming (Dunbar, 1991). Individuals can spend considerable amounts of time meticulously picking off parasites from themselves and group members. While there are questions by some about whether allogrooming is primarily for social reasons (e.g., bonding of group members), clearly one major function and outcome is the removal of parasites (Dunbar, 1991, Mooring et al., 2004; Chancellor & Isbell, 2008).

The remaining tetrapods typically lack the skills to use their forelimbs for effective grooming. They must rely on other mechanisms, such as oral grooming. Specifically, they can remove parasites with their teeth. Grazing, wild herbivores (e.g., moose, elk, impala and deer) have to contend with ectoparasites such as ticks, and frequently employ both auto- and allogrooming by using their mouths and biting for removal (Welch et al., 1991; Mooring et al., 2004; Yamada & Urabe, 2007). Other mechanisms used by wild and domesticated grazers, such as cows, are muscle twitching (including ears), head shaking, stomping legs and tail thrashing (Hart, 1994). Additionally, cows have been observed brushing up against trees, fence posts, etc., to remove ectoparasites (Kohari et al., 2007), and they can also allogroom (Kohari et al., 2007; Val-Laillet et al., 2009). So many of these grazing mammals spend significant time budgets on grooming, although in the case of some deer species, allogrooming may play a social role primarily and cleaning role secondarily (Yamada & Urabe, 2007).

Cleaning behaviors in lower mammals, such as rodents, have also been well documented – especially for the removal of fleas. For example, squirrels employ similar auto- and allogrooming techniques mentioned above for the larger mammals. Additionally, field studies have confirmed that parasite loads can drastically decrease fecundity in some squirrel species (Hillegass et al., 2010). Again, while allogrooming plays a social role, grooming certainly has a parasite-removal function (Hawlena et al., 2007).

Birds are well known for their preening activity, as the feathers and wings must be kept in optimal condition for flight (in air or in water)(Gill, 2012). In addition to this primary function, the removal of ectoparasites is also important. Thus, birds engage in both auto- and allopreening (Lewis et al., 2007; Gill 2012). Penguins can also spend a considerable amount of time cleaning themselves of debris and parasites (Viblanc et al., 2011). Birds are also known to clean other animals. The Galapagos Islands have several examples of such ectoparasite removal by birds. The most famous involves two species of Darwin finches and the Galapogos turtle. In an amazing behavioral display, the turtle extends its legs fully to elevate the body maximally so that the finches can access soft body parts from underneath the plastron, including proximal portions of the legs inside the shell to remove ectoparasites (MacFarland & Reeder, 1974). Finches and other bird species on the island are also known to clean parasites from the surfaces of both land and marine iguanas (Christian, 1980). Finally, wild grazing animals in the savannahs of Africa are allogroomed by two species of bird known collectively as oxpeckers. They are voracious consumers of parasites, such as ticks. However, they may also trigger problems for the hosts, as their consumption can include opening wounds and drinking blood from the host. So in some cases, the associations may transition from mutualistic to parasitic (Weeks, 2000).

Finally, the most abundant of all animals – the insects – have more than their fair share of potential parasites to remove, including closely related members within the arthropod taxon. Social insects show some of the greatest adaptations for grooming. Specifically, allogrooming is highly important to many social insect species (Fefferman, et al., 2007). Honey bees worldwide have to contend with mites, which can completely destroy bee hives when infestations become extreme (Stanimirovic et al., 2010). Thus, both auto- and allogrooming are used by honey bees for defense against mites (Danka & Villa, 2003). Ants also have to contend with a particularly vicious species of fly that lays its eggs in the head of ants and the developing larvae consume the soft tissues, including the brain. Interestingly, these tropical flies have been used in the southeastern US as biological control agents to rid areas of fire ants (which are non-native)(cf., Porter, 1998).

Leaf-cutting ants in tropical areas can potentially be devastated by toxic species of fungi. So auto- and allogrooming are important to remove fungal spores and hyphae to prevent colony damage in the nest (Walker & Hughes, 2009). Fungal infections can have bizarre behavioral effects on ants, which will be discussed later in this chapter. Ironically, leaf-cutting ants "farm" and protect less virulent fungal species in their nests for food (Little et al., 2006).

The animals discussed above illustrate that terrestrial environments have many examples of cleaning. When an animal is limited by its anatomy, the removal of external parasites and necrotic tissues, etc., becomes more challenging. Thus, many species must rely on supplemental allogrooming by others, which can also serve as an important social activity. However, terrestrial animals are usually much better equipped for autogrooming than aquatic animals (Losey, 1987). This leads to a significant "self-cleaning" problem for many aquatic organisms, especially in marine environments. From aquatic mammals, to fishes, to invertebrates... the problem of keeping the body surface clean from ectosymbionts is magnified, because in marine ecosystems the most convenient hard surfaces for larval forms to settle and grow are usually the surfaces of other organisms. For example, consolidated benthic (i.e., bottom) areas, such as rock, are usually already covered by sponges, corals, etc. This is why artificial objects, such as pilings, jetties, boat bottoms, etc., are quickly colonized by benthic organisms. These settling organisms can create a huge burden, financially, as they can "foul" these structures by decreasing the structural integrity (e.g., wooden piers) and increase drag and lower fuel economy with ships that are heavily impacted (which can necessitate that boat bottoms be scraped clear either by divers or have the entire boat lifted out of the water periodically to be thoroughly cleaned and repainted).

Aquatic mammals also have the need to protect themselves from ectoparasites. Like humans, many aquatic mammals have reduced fur or hair levels, which can potentially minimize parasite attachments. Additionally, some aquatic mammals – such as dolphins – swim rapidly through the more dense medium of water (compared to air), which could theoretically make it more challenging for potential ectoparasites to settle. However, when comparing the morphology of primates - with their dexterous abilities - to aquatic mammals, the latter group has much fewer grooming adaptations.

Parasites of aquatic mammals are well documented, with over 300 species of macroscopic worm and arthropodan parasites known to affect marine mammals (Evans & Raga, 2001). Impacts of these parasites can be significant and dangerous for the host mammal (Geraci & Aubin, 1987). Photos of large marine mammals, like whales, typically show upon closer inspection encrusting organisms growing on the body surface. Barnacles are a particularly common ectoparasite; while they do not use the whale's skin for food, they do settle in high numbers potentially increasing fluid-dynamic drag and creating skin lesions. There is speculation that the commonly observed breaching and fin slapping at the water's surface might be related to attempts to dislodge such parasites (Felix et al., 2006). Additionally, older whales and whales that are sick, and thus slower moving, typically have greater barnacle parasite loads (Slijper, 1979; Fertl, 2002). A single humpback whale was observed with an estimated 454kg (or 1000 lbs) of barnacles on its skin (Slijper, 1979). Clearly, if whales could autogroom effectively, such instances of heavy parasite loads would be unusual. Dolphins have been observed rubbing fins with each other, which could potentially be used as a form of allogrooming. However, it is likely the primary role is social, as little evidence has been observed that these actions actually remove parasites (Dudinski et al., 2009).

Manatees can typically swim in both marine and freshwater ecosystems. One advantage in this ability is that ectoparasites picked up in saltwater may actually be killed when the manatee ventures into freshwater (cf., Olivera-Gomez & Mellink, 2005). Additionally, freshwater fishes have been observed eating parasites and vegetation growing on the surface of manatees (Nico et al., 2009).

While dolphins and whales are extremely limited in their abilities to auto- and allogroom, other aquatic mammals may be better equipped. For example, otters have significant amounts of fur that has to be groomed and aerated for thermal insulation purposes (Osterrieder & Davis, 2011). The meticulous care of the fur involves both auto- and allogrooming, which in addition to serving a social function (as observed in many terrestrial mammals), maintains the functional aspects for swimming and presumably helps maintain a low parasite load.

Aquatic reptiles, with their exceptionally hard exoskeletons, serve as an especially attractive settlement surface. One of the most typical, aquatic reptilian representatives is the sea turtle. Like aquatic mammals, they spend most of the time in the water; although both groups are air breathers, which necessitates visits to the surface periodically. In the case of sea turtles, females typically exit the water for extended periods of time to lay their eggs in the sand. Ectosymbionts, which are aquatic, could possibly be affected during these reproductive excursions by the turtle out of water. Additionally, some sea turtles bask out of water primarily for thermoregulatory purposes, but ectoparasite removal may be a secondary function (Jantzen et al., 1992). However, barnacle symbionts would likely be unaffected as these animals can simply close the opening in the shell as those barnacles do when exposed in the intertidal zone when tidal fluctuations create extremely low tide levels. Leeches are also a problem ectoparasite for many sea turtles (George, 1997). So sea turtles have restricted abilities to autogroom, with their limited appendage length and flexibility. However, observations made by researchers with captive sea turtles that have been marked, for example with finger nail polish, show that such marks may actually be removed by the turtle either by rubbing against a hard object in the lab or attempting to rub it off with their flippers (Natasha Warraich, personal communication). The efficacy of sea turtle autogrooming for fouling organisms in the field is unclear. However, sea turtles have been observed attending cleaning stations (which will be discussed below).

The remaining examples discussed below represent animals that typically are fully aquatic. Thus, settling ectoparasites would not typically have to deal with exposure to air as a possible means to facilitate removal. For example, fishes are one of the major groups of macroscopic organisms in any aquatic environment, and parasites are likely found in or on all species (Feist & Longshaw, 2008). Fishes are under constant attack by ectoparasites, some of which are merely present as an intermediate larval stage in a complex life history. In any case, ectoparasites – including flatworms and copepods - are a major problem for fishes (Feist & Longshaw, 2008). Fishes autogroom primarily by chafing, or rubbing up against a hard object to dislodge the parasite (Wisendon et al., 2009). This behavior could be followed by body shakes, rapid acceleration swimming and "coughing" motions (Baker & Smith, 1997). Those who have had fishes in aquaria have likely observed these same behaviors, as ectoparasites can be a significant problem in such closed systems. Because of the limited abilities of fishes to autogroom, and the tremendous biological presence of potential parasites in the aquatic environment, allogrooming of fishes is extremely pervasive.

One of the most famous examples of an ectobiont on fishes (and almost any other large aquatic organism, including sea turtles and whales) involves another fish call the remora. Although most data suggest these fishes are commensalistic, by being phoretic or simply "hitching a ride," some research has demonstrated a reduced load of ectoparasites with the presence of some remora/host combinations (Mucientes et al., 2008). Additionally, those fishes (and other large aquatic vertebrates) that possess numerous or relatively large remora individuals could experience additional drag while swimming or even skin irritation (cf., Brunnschweiler, 2006). Thus, one of the more ubiquitous, and seemingly innocuous, fish ectosymbionts may actually be a parasite under certain conditions with some hosts. Interspecific cleaning by fishes is also prevalent in aquatic systems, and can be extremely effective, too (see cleaning stations discussion below).

Fishes are also involved in one of the most extreme, non-lethal (typically) parasitic associations, in which an isopod attaches to and consumes the fish's tongue and resides permanently in the oral cavity (Brusca, 1981; Bariche & Trilles, 2006; Jones et al., 2008; Bowman et al., 2010). Specifically, the isopod causes the fish's tongue to atrophy by consuming its blood supply and then attaching to the muscles of the remaining tongue base with powerful appendages. This is likely the only parasite that can functionally replace an organ, as the parasitized fishes continue to consume food normally – albeit, sharing the meal with its "prosthetic, parasitic" isopod.

Invertebrates are also known to allogroom fishes. The best known examples involve shrimp, which are effective at removal of flatworms and other parasites (McCammon et al., 2010) - but they consume the mucus of fishes, too. Interestingly, some cleaner shrimps also reside in sea anemones (Gwaltney & Brooks, 1994) and receive protection (Mihalik & Brooks, 1997), which is interesting because cleaners typically are not consumed. The efficacy of cleaning by these shrimps may vary depending on the shrimp/host combinations (McCammon et al., 2010). Perhaps those shrimp that are less involved in cleaning fishes of ectoparasites, and more involved in consuming fish mucus are more susceptible to predation by the fish host.

A particularly fascinating behavioral event occurs in some aquatic environments in which cohorts of animal hosts switch from potential predators to symbiotic cleaning "clients." Such events are best known in marine coral reef environments where fishes, sea turtles and other large species aggregate and wait their turn to be cleaned. Smaller cleaner fishes (e.g., wrasses, gobies, damsels) and cleaner shrimps occupy a particular area on a reef and through behavioral posturing (e.g., angle of the fish in water or waving brightly colored antennae of shrimp) signal to the larger client fish that their cleaning services are currently available. Additionally, and most importantly, the behaviors signal to the clients that the cleaner should not be consumed. Thus, some amazing events have been documented wherein the cleaner can even be allowed to enter the wide-open mouth of the client to remove parasites. Even sharks attend cleaning stations (Sazima et al., 2000). The evolution of cleaning stations demonstrates the clear need by many fishes to be cleaned and even possibly receive physical tactile stimulation (Poulin & Grutter, 1996). However, assessments of such associations to determine ultimate, long-term benefits are difficult to do in the field (Cote, 2000). Thus, we cannot categorically call all such associations mutualistic, as in some cases there may be no clear benefit – especially in the case of cleaner mimics like the saber-toothed blenny that eats host fish flesh while visually similar to the cleaner wrasse which focuses on parasites (cf., Sazima, 2002). But there are clear cases where shrimp have been examined and found fish parasites in their guts (Becker & Grutter, 2004).

There are some invertebrates with high potential to remove parasites. For example, echinoderms have numerous tube feet which they typically use for locomotion and feeding. These same structures could potentially be efficient for autogrooming. The reality, however, is that echinoderms still have numerous ectoparasites that clearly are not effectively removed (Jangoux, 1984). Other invertebrates with significant potential to autogroom are cephalopods, which typically have multiple, highly dexterous arms and tentacles with strong suckers. They, too, are victims to external parasites, such as copepods, flatworms and protists, (Hochberg, 1983; Gonzalez et al., 2003).

In summary, organisms in both terrestrial and aquatic environments are potential targets to parasites attempting to attach and gain resources from these potential hosts. Therefore, the need for cleaning is high, as potential parasites are diverse and abundant in both terrestrial and aquatic ecosystems.

In spite of all the host defenses discussed previously, parasitism is one of the most successful evolutionary strategies. Essentially every organism on Earth is a potential host! Thus, what happens once a parasite inevitably establishes and maintains a presence on or in the host?From the host's perspective, it should attempt to minimize the damage the parasite causes (which in most cases would involve an immune response of some kind). From the parasite's perspective, damage to the host should be minimized; otherwise, the parasite risks a decrease in its own fitness and survival. That is why with the exception of parasitoids – which are typically wasp larvae that purposely consume the host for food (Vinson & Iwantsch, 1980) – a good parasite does not kill its host. Or can it? What if the behavior of the host reduces reproductive success of the parasite? Can the parasite do anything about host behaviors? If the host is expendable at some point in the life history of the parasite, then the paradigm of non-lethal damage by a parasite is irrelevant. This can occur when a host serves as a vector or intermediate host for a parasite that must somehow get a larval stage (sometimes there are multiple, morphologically distinct larvae) into another habitat or organism. Research has uncovered some fascinating examples of behavioral and physiological changes in hosts that are induced by the parasite, typically to facilitate completion of a complex life cycle. While there may be skepticism about the behavioral effects of parasites on hosts (cf., Poulin, 2000), more studies are illustrating the potential for such interactions with compelling observations and data.

A recent landmark study has shown that bacteria in the guts of mice can not only influence digestion, but also influence brain hormonal activity and subsequent adult behaviors (Heijtz et al, 2011). The specific interactions discovered were that microbes were critical in development of neural pathways for motor activities and anxiety behaviors. These results suggest that symbiotic microbes – some of which are helpful and others harmful - may be critical components of development, both physiologically and behaviorally, in mammals.

While examples of parasite-induced behavioral changes in mammals are rare, lucid examples are present in some remaining vertebrates and invertebrates. Fishes are commonly intermediate hosts for numerous aquatic parasites. Killifishes can be infected by a trematode that directly affects the brains (specifically, neurotransmitter activity) of these fishes by inducing behavioral changes that make the fishes up to 30 times more susceptible to bird predators, which serve as the primary or definitive host to these parasites. Specifically, those killifishes infected with trematodes engaged in conspicuous swimming patterns making them more visible and susceptible to bird predators. Typical swimming by uninfected fish were more cryptic; thereby, minimizing predation (Lafferty & Morris, 1996; Lafferty, 2008).

One antipredator behavioral strategy employed by many animals is the avoidance of areas or times when light is abundant. This avoidance could be called photophobia. There are examples where a parasite clearly converts such light avoidance behavior to photophilia, which enhances the prospects of detection of the host by a predator. For example, there is a snail species that is parasitized by a trematode that first positions itself in one or both antennae of the snail. The snail host normally moves on the inferior surfaces of leaves and other vegetation during the day. However, infected snails are induced to move on top of the leaves where the pulsating, brightly striped larvae are easily seen through the transparent tissue of the antennae by bird predators. Consumption of the antenna/larva complex completes the parasite's life history (Wesenberg-Lund, 1931; Dobson 1988). Examples also exist wherein crustaceans are converted from photophobia to photophilia by parasitic flatworms, which subsequently doubled the rate of consumption by predators of the host (Moore, 1995, 2002).

There are also examples where parasites alter behavior in other dramatic ways. For example, grasshoppers and crickets infected with a nematomorph hair worm jump from land into an aquatic system. The hosts typically drown, but the parasitic worms swim away in search of a potential reproductive partner (cf., Libersat et al., 2009). Even fungi can affect host behavior, as seen in parasitized ants that are forced to position themselves in such a way to maximize dispersal of spores, which have developed inside the body of the soon-to-be dead ant (cf., Libersat et al., 2009).

Parasites can also take advantage of relatively programmed behaviors of hosts. So instead of altering the behavior of the host, the stereotyped behavior of the host enables the parasite to redirect the behavior in a beneficial way. Clear examples of such strategies exist in social groups where brood parasitism occurs. Some female cuckoo birds are known to lay their eggs in the nests of other bird species. The host species apparently does not recognize the "foreign" egg (as shell mimicry may occur) or even the cuckoo hatchling. The cuckoo hatchling will typically kill all of the resident hatchlings thereby ensuring the "host" mother feeds it as if it were one of its own hatchlings (Payne, 1998; Kruger & Davies, 2002).

Another example of brood parasitism involving cowbirds may in some ways have a more positive outcome in that most cowbirds do not kill host hatchlings, and some species may actually remove ectoparasites from the nest and host hatchlings. In one study, nests without a cowbird hatchling were likely to have much higher infestation rates of parasitic flies (Rothstein & Robinson, 1994). Thus, while brood parasitic birds can have devastating effects on host bird populations, in the latter case of cowbirds where parasite removal is an added factor, the association may trend toward mutualism (i.e., less parasitic with mutual benefits) in some specific instances (Winfree, 1999; Paracer & Ahmadjian, 2000).

Conclusion

One significant outcome of looking at the thousands of studies on symbioses is that rarely is there adequate data to accurately assess the cost-benefits in such associations. Such analysis would require detailed knowledge about nearly every aspect of each symbiont's life history, including their ecological niches. Additionally, many symbioses will likely vary temporally. That is, depending on the specific conditions the characterization of the association might vary. For example, commensalistic bacteria present on and in all animals can turn problematic when disturbance affects either the abundance and/or composition of the biota (Tiwari et al., 2011), or viral conversions turn normally safe bacteria into virulent forms (Blaser, 1998; Brüssow et al., 2004), which can also lead to activation of oncogenes and cancer in digestive tracts of humans (Blaser & Parsonnet, 1994). Essentially, characterization of an association as commensalistic, mutualistic or parasitic depends on the conditions at a particular time.

Much of this paper was dedicated to discussing how potential hosts prevent initial contact by symbionts, or once contacted, how to get them off (either through autogrooming or allogrooming). I focused primarily on macroscopic ectoparasites, as these are typically easier to see and detect (unless the endoparasite makes its presence easily detected as in the pulsating worm in the antennae of snails). The detection of ectoparasites can also be directly connected to host behaviors induced by the parasites – whether these behaviors are beneficial to the host or parasite. Clearly, in many situations, potential hosts perceive attachment attempts by other species as threats. Based on this observation, one might think that most symbioses start off as an "arms race." Speidel (2000) discusses symbioses in the context of neodarwinism and states "…there is a tendency to see symbiotic associations as the end result of a previously parasitic relationship. In such cases, the argument could be made for the following symbiotic sequence: Parasitism → Commensalism → Mutualism. Specifically, if the host cannot remove or prevent infestation by a symbiont that immediately triggers deleterious effects, neutralization of such impacts should commence. Ultimately, the host – in some cases – can turn the association into a mutually beneficial outcome. One of the most amazing, classical examples of this sequence comes from studies by Jeon (1972)(Jeon & Jeon, 1976) in which a pathogenic parasitic bacterium had infected an amoeba species thereby killing most individuals. However, some amoebae survived with the bacteria also still alive inside the cell. Subsequent generations (within a decade) of amoebae, with the vertically-transmitted bacteria, resulted in an obligate, mutualistic association in which both the symbiont and host would die if isolated from each other.

While many associations may start off in an antagonistic way (for at least one partner) and follow the aforementioned linear sequence, commensalism and even mutualism may also be starting points. Hosts that offer an innate defense against a variety of potential predators could serve as potential "protective" hosts to any symbionts that could neutralize the defense or feeding system against them individually. This strategy would be useful to the symbiont only if the defensive mechanism of the host remained intact; otherwise, both the symbiont and host would vulnerable. Cnidarians, with their stinging nematocysts, represent potentially excellent hosts for the initiation of these types of associations. Sea anemones are known to host a number of symbionts, some of which just hide in the general vicinity without making contact with the virulent tentacles. Others, such as fishes (cf., Dunn, 1981; Fautin, 1991) and some crustaceans (cf., Gwaltney & Brooks, 1994; Mihalik & Brooks, 1997) have biochemically and physiologically masked themselves from stimulating nematocyst discharge thereby enabling these symbionts to move among the tentacles with impunity (Brooks & Mariscal, 1984; Elliot & Mariscal, 1996). In most cases, there is empirical evidence that such symbionts living with or near sea anemones, and other cnidarians, receive protection from predators (cf., Brooks & Mariscal, 1985; Brooks, 1989; Brooks & Gwaltney, 1993).

Although, in some cases, the symbionts may consume small amounts of anemone tissue (Brooks & Mariscal, 1984; Khan et al., 2003), the evidence is also accumulating that cnidarians can benefit from the symbiont's presence by receiving nitrogen metabolites for symbiotic zooxanthellae living symbiotically in the host's tissues (Spotte, 1996; Porat & Chadwick-Furman, 2005; Roopin & Chadwick, 2009), which can stimulate growth and reproductive rates (Holbrook & Schmitt, 2005). Direct observations have also confirmed that, in some cases, the symbiont can provide protection to the host (Brooks & Gwaltney, 1993;

Fautin, 1991; McCammon, personal observation). So there are cases where a symbiont initiates an association with a potentially virulent host, as a likely commensal, but eventually – and possibly initially - benefits to both partners are the outcome.

In closing, the study of symbioses can transcend determining the cost-benefit ratios that allow us to categorize these associations based on the outcomes. Speidel (2000) also states "Perhaps it would be better to see them not so much in terms of what each partner is getting out of the relationship, but in terms of how the structure as a whole is functioning." That leads to the discussion of organismal phenotypes. Dyer (1989) points out that the phenotype is a composite of all organisms in associations, and states "Organisms as separate, completely definable entities may not exist." For example, a termite without its microbial host can ingest but not digest wood. We could say the same for ungulates, too. Clearly, in these associations the phenotype is based on the genotypes of not only the termite (or cow) but also the microbial symbionts that provide the cellulose digestion. Arguably, all eukarotic cells are at least partially composites of bacterial genomes, given that mitochondria and chloroplasts were likely derived from bacterial endosymbiotic events (Margulis & Bermudes, 1985; de Duve, 1996). Future scientific inquiry about any biological systems should always acknowledge and account for the likelihood that symbioses have influenced behavioral, physiological and ecological aspects of the evolution of species.

References

Baker, R.L., B.P. Smith. 1997. Conflict between antipredator and antiparasite behaviour in larval damselflies. *Oecologia* 109: 622-628.

Bariche, M., J-P Trilles. 2006. First record of the Indo-Pacific Cymothoa indica (Crustacea, Isopoda, Cymothoidae), a Lessepsian species in the Mediterranean Sea. *Acta Parasitologica.* 51(3): 223.

Becker, J. H., A.S. Grutter. 2004. Cleaner shrimp do clean. *Coral Reefs.* 23(4): 515-520.

Blaser, M. 1998. Helicobacters are indigenous to the human stomach: duodenal ulceration is due to changes in gastric microecology in the modern era. *Gut.* 43: 721-727.

Blaser, M., J. Parsonnet. 1994. Parasitism by the "Slow" bacterium Helicobacter pylori leads to altered gastric homeostasis and neoplasia. *Journal of Clinical Investigation.* 94: 4-8.

Bowman, T.E., S.A. Grabe, J.H. Hecht. 2010. Range extension and new hosts for the cymothoid isopod Anilocra acuta. *Journal of Chesapeake Science Earth and Environmental Science .* 18(4): 390-393.

Brooks, W.R. 1989. Hermit crabs alter sea anemone placement patterns for shell balance and reduced predation. *Journal of Experimental Marine Biology and Ecology.* 132: 109-121.

Brooks, W.R., R.N. Mariscal. 1984. The acclimation of anemone fish to sea anemones: Protection by changes in the fish's mucous coat. *Journal of Experimental Marine Biology and Ecology.* 81: 277-285.

Brooks, W.R., R.N. Mariscal. 1985. Protection of the hermit crab Pagurus pollicaris Say from predators by hydroid-colonized shells. *Journal of Experimental Marine Biology and Ecology.* 87: 111-118.

Brooks, W.R., C.L. Gwaltney. 1993. Protection of symbiotic cnidarians by their hermit crab hosts: Evidence for mutualism. *Symbiosis.* 15: 1-13.

Brunnschweiler, J.M. 2006. Sharksucker–shark interaction in two carcharhinid species. *Marine Ecology.* 27(1): 89–94.

Brusca, R.C. 1981. A monograph on the Isopoda Cymothoidae (Crustacea) of the Eastern Pacific. *Zoological Journal of the Linnean Society* 73 (2): 117–199.

Brüssow, H., C. Canchaya, W-D. Hardt. 2004. Phages and the evolution of bacterial pathogens: from genomic rearrangements to lysogenic conversion. *Microbiology and Molecular Biology Reviews.* 68(3): 560–602.

Chancellor, R.L., L.A. Isbell. 2008. Female grooming markets in a population of gray-cheeked mangabeys (*Lophocebus albigena*). *Behavioral Ecology.* Advance Access publication 26 September 2008.

Christian, K.A. 1980. Cleaning/feeding symbiosis between birds and reptiles of the Galapagos Islands: New observations of inter-island variability. *The Auk,* 979(4):887-889.

Combes, C. 2001. *Parasitism the Ecology and Evolution of Intimate Interactions.* The University of Chicago Press, Chicago.

Cote, I.M . 2000. Evolution and ecology of cleaning symbioses in the sea. *Oceanography and Marine Biology Annual Review.* 38: 311–355.

Danka, R.G., J. D. Villa. 2003. Autogrooming by resistant honey bees challenged with individual tracheal mites. *Apidologie* 34:591–596.

de Duve, C. 1996. The birth of complex cells. *Scientific American.* April: 50-57.

Dobson, A.P. 1988. The population biology of parasite-induced changes in host behavior. *Quarterly Review of Biology*, 63: 139–165.

Dudzinski, K.M., J.D. Gregga, C. A. Ribicc, S.A. Kuczaj. 2009. A comparison of pectoral fin contact between two different wild dolphin populations. *Behavioural Processes.* 80: 182–190.

Dunbar, R.I.M., 1991. Functional Significance of Social Grooming in Primates. *Folia Primatologica.* 57(3): 121-131.

Dunn, D.F. 1981. The clownfish sea anemones: Stichodactylidae (Coelenterata:Actiniaria) and other sea anemones symbiotic with pomacentrid fishes. *Transactions of the American Philosophical Society.* 71(1): 3-115.

Dyer, B.D. 1989. Symbiosis and organismal boundaries. *American Zoologist.* 29: 1085-1093.

Evans, P.G., J.A. Raga. 2001. *Marine Mammals: Biology and Conservation.* Plenum Publishers, New York.

Fautin, D.G. 1991. *The anemonefish symbiosis: what is known and what is not. Symbiosis.* 10:23-46.

Fefferman, N.H., J.F.A. Traniello, R.B. Rosengaus, D.V. Calleri II. 2007. Disease prevention and resistance in social insects: modeling the survival consequences of immunity, hygienic behavior, and colony organization. *Behavioral Ecology and Sociobiology.* 61(4): 565-577.

Feist, S.W., M. Longshaw. 2008. Histopathology of fish parasite infections – importance for populations. *Journal of Fish Biology.* 73(9): 2143–2160.

Felix, F., B. Bearson, J. Falconi. 2006. Epizoic barnacles removed from the skin of a humpback whale after a period of intense surface activity. *Marine Mammal Science.* 22(4): 979– 984.

Fertl, D. 2002. Barnacles. In: *Encyclopedia of marine mammals* (W.F. Perrin, B.G. Wursig, J.G.M. Thewissen, eds). Academic Press. San Diego, CA.

George RH. 1997. Health problems and diseases of sea turtles. In: *The Biology of Sea Turtles* (Lutz PL, Musick JA, eds). Boca Raton, FL:CRC Press, 363–385.

Geraci, J.R., St. Aubin, D.J., 1987. Effects of parasites on marine mammals. International. *Journal of Parasitology.* 17: 407–414.

Gill, S.A. 2012. Strategic use of allopreening in family-living wrens. *Behavioral Ecology and Sociobiology,* DOI: 10.1007/s00265-012-1323-6, Online First.

González, A.F., S. Pascual, C. Gestala, E. Abollob, A. Guerra. 2003. What makes a cephalopod a suitable host for parasite? The case of Galician waters. *Fisheries Research.* 60(1): 177– 183.

Gwaltney, C.L., W.R. Brooks. 1994. Host Specificity of the anemoneshrimp Periclimenes pedersoni and P. yucatanicus in the Florida Keys. *Symbiosis* 16: 83-93.

Hart, B.L. 1994. Behavioural defense against parasites: interaction with parasite invasiveness. *Parasitology,* 109: 139-151.

Hawlena, H., D. Bashary, Z. Abramsky, B.R. Krasnov, 2007. Benefits, costs and constraints of anti-parasitic grooming in adult and juvenile rodents. *Ethology.* 113(4):394–402.

Heijtz, R.D., S. Wang, F. Anuard, Y. Qiana,, B. Björkholm, A. Samuelsson, M.L. Hibberd, H. Forssberg, S. Pettersson. 2011. Normal gut microbiota modulates brain development and behavior. *Proceedings of the National Academy of Sciences.* 108(7): 3047–3052.

Hillegass, M.A., J.M. Waterman, J.D. Roth. 2010. Parasite removal increases reproductive success in a social African ground squirrel. *Behavioral Ecology,* Advance Access publication.

Hochberg F.G. 1983. The parasites of cephalopods: a review. *Memoirs of the National Museum of Victoria.* 44: 108-146.

Holbrook, S.J., R.J. Schmitt. 2005. Growth, reproduction and survival of a tropical sea anemone (Aciniaria): Benefits of hosting anemonefish. *Coral Reefs.* 24: 67-73.

Hopla, C.E., L.A. Durden, J.E., Keirans. 1994. Ectoparasites and classification. *Revue Scientifique Et Technique* (4):985-1017.

Jangoux, M. 1984. Diseases of Echinodermata. I. Agents microorganisms and protistans. *Diseases of Aquatic Organisms.* 2: 147-162.

Jantzen, F. J., G.L. Paukstis, E. D. Brodie III. 1992. Observations on basking behavior of hatchling turtles in the wild. *Journal of Herpetology.* 26(2): 217-219.

Jeon, K.W. 1972. Development of cellular dependence on infective organisms: Micrurgical studies in amoebas. *Science,* 176(4039): 1122-1123.

Jeon, K. W., M. S. Jeon, 1976. Endosymbiosis in amoebae: Recently established endosymbionts have become required cytoplasmic components. *Journal of Cellular Physiology.* 89(2): 337–344

Jones, C.M., T.L. Miller, A.S. Gruttera, T.H. Cribb. 2008. Natatory-stage cymothoid isopods: Description, molecular identification and evolution of attachment. *International Journal for Parasitology* 38: 3-4.

Khan, R.N., J.H.A. Becker, A.L. Crowther, I.D. Lawn. 2003. Sea anemone host selection by the symbiotic saddle cleaner shrimp *Periclimenes holthuisi. Marine and Freshwater Research.* 54: 653-656.

Kohari, D. T. Kosako, M. Fukasawa, H. Tsukada. 2007. Effect of environmental enrichment by providing trees as rubbing objects in grassland: Grazing cattle need tree-grooming. *Animal Science Journal.* 78(4):413–416.

Kruger, O., N.B. Davies. 2002. The evolution of cuckoo parasitism: a comparative analysis. *Proceedings of the Royal Society of London B*. 269(1489): 375-381.

Lafferty, K.D., 2008. Ecosystem consequences of fish parasites. *Journal of Fish Biology*. 73(9): 2083–2093.

Lafferty, K.D., A.K. Morris. 1996. Altered behavior of parasitized killifish increases susceptibility to predation by bird final hosts. *Ecology*. 77(5): 1390-1397.

Lewis, S., G. Roberts, M.P. Harris, C. Prigmore, S. Wanless. 2007. Fitness increases with partner and neighbour allopreening. *Biology Letters*. 3(4):386-389.

Libersat, F., A. Delago, R. Gal. 2009. Manipulation of host behavior by parasitic insects and insect parasites. *Annual Review of Entomology*. 54: 189-207.

Little, E.F., T. Murakami, U.G. Mueller, C.R. Currie. 2006. Defending against parasites: fungus- growing ants combine specialized behaviours and microbial symbionts to protect their fungus gardens. *Biology Letters*. 2(1): 12-16.

Losey, G.S. 1987. Cleaning symbiosis. Symbiosis 4, 229-258.

MacFarland, C.G., W. G. Reeder. 1974. Cleaning symbiosis involving Galápagos Tortoises and two species of Darwin's Finches. *Zeitschrift für Tierpsychologie*. 34(5):464–483.

Margulis, L., D. Bermudes. 1985. Symbiosis as a mechanism of evolution: Status of cell symbiosis theory. *Symbiosis*. 1: 101-124.

McCammon, A., P.C. Sikkel, D. Nemeth. 2010. Effects of three Caribbean cleaner shrimps on ectoparasitic monogeneans in a semi-natural environment. *Coral Reefs*. 29 (2): 419-426.

Mihalik. M.B., W.R. Brooks. 1997. Protection of the symbiotic shrimps Periclimenes pedersoni, P. yucatanicus, and Thor spp. from fish predators by their host sea anemones. 1997. *Proc. 6th International Conference on Coelenterate Biology*. 337-343.

Moore, J. 1995. *The behavior of parasitized animals. : When an ant... is not an ant. Bioscience*. 45(2): 89-96.

Moore, J. 2002. *Parasites and the Behavior of Animals*. Oxford University Press, New York.

Mooring, M. S., D. T. Blumstein, C. J. Stoner. 2004. The evolution of parasite-defense grooming in ungulates. *Biological Journal of the Linnean Society* 81:17–37.

Mucientes, G.R., N. Queiroz, S. J. Pierce, I. Sazima, J.M. Brunnschweiler. 2008. Is host ectoparasite load related to echeneid fish presence? *Research Letters in Ecology*. Volume 2008, Article ID 107576, 4 p.

Nico, L. G., W.F. Loftus, J. P. Reid. 2009. Interactions between non-native armored suckermouth catfish (Loricariidae: Pterygoplichthys) and native Florida manatee (Trichechus manatus latirostris) in artesian springs. *Aquatic Invasions*. 4(3) : 511-519.

Noë, R, Hammerstein P. 1995. Biological markets. *Trends in Ecology and Evolution*.10:336–340.

Olivera-Gomez, L.D., E. Mellink. 2005. Distribution of the Antillean manatee (Trichechus manatus manatus) as a function of habitat characteristics in Bahıa de Chetumal, Mexico. *Biological Conservation*. 121: 127–133.

Osterrieder, S.K., R.W. Davis. 2011. Sea otter female and pup activity budgets, Prince William Sound, Alaska. *Journal of the Marine Biological Association of the UK*. 91: 883-892.

Pagel, M., W. Bodmer. 2003. A naked ape would have fewer parasites. *Biology Letters*. 270 (1), 117-119.

Paracer, S., V. Ahmadjian, 2000. *Symbiosis; An Introduction to Biological Associations*. Oxford University Press, New York.

Payne, R.B. 1998. Brood parasitism in birds: Strangers in the nest. *Bioscience.* 48: 377-386.

Porat, D. N.E. Chadwick-Furman. 2005. Effects of anemonefish on giant sea anemones: Ammonium uptake, zooxanthella content and tissue regeneration. *Marine and Freshwater Behaviour and Physiology.* 38(1): 43-51.

Porter, S.D., 1998. Biology and behavior of pseudacteon decapitating flies (Diptera: Phoridae) that parasitize Solenopsis fire ants (Hymenoptera: Formicidae), *The Florida Entomologist,* 81(3): 292-309.

Poulin. R. 2000. Manipulation of host behaviour by parasites: a weakening paradigm? *Proceedings of the Royal Society of London B.* 267(1445): 787-792.

Poulin, R., A.S. Grutter. 1996. Cleaning Symbioses: Proximate and Adaptive Explanations: What evolutionary pressures led to the evolution of cleaning symbioses? *Bioscience.* 46(7): 512-517.

Roopin, M. N.E. Chadwick. 2009. Benefits to host sea anemones from ammonia contributions of resident anemonefish. *Journal of Experimental Marine Biology and Ecology.* 370: 27-34.

Rothstein, S.I., S.K. Robinson. 1994. Conservation and coevolutionary implications of brood parasitism by cowbirds. *Trends in Ecology and Evolution.* 9: 162-164.

Sazima, I. 2002. Juvenile snooks (Centropomidae) as mimics of mojarras (Gerreidae), with a review of aggressive mimicry in fishes. *Environmental Biology of Fishes.* 65(1): 37-45.

Sazima, I., R.L. Moura, S.T. Ross. 2000. Shark (Carcharhinus perezi), cleaned by the goby (Elacatinus randalli), at Fernando de Noronha Archipelago, Western South Atlantic. *Copeia:* 2000 (1): 297-299.

Slijper, E.J. 1979. *Whales.* Cornell University Press, Ithaca, NY.

Speidel, M. 2000. The parasitic host: Symbiosis contra Neo-Darwinism. *Pli.* 9: 119-38.

Spotte, S. 1996. Supply of regenerated nitrogen to sea anemones by their symbiotic shrimp. *Journal of Experimental Marine Biology and Ecology.* 198: 27-36.

Stanimirović, Z., S. Jevrosima, A. Nevenka, V. Stojić. 2010. Heritability of grooming behaviour in grey honey bees (Apis mellifera Carnica). *Acta veterinaria.* 60(2-3): 313-323.

Thompson, K.P.J. 2010. Grooming the naked ape: Do perceptions of disease and aggression vulnerability influence grooming behaviors in humans? A comparative ethological perspective. *Current Psychology.* 29:288-296.

Tiwari, S.K., A.A. Khan, P. Nallari. 2011. Helicobacter pylori: A benign fellow traveler or an unwanted inhabitant. *Journal of Medical and Allied Sciences.* 1(1): 2-6.

Val-Laillet, D.,V. Guesdon, M.A.G. von Keyserlingk, A. M. de Passillé, J. Rushen. 2009. Allogrooming in cattle: Relationships between social preferences, feeding displacements and social dominance. *Applied Animal Behaviour Science.* 116(2): 141-149.

Viblanc, V. A., A. Mathien, C. Saraux, V. M. Viera, R. Groscolas. 2011. It costs to be clean and fit: Energetics of comfort behavior in breeding-fasting. *Penguins.* PLoS ONE 6(7): e21110. doi:10.1371/journal. pone.0021110.

Vinson, S.B., G.F. Iwantsch. 1980. Host suitability for insect parasitoids. *Annual Review of Entomology.* 25: 397-419.

Walker, T.N., W.O.H. Hughes. 2009. Adaptive social immunity in leaf-cutting ants. *Biology Letters* 5: 446–448.

Weeks, P. 2000. Red-billed oxpeckers: vampires or tickbirds? *Behavioral Ecology.* 11(2):154- 160.

Welch, D. A., W. M. Samuel, and C. J. Wilkie. 1991. Suitability of moose, elk, mule deer, and white-tailed deer as hosts for winter ticks (Dermacentor albipictus). *Canadian Journal of Zoology* 69:2300–2305.

Wesenberg-Lund, C. 1931. Contributions to the Development of the Trematoda Digenea: The Biology of *Leucechloridium Paradoxum*. *Memoires de l'Academie Royale des Sciences et des Lettres de Danemark, Section des Sciences* (series 9) 4: 90142.

Winfree, R. 1999. Cuckoos, cowbirds and the persistence of brood parasitism. *Trends in Ecology and Evolution.* 14(9): 338-343.

Wisendon, B.D., C.P. Goater, C. T. James. 2009. Behavioral defenses against parasites and pathogens. In: *Fish Defenses* (Zaccone, C., Perriere A., Mathis, A., Kapoor, G., eds.). Vol. 2. Enfield: Science Publishers, 151–168.

Yamada, M., M. Urabe. 2007. Relationship between grooming and tick threat in sika deer Cervus nippon in habitats with different feeding conditions and tick densities. *Mammal Study.* 32(3):105-114.

In: Symbiosis: Evolution, Biology and Ecological Effects   ISBN: 978-1-62257-211-3
Editors: A. F. Camisão and C. C. Pedroso   © 2013 Nova Science Publishers, Inc.

Chapter 7

New Insights in the Actinorhizal Symbiosis

Didier Bogusz[*] and Claudine Franche[†]

Equipe Rhizogenèse, UMR DIADE,
Institut de Recherche pour le Développement, Montpellier, France

Abstract

More than 200 species of non-legume dicotyledonous plants, mostly trees and schrubs, belonging to eight different families and 24 genera can enter actinorhizal symbioses with the nitrogen-fixing actinomycete Frankia. Actinorhizal nodules arise from divisions occurring in the pericycle and display a lateral root structure with a central vascular system and infected cells in the expanded cortex. Recently, the development of genomics both in Frankia and in actinorhizal plants, together with the possibility to obtain transgenic actinorhizal plants following Agrobacterium gene transfer, offered valuable tools to achieve significant progress in the understanding of actinorhizal symbiosis. Molecular data indicates that, like in legumes, actinorhizal symbiosis is a highly controlled process, involving plant signals perceived by the actinomycete and specific Frankia factors perceived by plant roots, that activates a symbiotic signalling pathway. Current knowledge on the actinorhizal symbiosis will be discussed in the context of plant root endosymbioses evolution.

1. Introduction

The actinorhizal symbiosis is the result of a complex interaction between the nitrogen-fixing actinomycete *Frankia* and a diverse group of perennial dicotyledonous angiosperms. The symbiotic process occurs in conditions of nitrogen deficiency and results in the formation of specialized root organs, the actinorhizal nodules, inside which *Frankia* reduces atmospheric nitrogen to ammonium for the benefit of the host plant (Benson and Silvester, 1993; Franche et al., 1998; Wall, 2000; Chaia et al., 2010; Pawlowski and Demchenko, 2012). Actinorhizal nodules consist of multiple lobes, each of them possess a central vascular system and *Frankia* infected cells in the expanded cortex.

---

[*] E-mail address: didier.bogusz@ird.fr, Tel:+33-467416281
[†] E-mail address: claudine.franche@ird.fr, Tel:+33-467416260

With the exception of *Datisca* species which have herbaceous shoots (Davidson, 1973), actinorhizal plants are perennial dicots and include woody shrubs and trees that are major components of plant comunities world-wide, specially in nutrient-poor soils. They represent about 200 species encompassing 25 genera in 8 different angiosperm families, in three different orders: the *Betulaceae*, *Casuarinaceae* and *Myricaceae* of the order Fagales; the *Rosaceae*, *Rhamnaceae* and *Elaeagnaceae* of the order Rosales; and the *Coriariaceae* and *Datiscaceae* of the order Cucurbitales (Huss-Danell, 1997; Gualtieri and Bisseling, 2000; Vessey et al., 2005; Pawlowski et al., 2011). Examples of well-known genera include *Alnus* (alder), *Elaeagnus* (autumn olive), *Hippophae* (sea buckthorn) and *Casuarina* (she oak). Most of these plants are found in temperate zones, with only a few members growing in tropical or Arctic environments (Baker and Schwintzer, 1990; Sprent and Parsons, 2000).

Due to their high rates of nitrogen fixation comparable to those found in Legumes, they are important contributors to the global biological nitrogen fixation process (Torrey, 1976). Some species have a great economic value as timber and fuel wood, they also contribute to land reclamation and protection of coastal areas in tropical countries (Diouf et al., 2008; Dawson, 2008; Zhong et al., 2010). Compared to Rhizobium, *Frankia* exibits a number of unique features that make the actinorhizal symbiosis an original system to study (Pawlowski and Bisseling, 1996; Laplaze et al., 2008; Pawlowski and Sprent, 2008; Franche and Bogusz, 2011; Pawlowski and Demchenko, 2012). *Frankia* is a Gram positive filamentous actinobacteria that belongs to the family *Frankiaceae*. It is a facultative symbiont that differentiates three cell types in pure nitrogen-free cultures, vegetative hyphae, vesicles that are the site for nitrogen fixation and spores that are the reproductive structures. The recent sequencing of several *Frankia* genomes revealed a wide variation in genome size (Normand et al., 2007a; 2007b; Rawnsley and Tisa, 2007). Sizes varied from 5.43 Mb for *Frankia* CcI3 isolated from *Casuarina cunninghamiana* nodules, to 8.98 Mb for *Frankia* EAN1pec isolated from *Eleagnus angustifolia* nodules. Whereas some putative nodulation-(*nod*)-like genes were found in the actinomycetal genomes, these sequences were not organized in clusters and there was no homologous sequence to *nodA*, one of the common *nod* genes necessary for the synthesis of the Nod factor core in Rhizobium strains (Lerouge, 1994). These data suggest that signals molecules from *Frankia* involved in the initial plant-actinobacterial recognition could be different from the lipochitooligosacharides produced by Rhizobium in response to specific flavonoids excreted by Legume roots (Normand et al., 2007a).

So far, the lack of a genetic transformation system together with the slow growth rate and the filamentous nature of *Frankia* remain major constraints to the molecular understanding of regulatory events in actinorhizal nodulation (Benson and Silvester, 1993; Simonet et al., 1990; Lavire and Cournoyer, 2003; Bassy and Benson, 2007). However, major progress has been achieved in the functional analysis of several plant genes that play a key role in the establishment of the symbiotic process (Gherbi et al., 2008a; Laplaze et al., 2008; Pawlowski et al., 2011). Like in the Rhizobium-Legume symbiosis, actinorhizal root infection and nodule organogenesis are regulated by a sophisticated molecular dialogue between *Frankia* and the actinorhizal host plants (Wall and Berry, 2008; Perrine-Walker et al., 2010; Bogusz and Franche, 2011; Hocher et al., 2011a). The aim of this review is to report on the recent advances in the understanding of this original symbiotic system.

2. Infection Process and Nodule Ontogenesis in Actinorhizal Plants

Frankia can infect actinorhizal plants in two different ways, intracellular infection via root hair (Fagales) or intercellular penetration (Rosales and Cucurbitales) (Berry and Sunnel, 1990). The two infection pathways are determined by the host, and exhibit several differences presented below.

2.1. Intracellular Infection

The intracellular infection starts with root hair curling observed several hours after inoculation and induced by still unknown *Frankia* signals. While fully elongated root hairs remain straight, the infection zone corresponding to the deformed root hairs is restricted to emerging young root hairs. According to the host, all young root hairs (ex. *Casuarina*; Torrey, 1976) or only some of them (ex. *Comptonia*; Callaham *et al.*, 1979) are curled. Then *Frankia* hyphae penetrate some of the deformed root hairs and infection proceeds intracellularly in the root cortex. In the host cell, *Frankia* is surrounded by an invaginated cell membrane, and between the actinomycete cell wall and the host cell, there is an accumulation of host derived wall material (Huss-Danell, 1990; Berg, 1999). This structure is analogous to the infection threads formed during the intracellular infection of legumes by rhizobia (Gage, 2002; 2004).

Following the penetration of the actinomycete, a limited proliferation of the cortex is observed underneath the infected root hairs, leading to the formation of a small external protuberance on the root surface, called the prenodule. Most of prenodule cells are infected with *Frankia*. Furthermore, the prenodule cells exhibit the same characteristics as the nodule cells, suggesting that the prenodule represents a primitive symbiotic nitrogen-fixing organ (Laplaze et al., 2000). But, while cortical cells divisions lead to the formation of a nodule primordium in legumes, actinorhizal prenodules do not evolve in nodules. Concomitant with prenodule development, mitotic activity occurs in pericycle cells opposite to a protoxylem pole, giving rise to an actinorhizal lobe primordium (Duhoux et al., 1996). This nodule primordium will grow and become infected by *Frankia* hyphae progressing from the infected cells of the prenodule. The mature actinorhizal nodule consists of multiple lobes, each of which is a modified lateral root. In each lobe there is a central vascular bundle and *Frankia* is restricted to the cortical cells. The hyphae remain encapsulated in infection thread-like structures during the whole symbiotic process. At the evolutionary level, it is believed that it is a less specialized symbiotic state than the differentiation and release of bacteroids observed in Legumes.

2.2. Intercellular Infection

During intercellular infection, *Frankia* hyphae enter the root between epidermal cells, and colonize the root cortex intercellularly (Miller and Baker, 1985; Racette and Torrey, 1989). In contrast to rhizobia, *Frankia* does not depend on gaps in the root epidermis for entering the root. During the colonization of the cortex, root cortical cells secrete an electron-dense pectin and protein rich material into the intercellular spaces, and the formation of a nodule primordium is induced in the root pericycle (Liu and Berry, 1991; Valverde and Wall, 1999). *Frankia* hyphae infect primordium cells from the apoplast by intense branching of hyphae, concomitant with continuous invagination of the plant plasma membrane.

Intercellular infection takes place in host plants of the *Rhamnaceae*, *Elaeagnaceae* and *Rosaceae* families. In host plants of the actinorhizal Cucurbitales (*Datisca* and *Coriaria*), the infection mechanism has not been examined yet, but since no prenodule or infection thread are found in these plants, infection is assumed to follow the intercellular pathway. For some *Frankia* strains that can infect both *Elaeagnus* and *Myrica (intracellular infection),* or both *Shepherdia* and *Gymnostoma (root hair infection),* they enter via the route dictated by the host (Miller and Baker, 1986; Racette and Torrey, 1989).

2.3. The Actinorhizal Nodule

Mature actinorhizal nodules are indeterminate and multilobed structures. Each nodule lobe has a central vascular bundle surrounded by an endoderm, an expanded cortex, and a periderm. Due to activity of the apical meristem, actinorhizal nodule lobes show an indeterminate growth pattern and have a developmental zonation where specific patterns of gene expression are observed (Duhoux et al., 1996; Obertello et al., 2003). In zone 1, the apical meristem is free of *Frankia*. Adjacent to the meristem is the infection zone 2 where some of the young cortical cells resulting from the meristem activity are infected by *Frankia*. The subsequent nitrogen fixation zone 3 contains both hypertrophied infected cells and small uninfected cortical cells. However, in *D. glomerata* the infected cells form a continuous zone in the cortex (Pawlowski and Demchenko, 2012). Finally a basal senescence zone 4 is observed in old nodules; plant cells and bacteria degenerate and nitrogen fixation is switched off. Although all nodules lobes have an apical meristem, the growth of individual lobes is limited.

Among actinorhizal root nodules, there is a diversity in structural organization. The «Myrica type» nodules exhibit a so-called nodular root at the apex of each lobe; the «Alnus type» nodules do not have this feature (Duhoux et al., 1996). Nodular roots lack root hairs, have a smaller root cap, and are not infected by *Frankia*. They show a negative geotropism, have an extensive aerenchyma, and are believed to facilitate the diffusion of gazes in and out the nodule lobe (Callaham and Torrey, 1977; Schwintzer and Lancelle, 1983; Tjepkema, 1978). In Myrica-type nodules, aeration is provided by the presence of lenticels in the nodule periderm.

As the actinomycete colonize cortical cells of the actinorhizal nodule, it differentiates morphologically distinct structures. During the infection process, *Frankia* first grows as filamentous hyphae that proliferate in the newly infected host cell, and when the infected cell matures, the tips of the hypahe differentiate vesicles that will fix nitrogen (Newcomb and Wood, 1987). The thick-walled, multilaminated envelope of the nitrogen-fixing vesicles is believed to be the most important diffusion barrier that contributes to prevent inhibition of the nitrogenase complex (Parson et al., 1987; Berry et al., 1993; Dobritsa et al., 2001). It should be noted that the differentiation of vesicles is not observed in actinorhizal nodules of *Casurinaceae* hosts (Berg and McDowell, 1987). Oxygen protection involves different mechnisms including lignification of the infected cells walls (Silvester et al., 2008).

3. Signals Contributing to the Molecular Dialogue

3.1. Plant Molecules

In actinorhizal plants, although the involvement of flavonoids in symbiosis is still poorly understood, some evidence of chemo-attraction and proliferation of *Frankia* has been reported in the rhizosphere of several species (Smolander and Sarsa, 1990; Vessey et al., 2005). Benoit and Berry (1997) showed that flavonoid-containing preparations from seed washes of red alder (*Alnus rubra*) enhanced actinorhizal nodulation. These results were confirmed by Hughes et al. (1999) who observed that flavonols (quercetin and kaempferol) were the major compounds in *A. glutinosa* root exudates and that they could enhance the level of nodulation. Additional studies performed with *A. glutinosa* demonstrated that early stages in the infection process such as root hair curling, required exposure of *Frankia* culture to root culture filtrates before inoculation (Prin and Rougier, 1987; Van Ghelue, 1994; Van Ghelue et al., 1997). Data obtained by Popovici et al. (2010) further suggested that the interaction between root favonoids and *Frankia* was strain-specific. A set of compatible and non-compatible *Frankia* strains were treated with phenolic exudates extracted from *Myrica gale* seeds. Bacterial growth and nitrogen fixation were enhanced in compatible strains, and inhibited in incompatible ones.

The role of flavonoids in later stages of actinorhizal nodule development was suggested by some studies performed on *C. glauca* and *Elaeagnus umbellata*. Laplaze et al. (1999) revealed a cell-specific accumulation of flavans in mature nodule lobes of *C. glauca*, together with an accumulation of chalcone synthase transcripts. The accumulation of flavans was found to create a compartmentation in the nodule cortical tissue, with continuous layers of phenolic-containing cells separating layers containing both *Frankia*-infected and uninfected cells. It was suggested that this cell specific flavan accumulation might restrict endophyte invasion during nodule ontogenesis. In *E. umbellata*, Kim et al. (2003) isolated from a root nodule cDNA library a clone encoding a chalcone isomerase (CHI). This gene was found highly expressed in root nodules, with transcripts levels increasing during nodule development. In further experiments, *CHI* expression was localized in infected cells of *E. umbellata* root nodules (Kim et al., 2007). The high CHI expression in root nodules was proposed to be associated with defense mechanism against infection by *Frankia*.

Recent transcript analysis in the early stages of the infection process has shown that several genes linked to the flavonoid biosynthesis pathway are upregulated in *C. glauca* roots following infection by *Frankia* (Auguy et al., 2011; Hocher et al., 2011a). In this study, quantitative real-time PCR (qRT-PCR) was applied to monitor the expression of eight *C. glauca* genes encoding enzymes involved in flavonoid biosynthesis: chalcone synthase (*CHS*), chalcone isomerase (*CHI*), isoflavone reductase (*IFR*), flavonoid-3-hydroxylase (*F3H*), flavonoid 3'-hydroxylase (*F3'H*), flavonoid 3',5' hydroxylase (*F3'5'H*), dihydroflavonol 4-reductase (*DFR*) and flavonol synthase (*FLS*). Results showed that *FLS* and *F3'5'H* transcripts accumulated in mature nodules, whereas *CHI* and *IFR* transcripts accumulated preferentially early after inoculation with *Frankia*. Comparison of *IFR* and *CHI* expression in inoculated plants and in control plants cultivated with or without nitrogen confirmed that early expression of *IFR* was specifically linked to symbiosis. The role of flavonoids during actinorhizal symbiosis is currently under study in plants genetically transformed with RNAi-*CHS* constructs (Rhizogenesis, IRD Montpellier, France).

Interesting data were reported by Beauchemin et al. (2012), showning that Casuarina root exudates containing flavonoids alter the physiology, surface properties and plant infectivity of *Frankia* strain CcI3. Specific changes in fatty acids, proteins and carbohydrates spectral pattern were observed in *Frankia* cell wall, thus indicating that surface properties are altered in response to plant host signaling compounds in the context of infection. Furthermore, Beauchemin et al. (2012) correlated changes in *Frankia* surface with effects on plant-microbe interactions. The authors showed that pre-exposure to root exudates allowed *Frankia* CcI3 to nodulate host plants earlier than untreated actinobacterial cells, thus suggesting that this treatment was beneficial to the infection and nodulation process.

All these data support a role for the flavonoid compounds during different stages of actinorhizal symbiosis including pre-infection events, nodule initiation and differentiation.

3.2. *Frankia* Factors

Direct contact between *Frankia* hyphae and host root hairs is not necessary to obtain root hair deformation in actinorhizal plants infected via the intracellular process (Van Ghelue et al., 1997; Bhuvaneswari and Solheim, 2000). Bacteria-free filtrates from exponentially growing cultures of the actinomycete contain a factor(s) that induces root hair deformation. The susceptible zone remains confined to a region of the root just behind the root tip, from the location of young emerging root hairs to the location of mature root hairs. In the interaction between *A. glutinosa* and the strain ACoN24d, a 10^{-3} dilution of the supernatant was found to induce root hair deformation, and this root hair reaction was still observed for a 10^{-5} dilution (Cérémonie et al., 1999).

So far, strategies to purify and characterize symbiotic molecules from *Frankia* have relied on a bioassay based on root hair deformation with supernatants from *Frankia* cultures (McEwan et al., 1992; Van Ghelue et al., 1997; Cérémonie et al., 1999). The RHD factor(s) has a molecular weight of about 5000 Da (van Ghelue, 1994) and has properties that are different from those of the Nod factors. Whereas in legumes Nod factors are heat-stable, amphiphilic and chitinase sensitive (Dénarié et al., 1996), *Frankia* factors from the symbiotic strain ACoN24d of *Alnus* were found stable to heating up to 100°C, hydrophilic, and resistant to endochitinase and exochitinase from *Streptomyces griseus*. However, N-acetyl-glucosamine, the subunit of the Nod factor backbone, has been detected in the *Alnus* root hair deforming fraction, leaving open the existence of a Nod-factor-related *Frankia* compound (Cérémonie et al., 1999). Additional experiments showed that the *Frankia* supernatant did not induce any mitotic activity in actinorhizal root cortex, opposite to legumes where Nod factors induce cell divisions in cortical cells.

Recent data resulting from the characterization of symbiotic genes from *C. glauca* have led to the isolation of two plant promoters that are induced 24 h after contact with *Frankia* CcI3, before root hair infection (Rhizogenesis, IRD Montpellier, unpublished data). In transgenic *C. glauca* plants expressing transcriptional fusions between these promoters and the reporter gene ß-glucuronidase, *Frankia*-induced expression was found in the area of root hair emergence where further nodulation is expected to occur. Similar patterns of expression were observed when transgenic roots were incubated with supernatants of the strain CcI3 preinduced with *C. glauca* root exsudates from nitrogen-deprived plants. These transgenic plants will undoubtly provide very valuable tools to purify the RHD signal molecules from *Frankia*.

4. Plant Genes Involved in the Symbiotic Process

4.1. Signal Transduction Pathway

The development of genetic and genomic tools for the model legumes *M. truncatula* and *Lotus japonicus* has greatly facilitated the cloning of genes required for root symbiosis (Geurts et al., 2005; Oldroyd and Downie, 2006; Riely et al., 2006; Madsen et al., 2010). Some of these genes were found to be involved in the establishment of both rhizobia and mycorrhiza symbioses, and designated as common SYM genes constituting the «common SYM pathway». They include genes encoding a leucine-rich-repeat (LRR) receptor kinase (SymRK), a putative cation channel (DMI1/Castor andPollux), a Leucine-rich Repeat Receptor like Kinase (SymRK/DMI2/NORK), a Calcium and Calmodulin-dependant kinase (CCaMK/DMI3) and nucleoporins (Nup85 and Nup133). The question was raised whether some of these symbiotic genes were shared in the signal transduction pathway in response to *Frankia* factors and rhizobial Nod factors.

To answer this question, a functional study of *CgSymRK*, a gene isolated from *C. glauca*, homologous to the receptor-like kinase gene *SymRK* required for nodulation and mycorrhization in legumes, was undertaken. Downregulation of *CgSymRK* resulting from a RNA interference approach (Gherbi et al., 2008b) revealed that the frequency of nodulated RNAi-*CgSymRK* plants was reduced 2-fold compared to control *C. glauca* plants (Gherbi et al., 2008a). In addition, a range of morphological alterations was observed in the down-regulated *CgSymRK*-nodules (Gherbi et al., 2008b; Benabdoun et al., 2011). Additional experiments revealed that CgSymRK was also necessary for the establishment of the symbiosis with the arbuscular mycorrhiza *Glomus intraradices*. The knockdown of *CgSymRK* was seen to strongly affect penetration of the fungal hyphae into the root cortex, thus revealing the key role of CgSymRK in root endosymbioses in *Casuarina*, and the conservation of SymRK function between legumes and actinorhizal plants (Gherbi et al., 2008a). Similar experiments were undertaken with *DtSymRK, a SYMRK* sequence isolated from *D. glomerata* (Markman et al., 2008). Reduction of root mRNA levels of *DgSYMRK* was achieved via RNA-interference in composite plants resulting from the genetic transformation by *Agrobacterium rhizogenes*. In down-regulated *DgSYMRK* RNAi roots, no nodules were detected after inoculation by *Frankia*, except for small, primordial swellings on 16% of independent transformed roots. This result confirms that *SYMRK* is essential for both infection by *Frankia* and further actinorhizal nodule development.

In 2011, Hocher et al. were able to highlight the fact that, beyond SymRK, the whole array of genes from the Nod factor signal transduction pathway is shared between root nodule symbioses (RNS) in actinorhizal plants and legumes. The fact that a series of well-characterized symbiotic genes in legumes exhibit similar expression patterns in actinorhizal plants lends credibility to a common "SYM" pathway for endosymbioses and, for the first time, points to the possibility of a similar "NOD" pathway between RNS. This overlapping of legume and actinorhizal RNS reinforces the hypothesis of a common genetic ancestor of the nodulating clade with a genetic predisposition for nodulation (Soltis et al., 1995; Swensen and Mullin, 1997; Swensen and Benson, 2008).

4.2. Defense Genes

In plants, the spread of pathogens can be limited by the generation of defensive activities in the cells located in the vicinity of the pathogen. These defense reactions include the generation of reactive oxygen species (ROS) and reactive nitrogen species, the synthesis of phenolic compounds and protective proteins, the alteration of cell wall composition, and the induction of programmed cell death. Some of these activities have been observed during the establishment of symbioses between legumes and rhizobial bacteria, thus suggesting that nodulation may have evolved from a pathogenic interaction (Vasse et al., 1993; Baron and Zambrisky, 1995; Samac and Graham, 2007; Sprent and James, 2007).

In actinorhizal plants, data coming from the comparison of root and nodule transcriptomes in *C. glauca* and *A. glutinosa* has provided a global view of the genes that are differentially expressed (Hocher et al., 2006; Hocher et al., 2011b). For both hosts, the level of expression of several defence and stress genes was found to increase in a similar way to the one observed in Legumes. Several candidate genes from different actinorhizal plants have been chosen for functional analyses. These genes that are induced during the symbiotic process include sequences encoding chitinases, cysteine and subtilisin-like proteases, chalcone synthase and chalcone isomerase, Glycine- and histine-rich proteins, peroxydase and metallothionein-like proteins. The putative role of these sequences has been recently reviewed (Ribeiro et al., 2011). Functional analyses based on RNA interference will provide in the coming years complementary data on the putative role of these defense-related genes during infection and nodule ontogenesis in actinorhizal plants.

The mechanism of regulation of defense reactions by symbiotic signaling gene cascade is one of the subjects that requires study in the future to elucidate the communication that is necessary to establish a peaceful symbiotic relationship between *Frankia* and its actinorhizal host.

5. Genomics of *Frankia*

5.1. *Frankia* Genomes

The genomic resources available are expanding significantly since the 2007 publication of three *Frankia* genome sequences belonging to different host-compatibility groups, *Frankia alni* ACN14a that infects *Alnus, Frankia* CcI3 that infects *Casuarina* and *Frankia* EAN1pec that infects *Elaeagnus* (Normand et al., 2007a; Ransley and Tisa, 2007). A JGI program was later accepted to sequence the genome of a *Frankia* sp. that is an obligate endosymbiont of *D. glomerata* and the sequence is now available (Persson et al., 2011). There are other genomes underway with EuI1c and EUN1f, both infective on *Elaeagnus* and several others not yet completed. Interestingly, the size divergence in *Frankia* genomes is the largest yet reported for such closely related soil bacteria, ranging from 5.43 Mbp for the narrow host range strain CcI3 to 9.04 Mbp for the broad host range strain EAN1pec. In Dg1, the uncultured symbiont of *D. glomerata*, the genome sequence is 5.32 Mbp. These data suggest that gene deletion and duplication have occurred to different extents in the genomes during adaptation to host plants and their environments (Normand et al., 2007a).

Frankia genome analysis revealed the absence of canonical *nod* genes. Only a few, low similarity *nodB* and *nodC* homologs were detected. Moreover genes known to be involved in symbiosis such as *nif* (nitrogenase), *shc* (squalene hopene cyclase), *hup* (hydrogenase uptake) and *suf* (sulfur-iron cofactor synthesis) are scattered over the genomes away from the distant putative *nod* homologs (Normand et al., 2007b). This lack of a symbiotic island is in sharp contrast to the situation of rhizobia where *nif* genes are clustered with *nod* and ancillary genes, as a likely sequel of recent lateral transfers (Pappas and Cevallos, 2011). A consequence of this absence of symbiotic island is that the proximity of genes to the *nif* genes cannot be construed as an indication that they are involved in symbiosis.

5.2. *Frankia* Transcriptome

A whole genome *Frankia alni* microarray was set up and used for transcriptional analyses of symbiotic and free-living cells (Alloisio et al., 2010). These studies higlighted the following gene expression patterns: (i) nodule-induced genes were distributed mostly over several regions with high synteny between the three sequenced *Frankia* genomes; (ii) as expected, genes related with nitrogen fixation (*nif, suf, hup2, ispG, shc1,...*) were upregulated under symbiotic conditions; besides, a large number of genes with a putative role in transcription regulation, signaling processes, protein secretion, lipopolysaccharide and peptidoglycan biosynthesis were induced. It was also shown that *Frankia* symbiotic transcriptome was highly similar among the phylogenetically distant plant families *Betulaceae* and *Myricaceae*.

In 2011, Bickhart and Benson studied the behavior of the *Frankia Casuarina* sp. strain CcI3 transcriptome as a function of nitrogen source and culture age. To study global transcription in *Frankia* sp. CcI3 grown under different conditions, complete transcriptomes were determined using high throuput RNA deep sequencing. Samples varied by time (five days vs. three days) and by culture conditions (NH_4^+ added vs. N_2 fixing). Assembly of millions of reads revealed more diversity of gene expression between five-day- and three-day-old cultures than between three-day-old cultures differing in nitrogen sources. Heat map analysis organized genes into groups that were expressed or repressed under the various conditions compared to median expression values. Twenty-one SNPs common to all three transcriptome samples were detected indicating culture heterogeneity in this slow-growing organism. Significantly higher expression of transposase ORFs was found in the five-day and N_2-fixing cultures, suggesting that N starvation and culture aging provide conditions for on-going genome modification. The overall pattern of gene expression in aging cultures of CcI3 suggests significant cell heterogeneity even during normal growth on ammonia. These studies also sound a cautionary note when comparing the transcriptomes of *Frankia* grown in root nodules, where cell heterogeneity would be expected to be quite high.

6. Actinorhizal Plant Transcriptome

On the plant side, Expressed Sequence Tags (ESTs) libraries were obtained from roots (inoculated or not with *Frankia*) and nodules of *C. glauca* and *A. glutinosa*, and two arrays representing approximately 14 000 unigenes were developed (Hocher et al., 2011b). Using these tools, it has been found several homologs of genes involved in the Nod signal transduction pathway and shown their regulation during the symbiotic process. This transcriptome study also revealed that genes encoding enzymes involved in defence response and secondary metabolism were upregulated in young nodules. A similar observation was made for rhizobial infection at early stages of nodule formation in the model legume *L. japonicus* (Kouchi et al., 2004). Sequence analyses have indeed revealed the expression of numerous genes potentially implicated in actinorhizal nodulation.

In the coming years, it is expected that the Next Generation Sequencing technologies applied to several actinorhizal plants from different genera will bring a wide spectrum of new data and help understanding the key components of this original root endosymbiosis.

7. Evolutionary Aspects

In the families *Coriariaceae*, *Elaeagnaceae*, *Datiscaceae* and *Casuarinaceae*, all members are nodulated, whereas in *Betulaceae*, *Myricaceae*, *Rhamnaceae* and the *Rosaceae*, only a portion of the genera are nodulated. In at least one case (*Dryas*), nodulation apparently does not extend to all members of a single genus. These observations have led to the conclusion that, while the predisposition, or *potential*, to form the nitrogen-fixing symbiosis may have evolved only once in the Rosid I clade (Soltis et al.,1995), the realization of that potential has occurred and/or been lost multiple times in actinorhizal plants (Benson and Clawson, 2000; Swensen and Mullin, 1997 and 2008).

The numerous common features shared by rhizobial and actinorhizal nitrogen-fixing symbioses during the infection process are in agreement with a single origin of the predisposition for symbiotic nitrogen fixation in Angiosperms. Whereas it is not possible yet to compare the signal molecules exchanged by the two partners, a common symbiotic signaling pathway has been shown in actinorhizal, legume and arbuscular mycorrhization (Gherbi et al., 2008a; Markmann et al., 2008; Hocher et al., 2011b).

Besides the common SYM pathway, promoter studies have revealed a conservation of the «infection regulatory pathway» between actinorhizal and legume/rhizobium symbioses (Svistoonoff et al., 2010; Imanishi et al., 2011). A gene construct containing the *MtEnod11* promoter from *M. truncatula* fused to the reporter gene *gus* was introduced into *C. glauca*, and its pattern of expression was studied during *Frankia* infection. In legumes, Pro*MtEnod11* is widely used as an early infection marker for endosymbiotic associations involving both rhizobia and AM fungi (Journet et al., 2001; Charron et al., 2004). In *C. glauca*, Pro*MtEnod11::gus* expression was observed during *Frankia* infection in transgenic actinorhizal root hairs, prenodules and nodules. However, no activation of the *gus* reporter gene was detected prior to infection, nor in response to either rhizobial Nod factors or the amphiphilic peptide agonist MAS-7. Equally, Pro*MtEnod11::gus* expression was not elicited during the symbiotic associations with either ecto- or endomycorrhizal fungi. These observations suggest that there is a conservation of gene regulatory pathways between legumes and actinorhizal plants in cells accommodating endosymbiotic N2-fixing bacteria.

Conclusion

Progression in the molecular knowledge of the actinorhizal symbioses has been slow compared to Legumes. However, the development of new generation sequencing tools (Bickhart and Benson, 2011; Benson et al., 2011), new techniques for image analysis of plant cells, together with the functional tools provided by transgenic actinorhizal plants (Svistoonoff et al., 2010; Benadoun et al., 2011), are offering new avenues for investigating the symbiotic dialogue at different stages of nodule development. Major issues will be addressed in the coming years: 1) what is the biochemical structure of *Frankia* signals perceived by the root system in actinorhizal plants and do these signals share some common biochemical features with the Nod and the recently characterized Myc factors? 2) What are the plant genes contributing to the perception of the still unknown *Frankia* factors? 3) What are the common and specific features of the endosymbiotic actinorhizal, rhizobial and mycorrhizal processes? This will contribute to elucidate evolutionary relationships between plant root endosymbioses resulting from the symbiotic interactions with Frankia, rhizobia and endomycorrhizal fungi.

Acknowledgments

Work on actinorhizal plants in the Rhizogenesis Laboratory is supported by the *Institut de Recherche pour le Développement* (IRD) and by the *Agence Nationale de la Recherche* (ANR) Blanc project NewNod (ANR-06-BLAN-0095) and SESAM (BLAN-1708-01).

References

Alloisio, N., Queiroux, C., Fournier, P., Pujic, P., Normand, P., Vallenet, D., Medigue, C., Yamaura, M., Kakoi, K., Kucho, K. 2010. The Frankia alni symbiotic transcriptome. *Mol. Plant Microbe Interact.*, 23: 593-607.

Auguy, F., Abdel-Lateif, K., Doumas, P., Badin, P., Guerin, V., Bogusz, D., Hocher, V. 2011. Isoflavonoids pathway activation in actinorhizal symbioses. *Funct. Plant Biol.*, 38: 690-696.

Baker, D.D., Schwintzer, C. R. 1990. Introduction. In: *The Biology of Frankia and Actinorhizal Plants*. Schwintzer, C.R., Tjepkema, J.D. (eds.). Academic Press, New York, pp. 157-176.

Baron, C., Zambryski, P.C. 1995. The plant response in pathogenesis, symbiosis, and wounding: variations on a common theme? *Annu. Rev. Genet.*, 29: 107-129.

Bassi, C. A., Benson, D.R. 2007. Growth characteristics of the slow-growing actinobacterium Frankia sp. strain CcI3 on solid media. *Physiol. Plant.*, 130: 391-399.

Beauchemin, N.J., Furnholm, T., Lavenus, J., Svistoonoff, S., Doumas, P., Bogusz, D., Laplaze, L., Tisa, L. 2012. Casuarina root exudates alter the physiology, surface properties, and plant infectivity of *Frankia* sp. strain CcI3. *Appl. Environ. Microbiol.*, 78: 575-580.

Benabdoun, F.M., Nambiar-Veetil, M., Imanishi, L., Svistoonoff, S., Ykhlef, N., Gherbi, H. Franche, C. 2011. Composite actinorhizal plants with transgenic roots for the study of symbiotic associations with *Frankia*. *J. Bot.*, ID 702947, 8 p.

Benoit, L.F., Berry, A.M. 1997. Flavonoid-like compounds from red alder (Alnus rubra) influence nodulation *by Frankia* (Actinomycetales). *Physiol. Plant.*, 99: 588-93.

Benson, D.R., Brooks, J.M., Huang, Y., Bickhart, D.M., Mastronunzio, J.E. 2011. The biology of Frankia sp. strains in the post-genome era. *Mol. Plant-Microbe Interact.*, 24: 1310-1316.

Benson, D.R., Clawson, M.L. 2000. Evolution of the actinorhizal plant symbioses. In: Prokaryotic nitrogen fixation: A model system for analysis of biological process. Triplett E.W. (ed.). Horizon Scientific Press, Wymondham, UK, pp. 207-224.

Benson, D.R., Silvester, W.B. 1993. Biology of Frankia strains, actinomycete symbionts of actinorhizal plants. *Microbiol. Rev.*, 57: 293-319.

Berg, R.H. 1999. Frankia forms infection threads. *Can. J. Bot.*, 77: 1327-33.

Berg, R.H., McDowell, L. 1987. Endophyte differentiation in *Casuarina* actinorhizae. *Protoplasma,* 136: 104-117.

Berry, A.M., Harriott, O.T., Moreau, R.A., Osman, S.F., Benson, D.R., Jones, A.D. 1993. Hopanoid lipids compose the Frankia vesicle envelope, presumptive barrier of oxygen diffusion to nitrogenase. *Proc. Natl. Acad. Sci. U.S.A.,* 90: 6091-6094.

Berry, A.M., Sunnel, L.A. 1990. The infection process and nodule development. In: The biology of Frankia and Actinorhizal plants. Schwintzer, C.R., Tjepkema J.D. (eds). Academic Press, New York, pp. 61-81.

Bhuvaneswari, T.V., Solheim, B. 2000. Root-hair interactions in actinorhizal symbioses. In: *Root hairs - Cell and molecular biology*. Ridge, R.W., Emons, A.M.C. (eds.). Springer, pp. 311-327.

Bickhart, D.M., Benson, D.R. 2011. Transcriptomes of Frankia sp. strain CcI3 in growth transitions. *BMC Microbiol.*, 11: 192. doi:10.1186/1471-2180-11-192.

Callaham, D., Newcomb, W., Torrey, J.G., Peterson, R.L. 1979. Root hair infection in actinomycete-induced root nodule initiation in *Casuarina*, *Myrica* and *Comptonia*. *Bot. Gaz.*, 140: S1-9.

Callaham, D., Torrey, J.G. 1977. Prenodule formation and primary nodule development in roots of *Comptonia* (Myricaceae). *Can. J. Bot.*, 51: 2306-18.

Cérémonie, H., Debellé, F., Fernandez, M.P. 1999. Structural and functional comparison of Frankia root hair deforming factor and rhizobia Nod factor. *Can. J. Bot.*, 77: 1293-301.

Chaia, E. E., Wall, L. G., Huss-Danell, K. 2010. Life in soil by the actinorhizal root nodule endophyte *Frankia*. A review. *Symbiosis,* 51: 201-226.

Charron, D., Pingret, J.L., Chabaud, M., Journet, E.P. Barker, D.G. 2004. Pharmacological evidence that multiple phospholipid signaling pathways link Rhizobium nodulation factor perception in *Medicago truncatula* root hairs to intracellular responses, including Ca2+ spiking and specific ENOD gene expression. *Plant Physiol.*, 136: 3582-3593.

Davidson, C. 1973. An anatomical and morphological study of Datiscaeae. *Aliso*, 8: 49-110.

Dawson, J.O. 2008. Ecology of actinorhizal plants. In: *Nitrogen fixation research: origins and progress»*, vol. VI: Nitrogen-fixing Actinorhizal symbioses. Pawlowski, K., Newton, W.E. (eds). Springer, pp. 199-234.

Dénarié, J., Debellé, F., Promé, J.C. 1996. Rhizobium lipochitooligosaccharide nodulation factor: signaling molecules mediating recognition and morphogenesis. *An. Rev. Biochem.*, 65: 503-535.

Diouf, D., Sy, M-O., Gherbi, H., Bogusz, D., Franche, C. 2008. Casuarinaceae. In: *Compendium of Transgenic Crop Plants: Transgenic Forest Tree Species*, vol. 9. Kole, C.R., Scorza, R., Hall, T.C. (eds). Blackwell Publishing, Oxford, UK, pp. 279-292.

Dobritsa, S.V., Potter, D., Gookin, T.E., Berry, A.M. 2001. Hopanoid lipids in *Frankia*: identification of squalene-hopene cyclase gene sequences. *Can. J. Microbiol.*, 47: 535-540.

Duhoux, E., Diouf, D., Gherbi, H., Franche, C., Ahée, J., Bogusz, D. 1996. Le nodule actinorhizien. *Acta bot. Gallica*, 143: 593-608.

Franche, C., Laplaze, L., Duhoux, E., Bogusz, D. 1998. Actinorhizal symbioses: recent advances in plant molecular and genetic transformation studies. *Crit. Rev. Plant Sci.*, 17: 1-28.

Franche, C., Bogusz, D. 2011. Signaling and communication in the actinorhizal symbiosis. In: *Signaling and communication in plant symbioses*. S. Perotto and F. Baluska (eds). Springer, pp. 73-92.

Gage, D.J. 2002. Analysis of infection thread development using Gfp- and DsRed-expressing *Sinorhizobium meliloti*. *J. Bacteriol.*, 184: 7042-7046.

Gage, D.J. 2004. Infection and Invasion of Roots by Symbiotic, Nitrogen-Fixing Rhizobia during Nodulation of Temperate Legumes. *Microbiol. Mol. Biol. Rev.*, 68: 280-300.

Geurts, R., Federova, E., Bisseling, T. 2005. Nod factor signaling genes and their function in the early stages of Rhizobium infection. *Cur. Opin. Plant Biol.*, 8: 346-352.

Gherbi, H., Markmann, K., Svistoonoff, S., Estevan, J., Autran, D., Giczey, G., Auguy, F., Péret, B., Laplaze, L., Franche, C., Parniske, M., Bogusz, D. 2008a. SymRK defines a common genetic basis for plant root endosymbioses with AM fungi, rhizobia and Frankia bacteria. 2008. *Proc. Nat. Acad. Sci. U.S.A.*, 105: 4928-4932.

Gherbi, H., Nambiar-Veetil, M., Zhong, C., Félix, J., Autran, D., Girardin, R., Vaissayre, V., Auguy, F., Bogusz, D., Franche, C. 2008b. Post-transcriptional gene silencing in the root system of the actinorhizal tree *Allocasuarina verticillata*. *Mol. Plant-Microbe Interact.*, 21: 518-524.

Gualtieri, G., Bisseling, T. 2000. The evolution of nodulation. *Plant Mol. Biol.*, 42: 181-194.

Hocher, V., Auguy, F., Argout, X., Laplaze, L., Franche, C., Bogusz, D. 2006. Expressed sequence-tag analysis in *Casuarina glauca* actinorhizal nodule and root. *New Phytol.*, 169: 681-688.

Hocher, V., Alloisio, N., Bogusz, D., Normand, P. 2011a. Early signaling in actinorhizal symbioses. *Plant Signal. Behav.*, 6: 1377-1379.

Hocher, V., Alloisio, N., Auguy, F., Fournier, P., Doumas, P., Pujic, P., Gherbi, H., Queiroux, C., Da Silva, C., Wincker, P., Normand, P., Bogusz, D. 2011b. Transcriptomics of actinorhizal symbioses reveals homologs of the whole common symbiotic signaling cascade. *Plant Physiol.*, 156: 700-711.

Hughes, M., Donnelly, C., Crozier, A., Wheeler, C.T. 1999. Effects of the exposure of roots of *Alnus glutinosa* to light on flavonoids and nodulation. *Can. J. Bot.*, 77: 1311-5.

Huss-Danell, K. 1990. The physiology of actinorhizal nodules. In: *The biology of Frankia and actinorhizal plants*. Schwintzer, C.R., Tjepkema, J.D. (eds). Academic Press, Inc., New York, pp. 129-156.

Huss-Danell, K. 1997. Actinorhizal symbioses and their N2 fixation. *New Phytol.*, 136: 375-405.

Imanishi, L., Vayssières, A., Franche, C.,Bogusz, D., Wall, L., Svistoonoff, S. 2011. Transformed Hairy Roots *of Discaria trinervis*: A valuable tool for studying actinorhizal symbiosis in the context of intercellular infection. *Mol. Plant-Microbe Interact.*, 24: 1317-1324.

Journet, E.P., El-Gachtouli, N., Vernoud, V., de Billy, F., Pichon, M., Dedieu, A., Arnould, C., Morandi, D., Barker, D.G. and Gianinazzi-Pearson, V. 2001. Medicago truncatula *ENOD11*: A novel RPRP-encoding early nodulin gene expressed during mycorrhization in arbuscule-containing cells. *Mol. Plant-Microbe Interact.*, 14: 737-748.

Kim, H.B., Bae, J.H., Lim, J.D., Yu, C.Y., An, C.S. 2007. Expression of a functional type-I chalcone isomerase gene is localized to the infected cells of root nodules of *Elaeagnus umbellata. Mol. Cells*, 23: 405-409.

Kim, H.B., Oh, C.J., Lee, H., An, C.S. 2003. A type-I chalcone isomerase mRNA is highly expressed in the root nodules of *Elaeagnus umbellata. J. Plant Biol.*, 46: 263-270.

Kouchi, H., Shimomura, K., Hata, S., Hirota, A., Wu, G.J., Kumagai, H., Tajima, S., Suganuma, N., Suzuki, A., Aoki, T., Hayashi, M., Yokoyama, T., Ohyama, T., Asamizu, E., Kuwata, C., Shibata, D., Tabata, S. 2004. Large-scale analysis of gene expression profiles during early stages of root nodule formation in a model legume, *Lotus japonicus. DNA Res.*, 31: 263-274.

Laplaze, L., Duhoux, E., Franche, C., Frutz, T., Svistoonoff, S., Bisseling, T., Bogusz, D., Pawlowski, K. 2000. Casuarina glauca prenodule cells display the same differentiation as the corresponding nodule cells. *Mol. Plant-Microbe Interact.*, 13: 107-12.

Laplaze, L., Gherbi, H., Frutz, T., Pawlowski, K., Franche, C., Macheix, J-J., Auguy, F., Bogusz, D., Duhoux, E. 1999. Flavan-containing cells delimit Frankia-infected compartments in *Casuarina glauca* nodules. *Plant Physiol.*, 121: 113-122.

Laplaze, L., Svistoonoff, S., Santi, C., Auguy, F., Franche, C., Bogusz, D. 2008. In: *Molecular biology of actinorhizal symbioses.* In: *Nitrogen fixation research: origins and progress*, vol. VI: Nitrogen-fixing Actinorhizal symbioses. Pawlowski, K., Newton, W.E. (eds). Springer, pp. 235-259.

Lavire, C., Cournoyer, B. 2003. Progress on the genetics of the N2-fixing actinorhizal symbiont Frankia. *Plant Soil*, 254: 125-137.

Lerouge, P. 1994. Symbiotic host specificity between leguminous plants and rhizobia is determined by substituted and acetylated glucosamine oligosaccharides signals. *Glycobiol.*, 4: 127-134.

Liu, Q., Berry, A.M. 1991. The infection process and nodule initiation in *the Frankia-Ceanothus root nodule symbiosis. Protoplasma*, 163: 82-92.

Madsen, L.H., Tirichine, L., Jurkiewicz, A., Sullivan, J.T., Heckmann, A.B., Beck, A.S., Ronson, C.W., James, E.K., Stougaard, J. 2010. The molecular network governing nodule organogenesis and infection in the model legume *Lotus japonicus. Nature Com.*, 1: 1-12.

Markmann, K., Giczey, G., Parniske, M. 2008. Functional adaptation of a plant receptor-kinase paved the way for the evolution of intracellular root symbioses with bacteria. *PLoS Biol.*, 6: 497-506.

McEwan, N.R., Green, D.C., Wheeler, C.T. 1992. Utilisation of the root hair curling reaction in *Alnus glutinosa* for the assay of nodulation signal compounds. *Acta Oecol.*, 13: 509-510.

Miller, I.M., Baker, D.D. 1985. The initiation, development and structure of root nodules in *Elaeagnus angustifolia* L. (Elaeagnaceae). *Protoplasma*, 128: 107-19.

Miller, I.M., Baker, D.D. 1986. Nodulation of actinorhizal plants by Frankia strains capable of both root hair infection and intercellular penetration. *Protoplasma*, 131: 82-91.

Newcomb, W.R., Wood, S. 1987. Morphogenesis and fine structure of Frankia (Actinomycetales): the microsymbiont of nitrogen-fixing actinorhizal root nodules. *Int. Rev. Cytol.*, 109: 1-88.

Normand, P., Lapierre, P., Tisa, L.S., Gogarten, J.P., Alloisio, N., Bagnarol, E., Bassi, C.A., Berry, A.M., Bickhart, D.M., Choisne, N., Couloux, A., Cournoyer, B., Cruveiller, S., Daubin, V., Demange, N., Francino, M.P., Goltsman, E., Huang, Y., Martinez, M., Mastronunzio, J.E., Mullin, B.C., Nieman, J., Pujic, P., Rawnsley, T., Rouy, Z., Schenowitz, C., Sellstedt, A., Tvares, F., Tomkins, J.P., Vallenet, D., Valverde, C., Wall, L., Wang, Y., Medigue, C., Benson, D.R. 2007a. Genome characteristics of facultatively symbiotic Frankia sp. strains reflect host range and host plant biogeography. *Genome Res.*, 17: 7-15.

Normand, P., Queiroux, C., Tisa, L.S., Benson, D.R., Rouy, Z., Cruveiller, S., Medigue, C. 2007b. Exploring the genomes of Frankia. *Physiol. Plant.*, 130: 331-343.

Obertello, M., Sy, M-O., Laplaze, L., Santi, C., Svistoonoff, S., Auguy, F., Bogusz, D., and Franche, C. 2003. Actinorhizal nitrogen fixing nodules: infection process, molecular biology and genomics. *Afr. J. Biotechnol.*, 2: 528-538.

Oldroyd, G.E.D., Downie, J.A. 2008. Coordinating nodule morphogenesis with rhizobial infection in legumes. *Annu. Rev. Plant Biol.*, 59: 519-546.

Pappas, K.M., Cevallos, M.A. 2011. Plasmids of the Rhizobiaceae and their role in interbacterial and transkingdom interactions. In: *Biocommunication in Soil Microorganisms, Soil Biology* vol. 23. Witzany, G. (ed). Springer-Verlag, Berlin Heidelberg, pp. 295-337.

Parsons, R., Silvester, W.B., Harris, S., Gruijters, W.T.M, Bullivant. S. 1987. Frankia vesicles provide inducible and absolute oxygen protection for nitrogenase. *Plant Physiol.*, 83: 728-731.

Pawlowski, K., Bisseling, T. 1996. Rhizobial and Actinorhizal Symbioses: What Are the Shared Features? *Plant Cell*, 8: 1899-913.

Pawlowski, K., Sprent, J. 2008. Comparison between actinorhizal and legume symbiosis. In: *Nitrogen-fixing actinorhizal symbioses*. Pawlowski, K., Newton, W.E., (eds). Dordrecht, Springer Netherlands, pp. 261-288.

Pawlowski, K., Demchenko, K.N. 2012. The diversity of actinorhizal symbiosis. *Protoplasma*, in press.

Pawlowski, K., Bogusz, D., Ribeiro, A., Santos, P. 2011. Progress on research on actinorhizal plants. *Funct. Plant Biol.*, 38: 633-638.

Perrine-Walker, F., Gherbi, H., Imanishi, L., Benabdoun, F., Hocher, V., Ghodhbane-Gtari, F., Lavenus, J., Nambiar-Veetil, M., Svistoonoff, S., Laplaze, L. 2010. Symbiotic signalling in actinorhizal symbioses. *Curr. Protein Pept. Sci.*, 12: 156-164.

Persson, T., Benson, D.R., Normand, P., Vanden Heuvel, B., Pujic, P., Chertkov, O., Teshima, H., Bruce, D.C., Detter, C., Tapia, R., Han, S., Woyke, T., Pitluck, S., Pennacchio, L., Nolan, M., Ivanova, N., Pati, A., Land, M.L., Pawlowski, K., Berry, A.M. 2011. Genome sequence of "*Candidatus Frankia datiscae*" Dg1, the uncultured microsymbiont from nitrogen-fixing root nodules of the dicot *Datisca glomerata*. *J. Bacteriol.*, 193: 7017-7018.

Popovici, J., Comte, G., Bagnarol, E., Fournier, P., Bellvert, F., Bertrand, C. 2010. Differential Effects of Rare Specific Flavonoids on Compatible and Incompatible Strains in the *Myrica gale-Frankia* Actinorhizal Symbiosis. *Appl. Environ. Microbiol.*, 76: 2451-2460.

Prin, Y., Rougier, M. 1987. Preinfection events in the establishment of Alnus-Frankia symbiosis: study of the root hair deformation step. *Plant Physiol. (Life Sci. Adv)*, 6: 99-106.

Racette, S., Torrey, J.G. 1989. Root nodule initiation in *Gymnostoma* (Casuarinaceae) and *Shepherdia* (Elaeagnaceae) induced by Frankia strain HFPGpI1. *Can. J. Bot.*, 67: 2873-2879.

Rawnsley, T., Tisa, L.S. 2007. Development of a physical map for three Frankia strains and a partial genetic map for *Frankia* EuI1c. *Physiol. Plant.*, 130: 427-439.

Ribeiro, A., Graca, I., Pawlowski, K., Santos, P. 2011. Actinorhizal plant defense-related genes in response to symbiotic Frankia. *Funct. Plant Biol.*, 38: 639-644.

Riely, B.K., Mun, J.H., Ané, J.M. 2006. Unraveling the molecular basis for symbiotic signal transduction in legumes. *Mol. Plant Pathol.*, 7: 197-207.

Samac, D.A., Graham, M.A. 2007. Recent advances in legume–microbe interactions: recognition, defense response, and symbiosis from a genomic perspective. *Plant Physiol.*, 144: 582-587.

Silvester, W.B., Berg, R.H., Schwintzer, C.R., Tjepkema, J.D. Oxygen responses, hemoglobin, and the structure and function of vesicles. In: Nitrogen fixation research: origins and progress», vol. VI: Nitrogen-fixing Actinorhizal symbioses. Pawlowski, K., Newton, W.E. (eds). Springer, pp. 105-146.

Simonet, P., Normand, P., Hirch, M., Akkermans, A.D.L. 1990. The genetics of the Frankia-actinorhizal symbiosis. In: *Molecular Biology of Symbiotic Nitrogen Fixation*. Gresshoff, P.M. CRC Press, Bocaraton, USA. pp. 77-109.

Smolander, A., Sarsa, M.-L. 1990. *Frankia* strains in soil under Betula pendula: behavior in soil and in pure culture. *Plant Soil*, 122: 129-136.

Soltis, D. E., Soltis, P.S., Morgan, D.R., Swensen,S.M., Mullin, B.C., Dowd, J.M., Martin, P.G. 1995. Chloroplast gene sequence data suggest a single origin of the predisposition for symbiotic nitrogen fixation in angiosperms. *Proc. Natl. Acad. Sci. U.S.A*, 92: 2647-2651.

Sprent, J.I., James, E.K. 2007. Legume evolution: Where do nodules and mycorrhizas fit in? *Plant Physiol.*, 144: 575-581.

Sprent, J.I., Parsons, R. 2000. Nitrogen fixation in legume and non-legume trees. *Field Crops Res.*, 65: 183-196.

Schwintzer, C.R., Lancelle, S.A. 1983. Effect of water-table depth on shoot growth, root growth, and nodulation of Myrica gale seedlings. *J. Ecol.*, 71: 489-501.

Svistoonoff, S., Sy, M.-O., Diagne, N., Barker, D., Bogusz, D., Franche, C. 2010. Infection-specific activation of the Medicago truncatula Enod11 early nodulin gene promoter during actinorhizal root nodulation. *Mol. Plant-Microbe Interact.*, 23: 740-747.

Swensen, S.M., Mullin, B.C. 1997. Phylogenetic relationships among actinorhizal plants: The impact of molecular systematics and implications for the evolution of actinorhizal symbioses. *Physiol. Plant.*, 99: 565-573.

Swensen, S.M., Benson, D.R. 2008. Evolution of actinorhizal host plants and Frankia endosymbionts. In: Nitrogen fixation research: origins and progress», vol. VI: Nitrogen-fixing Actinorhizal symbioses. Pawlowski, K., Newton, W.E. (eds). Springer, pp. 73-104.

Tjepkema, J. 1978. The role of oxygen diffusion from the shoots and nodule roots in nitrogen fixation by root nodules of *Myrica gale*. *Can. J. Bot.*, 56: 1365-1371.

Torrey, J.G. 1976. Initiation and development of root nodules of Casuarina (Casuarinaceae). *Amer. J. Bot.*, 63: 335-344.

Valverde, C., Wall, L.G. 1999. Time course of nodule development in *Discaria trinervis* (Rhamnaceae) -Frankia symbiosis. *New Phytol.*, 141: 345-354.

van Ghelue, M. 1994. Interactions in actinorhizal symbioses. Ph. D. thesis, The University of Tromsö, Tromnsö, Norway.

van Ghelue, M., Lovaas, E., Ringo, E., Solheim, B. 1997. Early interaction between Alnus glutinosa and Frankia strain Arl3. Production and specificity of root hair deformation factors. *Physiol. Plant.*, 99: 579-587.

Vasse, J., F. de Billy, F., Truchet, G. 1993. Abortion of infection during the *Rhizobium meliloti*-alfalfa symbiotic interaction is accompanied by a hypersensitive reaction. *Plant J.*, 4: 555-566.

Vessey, J.K., Pawlowski, K., Bergman, B. 2005. Root-based N2-fixing symbioses: Legumes, actinorhizal plants, Parasponia sp and cycads. *Plant Soil*, 274: 51-78.

Wall, L.G., Berry, A.M. 2008. Early interactions, infection and nodulation in actinorhizal symbiosis. In: *Nitrogen-fixing actinorhizal symbioses*. Pawlowski, K., Newton, W.E. (eds). Springer, Dordrecht, The Netherlands, pp. 147-166.

Wall, L.G. (2000). The actinorhizal symbiosis. *J. Plant Growth Reg.*, 19: 167-182.

Zhong, C., Zhang, Y., Chen, Y., Jiang, Q., Chen, Z., Liang, J., Pinyopusarerk, K., Franche, C., and Bogusz, D. Casuarina research in China. *Symbiosis,* 1: 107-114.

In: Symbiosis: Evolution, Biology and Ecological Effects   ISBN: 978-1-62257-211-3
Editors: A. F. Camisão and C. C. Pedroso   © 2013 Nova Science Publishers, Inc.

Chapter 8

THE ROLE OF LEGUMES FOR SUSTAINABILITY: RESEARCH NEEDS FOR FUTURE PROSPECTS

Corina Carranca[*]

Instituto Nacional de Investigação Agrária, Oeiras, Portugal
CEER, Instituto Superior de Agronomia, Lisboa, Portugal

Abstract

Improved nitrogen (N) management is needed to optimize economic returns to farmers and minimize environmental concerns associated with the N use. Symbiotically N2 fixed is of particular significance in sustainable agriculture as it allows reducing the use of chemical N in the production of field crops. Intercropping or crop rotation including legumes promises a more sustainable plant production in many agricultural systems through the N transfer and N release from legume residue. Sharing of N sources between the intercropped N2 fixing and non-fixing plants contribute to a better overall N use. In a crop rotation, the final contribution of fixed N2 to the soil depends upon the crop N balance, environmental conditions and agricultural practices.

Keywords: biodiverse pasture, crop rotation, fertilizer value, intercropping, Mediterranean conditions, mycorrhiza, N balance, N release, N transfer, N_2 fixation, natural sward, temperate and tropical climate

1. Introduction

The decline of soil fertility with loss of organic matter (OM), the excessive use of chemical fertilizers, the inappropriate use of water resources and the increase of soil acidity and salinity, particularly in dry regions, all pose real threats to both economic and environmental sustainability. Soil losses above 60 t ha^{-1} yr^{-1} have been reported in South Africa by burning crop residues in maize areas (Murungu et al., 2010). In Europe, particularly the south, soil erosion has been caused by the irregular and intense rainfall events (115

---

[*] E-mail address: c.carranca.ean@clix.pt, Tel: +351 214403517, Fax: +351 214416011

million ha) and wind (42 million ha), but also by wrong agricultural practices (tillage, irrigation, crop sequence, etc.) and fires. Portugal has the greatest burnt area in Europe (5%), five times greater than other southern European countries (IPCC, 2007). Combating soil erosion is thus a primary concern for agricultural producers in south Europe and Africa, but also in the United States of America (USA) and other American countries. Many countries have incorporated conservation agriculture, including conservation tillage systems in their efforts to maintain a profitable crop output. Legume cover crop or crop residue are important tools for this purpose, also supplementing the soil with nitrogen (N). Forage legumes have been used to recover degraded soils via their ability to improve the physical, chemical and biological properties of soils. Permanent biodiverse pastures (including more than twenty plant species, among legumes and non-legumes) are important for soil conservation in Mediterranean ecosystems, where most soils are degraded (51% of soils in European Member States). They are the central part of Mediterranean High Natural Value farming system and contribute to the decrease of soil erosion and desertification, and to the increase of soil fertility and biodiversity (Sequeira, 2008). White clover (*Trifolium repens* L.) and subclover (*T. subterraneaum* L.) are very suitable for soil conservation, but lucerne (*Medicago* sp.) is also used in drier areas.

2. Fertilizer Value of Legumes

Systems involving legumes represent a sustainable alternative to conventional agriculture by capture of atmospheric N_2. These systems include intercropping or crop rotation. Sharing of N sources between the intercropped N_2 fixing and non-fixing plants contribute to a better overall N use and reduces the post-harvest soil N availability, which in some agro-ecosystems may be easily leached to the groundwater or lost by denitrification or incomplete nitrification. Crop rotations including legumes also seek to balance the fertility demand of various sequential crops and avoid excessive depletion of soil nutrients, as happens by growing the same crop in the same place for several years (monoculture). By rotation, farmers can keep the field under continuous production, without the need of fallow, and reducing the need for mineral fertilizers, with a better control for pests and diseases.

2.1. Intercropping

Intercropping, also referred to as a mixed cropping or polyculture is the practice of growing simultaneously two or more compatible crops in close proximity to promote interaction between them and increase the productivity per unit of land. The component crops of an intercropping system do not necessarily have to be sown or harvested at the same time, but they should be grown simultaneously for a great part of their cycles. The options include row, strip, and mixed cropping systems, normally with a main crop of primary importance for economic or food production reasons, and one or more other crops. There are several plant species that can be used for intercropping such as annuals, e.g. cereals and legumes, perennials, including shrubs and trees, and a mixture of the two (annuals and perennials). In this case the term mostly used is agroforestry.

Intercropping is a widely distributed practice, because of its high and more stable productivity in a wide range of crop combinations, good pest and disease control, minimal use of inputs such as fertilizers and pesticides, ecological services, and economic profitability (Ouma and Jeruto, 2010; Yu et al., 2010; Lithourgidis et al., 2011).

Careful planning is required, taking into account the soil, climate, and crop species and variety. Intercrops should not compete each other for water, nutrients and sunlight. Strategies may include planting a deep-rooted crop with a shallow-rooted one, or planting a tall crop with a short crop that requires only partial shade, or planting one or more N_2 fixers with one or more non-fixers. Factors including plant height, size of the leaves, and orientation and distribution of leaves in the plant canopy are important influencing the successful of this cropping system. These variables affect the amount of sunlight that passes through the canopy (light interception) and influences the leaf photosynthetic rate. In intercropping systems with different canopy heights, the crop under the canopy needs to be shade tolerant to be productive (Ouma and Jeruto, 2010). Legumes generally suffer more from shading than do grasses (Mengel and Kirkby, 2001), since they need light for the photosynthetic process to produce a high level of ATPs for the N_2 fixation process.

Early- and slow-maturing crops are used combined in intercropping systems to ensure efficient utilization of the growing season length, and because intercrops are more productive when differing greatly in growth duration (Lithourgidis et al., 2011). For example, when a long-duration pigeon pea [*Cajanus cajan* (L.) Mills.] cultivar was grown with a mixture with three cereals of different growth durations, i.e. setaria (*Setaria*), pearl millet [*Pennisetum glaucum* (L.) R. Br.] and sorghum [*Sorghum bicolor* (L.) Moench], the Land Equivalent Ratio (LER=the relative land area under sole crops required to produce yields achieved in intercropping) was highest with the fast-maturing setaria and lowest with the slow-growth sorghum. When the LER>1, the intercropping favours the growth and yield of the species, whereas when the LER<1 the intercropping depresses the growth and yield of crop components (Ouma and Jeruto, 2010; Lithourgidis et al., 2011). Ouma and Jeruto (2010) reported that intercropping maize (*Zea mays* L.) with cowpea (*Vigna unguiculata* L.) and soybean [*Glycine max* (L.) Merr.] increased the productivity by 22-32% on the LER basis, and the productivity of wheat (*Triticum aestirum* L.) intercropped with chickpea (*Cicer arietinum* L.), or barley (*Hordeum vulgare* L.) intercropped with lentil (*Lens culinaris* L.) increased by 29% and 28%, respectively.

Intercropping legumes with non-legumes is the most common system in the tropics and has also been largely adopted in European countries. Here, legumes are grown in the inter-row spacing with fruit and olive (*Olea europaea* L.) trees, vineyards (*Vitis vinifera* L.) and forest trees, and mixed in swards. In tropical and subtropical regions, intercropping is mostly associated with food grain production, where maize, sorghum, millet (*Millet*), but less rice (*Oryza sativa* L.) are cereals primarily used together with cowpea, groundnut (*Arachis hypogaea* L.), soybean, chickpea, bean (*Phaseolus* sp.), and pigeon pea. In these regions, intercropping banana (*Musa* sp.) with beans has also been reported to reduce the incidence of weevils and nematodes, and in Kenya, young fruit trees are intercropped with all types of annual crops, including beans and peas (*Pisum sativa* L.), as a way of attaining food security and income before the trees mature (Ouma and Jeruto, 2010).

Table 1. Pastures composition by the end of winter (February 2011) and spring (April 2011) in Mediterranean conditions, as influenced by the pasture age and diversity, and tree (*Quercus ilex* L.) canopy (Carranca, unpublished data)

Source of variation	Legume aerial biomass (kg DM ha⁻¹)	Legume root biomass (kg DM ha⁻¹)	Non-legume aerial biomass (kg DM ha⁻¹)	Non-legume root biomass (kg DM ha⁻¹)
	Vegetative period			
Pasture type				
improved>30 years-old	784 a	389 a	299 a	216 b
improved 12 years-old	646 a	122 a	497 a	532 ab
improved 5 years-old	240 a	53 a	370 a	436 b
natural>25 years-old	254 a	99 a	549 a	1239 a
Tree canopy influence				
under	88 b	21 b	541 a	861 a
out	875 a	311 a	313 b	350 a
ANOVA				
Pasture type	ns	ns	ns	3.27*
Tree canopy influence	15.40***	8.18**	7.16*	ns
Interaction	ns	ns	ns	ns
	Bloom period			
Pasture type				
improved>30 years-old	710 b	52 c	6199 a	415 b
improved 12 years-old	2102 a	244 b	5557 a	1056 ab
improved 5 years-old	1059 ab	513 a	8769 a	750 ab
natural>25 years-old	907 b	120 c	7071 a	1252 a
Tree canopy influence				
under	933 a	267 a	7689 a	1100 a
out	1456 a	198 b	6109 a	636 b
ANOVA				
Pasture type	5.30**	53.39***	ns	4.45*
Tree canopy influence	ns	6.19*	ns	7.16*
Interaction	ns	30.95***	ns	ns

ANOVA=Analysis of variance; DM=dry matter; ns, *, **, *** = F-values not significant (P>0.05), and significant for P<0.05, 0.01 and 0.001, respectively; in each column and for each characteristic, means with the same letter are not significantly different (P<0.05), according to Bonferroni's test.

In legume-mixed swards, a major difficulty is their slow establishment, particularly in dry climates like the tropical and Mediterranean-type, made worse when sowing the seed too deep or too late. During establishment, legume development is frequently restricted by competition from non-legumes (mostly for light and water), often decreasing to <10% after 2-3 years. In south Europe, the aboveground biomass produced by the subclover during the autumn-winter in different rotational grazing pastures (natural and biodiverse), varying from 5, 12 and greater than 25 years-old was 482 kg dry weight (DW) ha⁻¹, i.e. 11% greater than associated non-legumes (429 kg DW ha⁻¹) (Table 1). In spring, subclover aerial material (1194 kg DW ha⁻¹) was 51% lower than non-legumes (2340 kg DW ha⁻1). These pastures with different age and composition did not differ in growth response. Pastures composition in the improved stands was based on *Lolium* and *Phalaris* as non-legumes, and composite plants and *Plantago* in the natural sward (Carranca, data not published). The above values agreed with Rochon et al. (2004) for Mediterranean environments, but did not respect the results obtained before the climate change in these Mediterranean regions. According to Rochon et al. (2004), annual legume-based pastures need to be grazed in order to maintain a total biomass averaging 1000 and 1500 kg DW ha^{-1}. They found higher forage availability under rotational grazing than continuous stocking, but swards were richer in annual ryegrass (*Lolium multiflorum* Lam.) and poorer in leguminous species, compared to continuous stocking.

Plant persistence in permanent stands is an imperative to livestock production. The legume-based performance is very dependent on sward management, particularly the palatability (governing the animal-litter pathway) and decomposability (addition to soil organic-mineral N) of the legume material. Timing, frequency and intensity of grazing induce botanical changes in (improved) pastures, and it is generally considered that rotational grazing is less damaging than continuous grazing to clover (Haggar, 1988; Cadish et al., 1994; Rochon et al., 2004). During flowering stage of annual legumes, livestock should be carefully managed in order to avoid the depletion of the seed bank caused by seed intake (Rochon et al., 2004).

Information on how to manage properly the white clover (temperate areas) and annual self-regenerating legumes (drier areas), either as a monoculture (temperate zones) or mixed with non-legumes is still limited.

This is particularly relevant in improved tropical-type pastures where the problem of a legume-based persistency is evident since grasses possess a C_4 photosynthetic pathway and are thus highly aggressive in competition with introduced legumes, which are C_3 plants (Giller and Cadisch, 1995).

2.1.1. N Transfer

The proportion of legumes required to balance the N cycle of grazed swards depends on the rate of pasture utilization, the efficiency of N_2 fixation, and the recycling of excreta, litter and belowground material. In grazed temperate pastures, legumes proportion has been estimated to be just over 10% (of herbage biomass production), while for mixed tropical pastures estimates range 11-45%, depending on pasture utilization, i.e. 60-117 kg fixed N_2 ha^{-1} (Cadish et al., 1994; Giller and Cadisch, 1995; Thomas, 1995; Carranca, unpublished). In Western Europe, the proportion of 1:1 for legume:non-legume seeds has been highly recommended.

A strategic use of mineral N at the end of winter (February) or in early spring has been advocated in north Europe, on the assumption that the increased non-fixer growth at this time is less competitive to the slow-growing legume. As the system becomes N limited, the legume (mostly white clover) becomes dominant due to its access to atmospheric N_2 (Tranin et al., 2000). In Mediterranean situations, improved pastures (mostly annual self-regenerating subclover and alfalfa) receive no mineral N and are sustainable swards. Only phosphorus (P) and potassium (K) fertilizers are added bi-annually.

The main pathways for N transfer from legumes to the soil in mixed swards are through the excreta by grazing livestock (75-95% of legume N is used by cattle and then excreted through feces and urine to the soil), and then by physical and microbial decomposition of unused leaves and stems of legumes, followed by decomposition of rhizodeposits (root exudates and products of nodule and root necrosis), root systems and senescent nodules (Rochon et al., 2004; De Varennes et al., 2007).

Tranin et al. (2000) reported that the main pathway of belowground N transfer occurs via decomposing roots (non-colonized roots decompose in few days) rather than via root exudates or direct mycorrhizal hyphae transfer. This statement is probably not the general rule since the exudates of nodulated plants are rich in easily decomposable N-compounds. Nitrogen fixing plants exude greater amounts of ammonium (NH_4^+) and amino acids into the mycorrhizosphere (rhizosphere with the influence of mycorrhized roots) than non-legumes by passive diffusion attributed to a concentration gradient from root to soil (Tranin et al., 2000; Jensen and Hauggaard-Nielsen, 2003; Paynel et al., 2008). Khan et al. (2000) and Unkovich and Pate (2000) found 20-45% of the total plant N allocated to roots of clover and alfalfa (Medicago sp.). Peoples et al. (1995) reported that N present in nodulated roots could reach 15-50 kg N ha^{-1}. If the carbon:nitrogen (C:N) ratio of exudates of nodulated plants is lower than that of non-nodulated plants, less soil N is immobilized in the nodulated rhizosphere and more N is taken up by the associated crops.

In natural meadows, the transfer of fixed N_2 from legumes to non-legumes was observed to increase with increasing plant species richness, but in a legume-based system, Carlsson et al. (2009) concluded that more important than species richness are species composition and functional traits regarding how neighbouring plant species influence N_2 fixation. These authors (2009) measured an increase of N_2 fixation by white clover or alsike clover (*T. hybridum* L.) in presence of grasses, but the competition for light from grasses limited the growth of clovers. They also observed that the tall grass (*Phalaris arundinacea* L.) was a particularly strong competitor for inorganic soil N, increasing the perennial clover dependence on N_2 fixation. When two grasses were grown together with clover, only ryegrass (*Festuca* L.) benefited from the presence of the legume, but the benefit decreased with increasing clover abundance (Marty et al., 2009). Haggar (1988) observed that the tetraploid perennial ryegrass (*Lolium perenne* L.) permitted a better associated white clover performance than did diploid cultivars or the early flowering ryegrasses. These data show the importance of evaluating the appropriate companion crop(s) to grow intercropped with legumes, especially in mixed swards.

In agricultural systems, Pal and Shehu (2001) reported a direct N transfer from nodulating soybean, lablab (*Lablab* sp.), mungbean (*Vigna radiata* L.) and black lentil (*Vigna mungo* L.) to the intercropped maize of the order of 27, 26, 21 and 18 kg N ha^{-1}, respectively. Substantial direct N transfer from N-donor soybean to N-receiver maize via the hyphae of *Glomus mossae* took place only when soybean was supplied with mineral N, but not when relying on N_2 fixation (He et al., 2003). In turn, Johansen and Jensen (1996) observed that N moved from the N-donor berseem (*T. alexandrinum* L.) to the N-receiver maize through the hyphae of *G. intraradices* without entering the soil solution, when the N concentration difference in the two plants constituted the driving force.

Similarly, N transfer in a pasture to the neighbouring grasses or other non-fixers plants can occur by the direct interconnection of both crop roots via mycorrhizal.

The possible bi-directional net N transfer, either as organic and/or inorganic N in the mycorrhizosphere from N-donor and N-receiver plants can be important in intercropping systems for sustainability (Cheng and Baumgartner, 2004; He et al., 2009). Nevertheless, Johansen and Jensen (1996) reported a very low N transfer (3 g N kg^{-1} DW) from a non-legume (barley) to a legume (pea) via arbuscular mycorrhizal (AM) hyphae. For such small N transfer, it was statistically difficult to show that N transfer had effectively occurred.

2.2. Crop Rotation

Crop rotation is the growing of succeeding crops of different genus, species or variety than the previous crop on a piece of land in a planned succession. The planned rotation may vary from two or three years, or a longer period. Longer rotations including pastures are preferable for sustainability and soil conservation. The aim of a rotation is to balance nutrient demands, improve soil quality, reduce the build-up of pests, diseases and weeds, and increase net profits. By this practice, yields can increase by 15-20% when compared to monoculture.

In the rotation, leguminous crops like pulses, green beans, or fodder crops can be used in-between the seasons of cereal or other cash crops [sugarbeet (*Beta vulgaris* L.), vegetables]. In Africa, the introduction of legumes such as chickpea, fababean (*Vicia faba* L.) and field bean in a wheat- or maize-based cropping rotation was a viable strategy for the reduction of inorganic fertilizer use (Danga et al., 2009; Li et al., in press).

In this cropping system is important to evaluate the impact of legumes on the performance of subsequent non-legume, in order to establish the most adequate crop sequence. It is not sufficient that a legume in a rotation provides an increase in soil N; this N must be available to the subsequent non-leguminous crop during the phases of higher N requirements. Since crop N is partitioned either into seed, vegetative parts or roots at crop maturity, not all of the N_2 fixed by the crop is available for return to the soil. The final contribution of fixed N_2 to the soil following harvest will depend upon the N balance at harvest (credit N). The N balance (eq.1) is determined by the difference between the amount of N2 fixed in total plant and removed N in the seed or feeding plant material:

N balance (kg ha^{-1}) = (N_2 fixed in total plant, kg ha^{-1}) – (seed N or N in feed material, kg ha^{-1}(eq.1)

If N balance is positive, there is a positive addition of N to the soil (N benefit or N input), otherwise there is a negative N credit (N deficit). Final N balance can range from as little as -132 kg ha^{-1} in non-nodulated soybeans to as much +135 N kg ha^{-1} in lupine (*Lupinus* sp.) (Peoples et al., 1995). Some legumes take soil N in addition with N_2 fixed and their contribution for soil fertility is not as high as those which fixation is around 100%. In a Mediterranean area, under rainfed conditions, Carranca et al. (1999, 2008, 2009a) estimated an N benefit by mature fababean and chickpea of the order of +19 and +5 kg N ha^{-1}, respectively, and +30 and +31 kg N ha^{-1} by white and yellow lupine (*Lupinus albus* L. and *L. luteus* L.), respectively. At pod-filling, yellow lupine showed a 100% fixation rate, with a credit N of +103 kg N ha-1. In Canada, Goss et al. (2002) measured an N input of +25 to +36 kg N ha^{-1} by soybean rooting system.

Peas can fix 20-70% N (Jensen, 1988; Carranca et al., 1999). The lowest fixation rate was obtained under drought stress and produced a negative N balance of -22 kg N ha^{-1} (Carranca et al., 1999), showing that this legume took much N from the soil which was allocated to the grain. According to Jensen (1988), this pulse takes up fertilizer N as efficiently as spring barley (Table 2). The short growth cycle of peas, particularly as spring crop in temperate regions fits well into rotations with cereals and other field crops, but the negative N balance after the legume harvest should be taken into account when deciding for the best crop sequence.

Table 2. Nitrogen balance for pea (*Pisum sativum* L.). Only N in aboveground plant part was considered. Data were from crops receiving no fertilizer N (Source: Jensen, 1988)

Parameter	Experiment				
	1980	1981	1982	1984	Mean
Crop N (kg N ha^{-1})	259	240	323	309	283
N in straw (kg N ha^{-1})	79	70	51	63	66
N in seeds (kg N ha^{-1})	180	170	272	246	217
Total N$_2$ fixed (kg N ha^{-1})	181	173	233	244	208
%N derived from atmosphere	70	72	72	79	73
N taken from soil (kg N ha^{-1})	78	67	90	65	75
Credit N[a] (kg N ha^{-1})	+1	+3	-39	-2	-9

[a]Credit N (kg N ha^{-1}) = Total N$_2$ fixed (kg N ha^{-1}) - N in seeds (kg N ha^{-1}).

Table 3. Dry matter (DM) yield (t ha^{-1}), N concentration (g kg^{-1} DM), C:N ratio, N at harvest (kg ha^{-1}), and insoluble and soluble lignin in pea (*Pisum sativum* L.) and white lupine (*Lupinus. albus* L.) mature residues. Means are from four experiments (1980-1983) for peas and two experiments (2002-2004) for lupine. (Sources: Jensen, 1988; Carranca et al., 2009c)

Plant part	Dry matter (t ha^{-1})	N (g kg^{-1} DM)	C:N	N (kg ha^{-1})	Insoluble lignin (g kg^{-1} DM)	Soluble lignin (g kg^{-1} DM)
Pea (*Pisum sativum* L.)						
Straw	3.1	17	24	52	-	-
Pod walls	1.1	12	33	13	-	-
Roots+visible nodules [a]	0.2	25	16	5	-	-
Total	4.4	18	25	70	-	-
White lupine (*Lupinus albus* L.)						
Straw	-	7	69	-	185	32
Pod walls	-	17	28	-	139	45
Roots+visible nodules [a]	-	7	67	-	252	29
Total	4.2	16	55	59	-	-

[a] = roots determined by simple excavation at harvest; -= not reported.

Nutrient release from legume residue plays an important role in crop productivity through the effect of timing on nutrient availability. When adequate soil water and temperature are available for microbial decomposition, a high quality legume residue decomposes quickly and immediately begins releasing N. The amount of N that is released from a productive legume can be substantial (depending on fixation rate and N harvest index), but the rate and duration of N release depends on environmental conditions, and the amount and quality of plant material (Table 3).

Generally, tillage hastens the breakdown of residues and raises the level of available N. Burying residues is important to reduce the ammonia (NH_3) loss by volatilization. Up to 14% of the added soil N in lentil (*Lens culinaris* L.) stove was lost within fourteen days by NH_3 volatilization (Giller and Cadisch, 1995). Residue conservation on the soil surface by soil conservation practice promotes the N immobilization (Jensen and Hauggaard-Nielsen, 2003; Carranca et al., 2009b,c). Depending on the situation, the choice between tillage and no-tillage should be well evaluated.

Litterbag studies suggested that 50-80% of the added legume residue-N was released during the first subsequent crop cycle under temperate conditions, and 70-95% was released under tropical-type climate (Giller and Cadisch, 1995; Carranca et al., 2009b,c). Recovery of N from legume residue in the second and further subsequent crops was very poor, and only 1-2% was recovered by the third crop (Jensen, 1994; Giller and Cadisch, 1995).

Five days after the addition of mature white lupine residue (4 t ha^{-1} of above- and belowground) to the soil in summer (June), Carranca et al. (2009b,c) found that little N was immobilized in the soil (0.3-2 kg N ha^{-1}), corresponding to less than 3% of added residue-N. They also observed that 86% of residue-N was nitrified after six months, producing 30 kg NO_3^--N ha^{-1}. The pattern of decomposition of lupine mature residue was represented by a two-step process: a first phase describing a relatively fast (exponential) decomposition rate of a labile (active) pool, followed by a much slower and longer evolution of a recalcitrant material. The decay constant for the lupine amended soil was high, i.e. $k=0.04$ (d^{-1}) at 18-19 °C, and $k=0.08$ (d^{-1}) for a longer period of decomposition at 25-26 °C (Carranca et al., 2009c).

The presence of factors such as lignin and polyphenol in the residue composition can reduce the rate of litter decomposition. The mixing of residues of different quality is one approach to managing rates of decomposition and N release which can help to optimize the efficient use of fixed N_2 (Gilles and Cadisch, 1995).

A release curve of residue-N by vetch (*Vicia villosa* Roth) fitted well the growth pattern for some vegetables such as sweet corn and watermelon (*Citrullus lanatus* cv. Mardigrass). These crops had their higher demand for N at the time when the residue was releasing maximum amounts of N (Segura, 2006). Carranca et al. (2009b) found that fodder oat (*Avena sativa* L.) following the white lupine in a single rotation produced 11 t DM ha^{-1} without mineral N fertilization. This yield was above the current oat production (9 t DM ha^{-1}) in Portuguese soils fertilized with 80 kg mineral N ha^{-1}.

Recent studies have shown that the belowground N contribution of legumes can be greater than previously thought (>10% of plant N). In case of pulses it can account for 22-68% of plant N, as supported by Yasmin et al. (2006), Mathieu et al. (2007) and McNeil and Fillery (2008). Also Khan et al. (2000) and Unkovich and Pate (2000) reported that roots of fababean, chickpea and lupine represented 26-43%, 40-68% and 28-40% of total plant N, respectively. At the end of a legume–oat rotation, De Varennes et al. (2007) found that 8 and 44 kg N ha^{-1} in fodder oat were derived from the white lupine roots in undisturbed or disturbed soils, respectively.

Nitrogen fixing crops are therefore an integral part of strategies to maintain high levels of soil organic N in temperate, Mediterranean and tropical agricultural systems. The choice of appropriate rotations is central for the optimization of N use efficiency by succeeding crops.

Conclusion

Improved N management is needed to optimize economic returns to farmers and minimize environmental concerns associated with the N use. Symbiotically N_2 fixed is of particular significance in sustainable agriculture as it allows reducing the use of chemical N in the production of field crops. Intercropping and crop rotation including legumes have advantages of risk minimization, reduction of soil erosion, increase food security and soil conservation, and should be practiced all over the world, especially in Africa. Considering the multiple advantages that can occur from intercropping for a sustainable agriculture, and given the possible numerous intercrop combinations and the different environmental conditions involved in each particular case, generalization to agronomic recommendations may not be feasible. It seems reasonable to continue research on the possibilities of growing more than one crop in a field at the same time.

Countries must replace mineral N fertilizers by planting legumes either as intercropping or in rotation and legumes must be generalized in human and animal diets.

References

Cadish, G.; Schunk, R.M. and Giller, K.E. (1994). Nitrogen cycling in a pure grass pasture and a grass-legume misture on a red latosol in Brazil. *Tropical Grasslands* 28:43-52.

Carlsson, G.; Palmorg, C.; Jumpponen, A.; Scherer-Lorenzen, M.; Högberg, P. and Huss-Danell, K. (2009). N_2 fixation in three perennial Trifolium species in experimental grasslands of varied plant species richness and composition. *Plant Ecology* 205:87-104.

Carranca, C.; De Varennes, A. and Rolston, D.E. (1999). Biological nitrogen fixation by fababean, pea, and chickpea under field conditions estimated by the ^{15}N isotope dilution technique. *European Journal of Agronomy* 10:49-56.

Carranca, C.; Madeira, M.; Torres, M.O.; Pina, J.P. and Marques, P. (2008). Variação sazonal da fixação simbiótica em duas espécies de Lupinus sujeitas a diferentes práticas de preparação do solo. Resumo das Comunicações do 1 Congresso Luso-Espanhol de Fixação de Azoto. *Estoril*: 14. (1-4 de Junho).

Carranca, C.; Torres, M.O. and Baeta, J. (2009a). White lupine as a beneficial crop in Southern Europe. I - Potential for N mineralization in lupine amended soil and yield and N_2 fixation by white lupine. *European Journal of Agronomy* 31:183-189.

Carranca, C.; Oliveira, A.; Pampulha, M.E. and Torres, M.O. (2009b). Temporal dynamics of soil nitrogen, carbon and microbial activity in conservative and disturbed fields amended with mature white lupine and oat residues. *Geoderma* 151:50-59.

Carranca, C.; Rocha, I.; De Varennes, A.; Oliveira, A.; Pampulha, M.E. and Torres, M.O. (2009c). Effect of tillage and temperature on potential nitrogen mineralization and microbial activity and microbial numbers of lupine amended soil. *Agrochimica* LIII (3), May-June:183-195.

Cheng, X. and Baumgartner, K. (2004). Arbuscular mycorrhizal fungi-mediated nitrogen transferfrom vineyard cover crops to grapevines. *Biology and Fertility of Soils* 40:406–412.

Danga, B.O.; Ouma, J.P.; Wakindiki, I.I.C. and Bar-Tal, A. (2009). Legume-wheat rotation effects on residual soil mixture, nitrogen and wheat yield in tropical regions. *Advances in Agronomy* 101:315-349.

De Varennes, A.; Torres, M.O.; Queda, C.; Goss, M.J. and Carranca, C. (2007). Nitrogen conservation in soil and crop residues as affected by crop rotation and soil disturbance under Mediterranean conditions. *Biology and Fertility of Soils* 44:49-58.

Giller, K.E. and Cadisch, G. (1995). Future benefits from biological nitrogen fixation: An ecological approach to agriculture. *Plant and Soil* 174:
255-277.

Goss, M.; De Varennes, A.; Smith, P.S. and Ferguson, J.A. (2002). N_2 fixation by soybeans grown with different levels of mineral nitrogen, and the fertilizer replacement value for a following crop. *Canadian Journal of Soil Science* 82:139-145.

Haggar, R.J. (1988). Agronomic limitations to production of forage legumes. In: P. Plancquart and R. Haggar (Eds.), *Legumes in farming systems* (pp. 64-69). Kluwer Acad. Publish., Commission of the European Communities.

He, X.-H., Critchley, C. and Bledsoe, C. (2003). Nitrogen transfer within and between plants through common mycorrhizae networks (CMNs). *Critical Reviews in Plant Sciences* 22:531-567.

He, X.; Xu, M.; Qiu, G.Y. and Zhou, J. (2009). Use of ^{15}N stable isotope to quantify nitrogen transfer between mycorrhizal plants. *Journal of Plant Ecology* 2:107-118.

IPCC (Intergovernmental Pannel for Climatic Conditions), 2007. 4th Assessment Report.

Jensen, E.S. (1988). The role of pea cultivation in the nitrogen economy of soils and succeeding crops. In: P. Plancquart and R. Haggar (Eds.), *Legumes in farming systems* (pp. 3-15). Kluwer Acad. Publish., Commission of the European Communities.

Jensen, E.S. (1994). Dynamics of mature pea residue nitrogen turnover in unplanted soil under field conditions. *Soil Biology and Biochemistry* 26:455-464.

Jensen, E.S.and Haugggaard-Nielsen, H. (2003). How can increased use of biological N_2 fixation in agriculture benefit the environment? *Plant and Soil* 252:177-186.

Johansen, A. and Jensen, E.S. (1996). Transfer of N and P from intact or decomposing roots of pea to barley interconnected by an arbuscular mycorrhizal fungus. *Soil Biology and Biochemistry* 28:73-81.

Khan, D.F.; Chen, D.; Herridge, D.F.; Schwenke, G.D. and Peoples, M.B. (2000). Contribution of below-ground legume N to crop rotations. *Proceedings of the International Symposium on Nuclear Techniques in Integrated Plant Nutrient Water and Soil Management*. Vienna, IAEA: 33-34.

Li, C.-J.; Li, Y.-Y.; Yu, C.-B.; Sun, J.-H.; Christie, P.; An, M.; Zhang, F.-S. and Li, L. Crop nitrogen use and soil mineral nitrogen accumulation under different crop combinations and patterns of strip intercropping in northwest China. *Plant and Soil* (doi: 10.1007/s11104-010-0686-6).

Lithourgidis, A.S.; Dordas, C.A.; Damalas, C.A. and Vlachostergios, D.N. (2011). Annual intercrops: An alternative pathway for sustainable agriculture. *Australian Journal of Crop Science* 5:396-410.

Marty, C.; Pornon, A.; Escaravage, N.; Winterton, P. and Lamaze, T. (2009). Complex interactions between a legume and two grasses in a subalpine meadow. *American Journal of Botany* 96:1814-1820.

Mathieu, S.; Fustec, J.; Faure, M.L.; Corre-Hellou, G. and Crozat, Y. (2007). Comparison of two ^{15}N labelling methods for assessing nitrogen rhizodeposition of pea. *Plant and Soil* 295:193-205.

McNeil, A.M. and Fillery, I.R.P. (2008). Field measurement of lupin belowground nitrogen accumulation and recovery in the subsequent cereal-soil system in a semi-arid Mediterranean-type climate. *Plant and Soil* 302:297–316.

Mengel, K. and Kirkby, E.A. (2001). *Principles of plant nutrition* (5th ed). Kluwer Academic Publishers, London.

Murungu, F.S.; Chiduza, C. and Muchaonyerwa, P. (2010). Biomass accumulation, weed dynamics and nitrogen uptake by winter cover crops in a warm-temperate region of South Africa. African *Journal of Agricultural Research* 5:1632-1642.

Ouma, G. and Jeruto, P. (2010). Sustainable horticultural crop production through intercropping: The case of fruits and vegetable crops: A review. *Agriculture and Biology Journal of North America.* (doi:10.5251/abjna.2010.1.5.1098.1105).

Pal, U.R. and Shehu, Y. (2001). Direct and residual contributions of symbiotic nitrogen fixation by legumes to the yield and nitrogen uptake of maize (*Zea mays* L.) in the Nigerian savannah. *Journal of Agronomy and Crop Science* 187:53-58.

Paynel, F.; Lesuffleur, F.; Bigot, J.; Diquélou, S. and Cliquet, J.B. (2008). A study of ^{15}N transfer between legumes and grasses. *Agronomy and Sustainable Development* 28:281-290.

Peoples, M.B.; Herridge, D.F. and Ladha, J.K. (1995). Biological nitrogen fixation: An efficient source of nitrogen for sustainable agricultural production? *Plant and Soil* 174:3-28.

Rochon, J.J.; Dayle, C.J.; Greef, J.M.; Hopkins, A.; Molle, G.; Sitzia, M.; Scholefield, D. and Smith, C.J. (2004). Grazing legumes in Europe: a review of their status, management, benefits and future prospects. *Grass and Forage Science* 59:197-214.

Segura, L.M.A. (2006). *Potential benefits of cover crop based systems for sustainable production of vegetables.* M. Sc. Thesis, University of Florida, USA.

Sequeira, E.M. (2008). Pasture and fodder crop as part of high natural value farm systems at Mediterranean dryland agroecosystems. *Option Méditerranéennes,* Serie A (79):1-22.

Thomas, R.J. (1995). Role of legumes in providing N for sustainable tropical pasture systems. *Plant and Soil* 174:103-118.

Trannin, W.S.; Urquiaga, S.; Guerra, G.; Ibijbijen, J. and Cadish, G. (2000). Interspecies competition and N transfer in a tropical grass-legume mixture. *Biology and Fertility of Soils* 32:441-448.

Unkovich, M.J. and Pate, J.S. (2000). An appraisal of recent field measurements of symbiotic N_2 fixation by annual legumes. *Field Crop Research* 65:211-228.

Yasmin, K.; Cadish, G. and Baggs, E.M. (2006). Comparing ^{15}N-labelling techniques for enriching above- and below-ground components of the plant-soil system. *Soil Biology and Biochemistry* 38:397-400.

Yu, C.-B.; Li, Y.-Y.; Li, C.-J.; Sun, J.-H.; He, H.; Zhang, F.-S. and Li, L. (2010). An improved nitrogen difference method for estimating biological nitrogen fixation in legume-based intercropping systems. *Biology and Fertility of Soils* 46:227-235.

In: Symbiosis: Evolution, Biology and Ecological Effects   ISBN: 978-1-62257-211-3
Editors: A. F. Camiso and C. C. Pedroso   © 2013 Nova Science Publishers, Inc.

Chapter 9

MODELS OF SYMBIOTIC ASSOCIATIONS IN FOOD CHAINS

Elena Caccherano[1], Samrat Chatterjee[2], Lucia Costa Giani[1],
Luigi Il Grande[1], Tiziana Romano[1], Giuseppe Visconti[1]
and Ezio Venturino[1]
[1] Dipartimento di Matematica "Giuseppe Peano",
Università di Torino, Torino, Italy
[2] Immunology Group, International Centre for Genetic
Engineering and Biotechnology, New Delhi, India

Abstract

In this study we investigate various ecosystems containing always a symbiotic pair. In the first model, the latter is subject to the action of a predator. In the second system, the mutualistic association at the trophic level is located above a common prey population. In the third model a three level food chain is investigated, with the symbiotic populations constituting the intermediate trophic level. These are situations that can well arise in nature.

Finally, we make considerations on a symbiotic system in which a population X predates on another population Z; a third population Y lives in symbiosis with X; finally Z predates on Y. Such an association is rather difficult to find in nature, but an interpretation of it could come from the world of finance, in which biological "populations" are substituted by corporations. At times they interact among each other in search for synergies, but at the same time it may occur that part of these associations compete with each other in other fields, giving rise to rather intricated situations as the one described above.

Keywords: predator-prey, trophic level, obligated mutualism, facultative mutualism

1. Introduction

This Chapter presents a mathematical study of simple symbiotic associations, which could be facultative or obligated, embedded in larger webs. Specifically, we want to consider symbiotic systems which, in more complex food chains, are prey for a top predator, or predate on a population at a lower trophic level, or lie at an intermediate level, below a predator population and at the same time feeding on a common prey; finally we consider a mixed symbiotic-competing type of model.

We do not consider specific biological examples, as the aim of the paper is rather of theoretical nature, and as it is usual for mathematical models, its applicability is wide. Without substantial changes, and with suitable interpretations, it could indeed be applied for any aquatic or terrestrial system.

The Chapter is organized as follows. In its main four parts, we consider a symbiotic system which appears in a food chain at different levels. In Section 1, the two symbiotic populations are both prey of a common predator. A variant of this situation consists in assuming that the symbiotic system is instead poisonous for its predators. This alternative is considered briefly in Section 3. In Section 2 instead the mutualistic association represents the predators of a prey at a lower trophic level. Section 3 contains a model in which the symbiotic system lies at an intermediate level, with predators on top and prey on which the symbiotic populations feed at the bottom. The final Section 4 presents a more complicated system, for which a biological interpretation is rather difficult but that can have an application for instance in the economics domain, as outlined below.

In the situation envisioned in Section 4, a symbiotic association has different types of interactions with a third population. The latter indeed is a prey for one of the symbiotic populations, while it predates on the other one. For it, we however provide an interpretation not of biological type, as that is kind of hard to find, but in the economic realm. In the financial world, in the stock market, it is not uncommon that companies buy stocks of other enterprises. In this way they may influence the latter's activities. So in principle it may happen that two societies that are allied in some way, could instead "compete" over a third one, in the sense that they may have opposite goals on it. Or, conversely, that the third enterprise might try to outcompete one of the two associated ones, while being subject to the same action on the part of the other one. Our model would give an outlook at the final result of such a situation.

Thoroughout the chapter, all parameters are assumed to be nonnegative.

2.   A Predator-Symbiotic Prey System

We initially consider an ecosystem in which three populations thrive, where a top predator hunts two populations that live in symbiosis. We will examine a few variants of this situation, considering both facultative as well as obligated mutualism.

2.1.   Facultative Mutualism

In the first model we suppose that the symbiotic populations S_1 and S_2 have an optional interaction, and are both hunted by the predators P. For the latter we assume that other food sources are available, thereby explaining the presence of the logistic term in the first equation. The remaining two terms express the different rewards predators obtain by hunting the two symbiotic populations. The symbiosis is facultative since in absence of one of the two populations and of predators, the other one would settle at a stable equilibrium, provided by the environment carrying capacity. Thus, once again, this fact is expressed by suitable logistic terms in the last two equations. In addition they contain the positive interaction due to mutualism and the negative effects caused by predators' hunting. The model's governing equations therefore are

$$
\begin{aligned}
\dot{P} &= r\left(1 - \frac{P}{H}\right)P + e_2 b_1 S_1 P + e_3 b_2 S_2 P, \\
\dot{S_1} &= r_1\left(1 - \frac{S_1}{K_1}\right)S_1 + aS_1 S_2 - b_1 S_1 P, \\
\dot{S_2} &= r_2\left(1 - \frac{S_2}{K_2}\right)S_2 + ae_1 S_1 S_2 - b_2 S_2 P,
\end{aligned}
\tag{1}
$$

where H, K_1, K_2 are the environment's carrying capacities for each population, r, r_1, r_2 are the reproductive rates, a represents the symbiotic interaction return rate, e_1 is a parameter that measures the relative benefit that population S_2 obtains from the symbiotic association, compared to the one of S_1. Finally b_1, b_2 are the predation rates and e_2, e_3 represent the conversion factors of the two types of prey into new predators.

2.1.1. Equilibria

To establish the system's dynamics, it is essential to determine the equilibria. The system (1) has all the eight possible equilibrium points arising by annihilating the right hand side of (1). We summarize them in the following Table:

	Population		
Equilibrium	P	S_1	S_2
E_1	0	0	0
E_2	0	0	K_2
E_3	0	K_1	0
E_4	0	+	+
E_5	H	0	0
E_6	+	0	+
E_7	+	+	0
E_8	$P^{(8)}$	$S_1^{(8)}$	$S_2^{(8)}$

Specifically, the population values at the boundary equilibria, i.e. those for which one or more populations vanish, are

$$
\begin{aligned}
S_1^{(4)} &= \frac{K_1 r_2(r_1 + aK_2)}{r_2 r_1 - a^2 e_1 K_1 K_2}, & S_2^{(4)} &= \frac{K_2 r_1(r_2 + ae_1 K_1)}{r_2 r_1 - a^2 e_1 K_1 K_2}, \\
P_6 &= \frac{H r_2(r + e_3 b_2 K_2)}{r_2 r + b_2^2 H K_2 e_3}, & S_2^{(6)} &= \frac{K_2 r(r_2 - b_2 H)}{r_2 r + b_2^2 H K_2 e_3}, \\
P_7 &= \frac{H r_1(r + e_2 b_1 K_1)}{r_1 r + b_1^2 H K_1 e_2}, & S_1^{(7)} &= \frac{K_1 r(r_1 - b_1 H)}{r_1 r + b_1^2 H K_1 e_2}.
\end{aligned}
$$

The system coexistence equilibrium E_8 has instead the following components

$$
S_1^{(8)} = \frac{1}{\Delta}\left[\left(e_3 b_2 - \frac{ar}{b_1 H}\right)\left(r_2 - \frac{b_2 r_1}{b_1}\right) - \left(\frac{rr_1}{b_1 H} - r\right)\left(\frac{b_2 a}{b_1} + \frac{r_2}{K_2}\right)\right],
\tag{2}
$$

$$S_2^{(8)} = \frac{1}{\Delta}\left[\left(\frac{rr_1}{b_1 H K_1} + e_2 b_1\right)\left(\frac{b_2 r_1}{b_1} - r_2\right) - \left(ae_1 - \frac{r_1 b_2}{b_1 K_1}\right)\left(\frac{rr_1}{b_1 H} - r\right)\right],$$

$$P^{(8)} = \frac{1}{b_1}\left[r_1\left(1 - \frac{S_1^{(8)}}{K_1}\right) + a S_2^{(8)}\right],$$

$$\Delta = \left(ae_1 + \frac{b_2 r_1}{b_1 K_1}\right)\left(\frac{ar}{b_1 H} - e_3 b_2\right) - \left(\frac{rr_1}{b_1 H K_1} + b_1 e_2\right)\left(\frac{b_2 a}{b_1} + \frac{r_2}{K_2}\right).$$

Further, we need to assess when these points are biologically feasible, which means that the populations levels do not attain negative values. In the following we thus summarize the restrictions imposed on the parameters to make the system's boundary equilibria feasible. While E_1, E_2, E_3 and E_5 are always feasible, respectively for E_4 we have

$$a^2 e_1 K_1 K_2 < r_1 r_2, \tag{3}$$

for E_6 instead

$$b_2 H \leq r_2, \tag{4}$$

and for E_7 we find

$$b_1 H \leq r_1. \tag{5}$$

The feasibilty conditions for the coexistence equilibrium E_8 are instead

$$r_1 K_1 + a K_1 S_2^{(8)} \geq r_1 S_1^{(8)} \tag{6}$$

together with

$$H\left[r_2 b_1(b_2 e_3 K_2 + r) + ab_2 r K_2\right] \geq rr_2(aK_2 + r_2) + H b_2^2 e_3 r_1 K_2, \tag{7}$$

$$2b_2 r_1^2 r + b_1^2 H K_1(b_2 e_2 r_1 + ae_1 r) \geq b_1(rr_1 r_2 + b_1^2 e_2 r_2 H K_1 + ae_1 rr_1 K_1 + b_2 rr_1 H),$$

$$a^2 b_1 e_1 r K_1 K_2 > b_1 rr_1 r_2 + b_1^3 e_2 r_2 H K_1 + ab_1^2 b_2 H K_1 K_2(e_2 + e_1 e_3) + b_1 b_2^2 e_3 r_1 H K_2,$$

or the converse of all the above inequalities.

2.2.2. Stability Analysis

To assess equilibria is not enough, since only those that are stable can ultimately be achieved by the system's trajectories. To study their stability, we need at first to compute the system's Jacobian matrix,

$$J = \begin{pmatrix} J_{11} & e_2 b_1 P & e_3 b_2 P \\ -b_1 S_1 & J_{22} & a S_1 \\ -b_2 S_2 & ae_1 S_2 & J_{33} \end{pmatrix} \tag{8}$$

where

$$J_{11} = -\frac{rP}{H} + r\left(1 - \frac{P}{H}\right) + e_2 b_1 S_1 + e_3 b_2 S_2,$$

$$J_{22} = -\frac{r_1}{K_1} S_1 + a S_2 - b_1 P + r_1\left(1 - \frac{S_1}{K_1}\right),$$

$$J_{33} = -\frac{r_2}{K_2} S_2 + ae_1 S_1 - b_2 P + r_2\left(1 - \frac{S_2}{K_2}\right).$$

Stability is achieved when all eigenvalues of J evaluated at the populations' equilibrium level are negative.

For the equilibrium point E_1, the Jacobian reduces to a diagonal matrix from which the eigenvalues are immediate. We find three positive eigenvalues, so that the origin is always unstable. This is a good result from the environmental point of view, since it indicates that the ecosystem never disappears. It is of course a consequence of the logistic assumptions we made for each population.

For the equilibrium point E_2, the Jacobian is a lower triangular matrix for which again the eigenvalues are easily read off from the diagonal values as $r + e_3 b_2 K_2 > 0$, $r_1 + a K_2 > 0$, $-r_2 < 0$. In view of the two positive eigenvalues, we conclude that E_2 is always unstable. Thus the second symbiotic population cannot survive alone in the ecosystem, in spite of the mutualism being facultative. This is an interesting, unexpected result, showing the power of mathematical modelling, that provides a quantitative tool, which is better than pure verbal arguments. With the latter indeed such a result would not have been easy to detect.

For the equilibrium point E_3, the Jacobian again provides easily and explicitly the eigenvalues: $r + e_2 b_1 K_1 > 0$, $-r_1 < 0$, $r_2 + a e_1 K_1 > 0$. Two eigenvalues are again positive so that E_3 is also always unstable. This is a symmetric result as for E_2, stating that also the other symbiotic population S_1 cannot survive alone. Note that both these results, however, are not particular of this system. They do hold in fact just in the symbiotic association of S_1 and S_2 without predators. Indeed, one of the eigenvalues of the subsystem obtained by deleting the first row and column of the Jacobian, i.e. excluding the predators, has still positive value. Thus for this result, predators are not instrumental, it is rather a consequence of the mutualism itself.

For the equilibrium point E_4, we obtain the Jacobian matrix $J_4 \equiv J(E_4)$. This gives a cubic characteristic equation $\Pi(\lambda) \equiv \sum_{k=0}^{3} A_k \lambda^k = 0$, with $A_3 = 1$, $A_0 = -\det(J_4)$. But

$$\det(J_4) = \frac{S_1 S_2}{K_1 K_2}(e_2 b_1 S_1 + e_3 b_2 S_2)(r_1 r_2 - a^2 e_1 K_1 K_2) > 0,$$

in view of the feasibility condition (3) for E_4. This ensures that $\Pi(\lambda)$ has always at least one positive root and this suffices to state that the equilibrium point E_4 is always unstable.

For the equilibrium point E_5, the Jacobian is an upper triangular matrix, so that the eigenvalues are found as $-r$, $r_1 - b_1 H$, $r_2 - b_2 H$. To achieve stability, we impose them to be negative. The stability condition is then

$$H > \max\left\{\frac{r_1}{b_1}, \frac{r_2}{b_2}\right\}. \tag{9}$$

Thus predators can survive in the environment only if their carrying capacity due to other food sources exceeds the ratio of prey reproduction rate and their mortality due to hunting by predators.

For the equilibrium points E_6 one eigenvalue of the Jacobian factors out, to give the stability condition

$$a S_2^{(6)} + r_1 < b_1 P_6. \tag{10}$$

The remaining monic quadratic characteristic equation $\sum_{k=0}^{2} A_k \lambda^k$, $A_2 = 1$ has positive coefficients, given by

$$A_1 = \frac{r}{H} P_6 + \frac{r_2}{K_2} S_2^{(6)} > 0, \quad A_0 = \frac{r}{H} P_6 \frac{r_2}{K_2} S_2^{(6)} + e_3 b_2^2 P_6 S_2^{(6)} > 0,$$

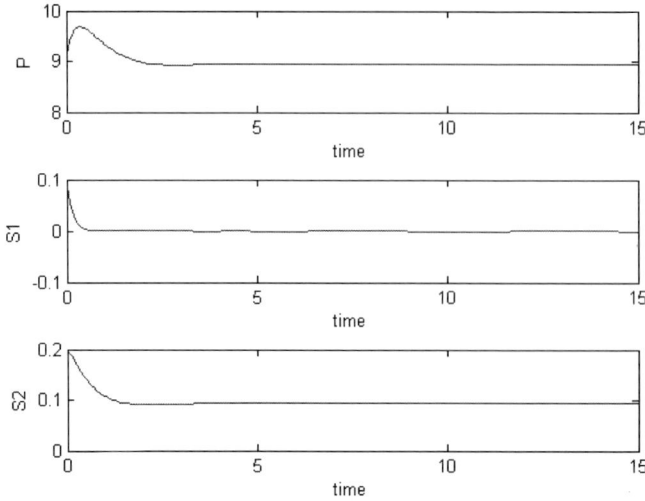

Figure 1. E_6 is stable for the following parameter values $r = 2, r_1 = 3, r_2 = 9, H = 8,$ $K_1 = 15, K_2 = 20, b_1 = 1, b_2 = 1, e_1 = 0.1, e_2 = 3, e_3 = 2.5, a = 0.02.$

so that its roots have always negative real parts. Stability is obtained just by satisfying (10).

At E_7 the analysis is quite similar, one eigenvalue gives

$$ae_1S_1^{(7)} + r_2 < b_2P_7, \tag{11}$$

and the coefficients of the remaining quadratic $\sum_{k=0}^{2} B_k\lambda^k$, $B_2 = 1$ satisfy the Routh-Hurwitz conditions,

$$B_1 = \frac{r}{H}P_7 + \frac{r_1}{K_1}S_1^{(7)} > 0, \quad B_0 = \frac{r}{H}P_7\frac{r_1}{K_1}S_1^{(7)} + e_2b_1^2P_7S_1^{(7)} > 0,$$

so that stability is given just by (11).

We performed also numerical simulations, see Figures 1-2, to verify these results. They are obtained respectively for the following sets of parameter values, $r = 2, r_1 = 3, r_2 = 9,$ $H = 8, K_1 = 15, K_2 = 20, b_1 = 1, b_2 = 1, e_1 = 0.1, e_2 = 3, e_3 = 2.5, a = 0.02$ for E_6 and $r = 2, r_1 = 9, r_2 = 1.5, H = 8, K_1 = 15, K_2 = 20, b_1 = 1, b_2 = 0.45, e_1 = 0.1,$ $e_2 = 3, e_3 = 2.5, a = 0.02$ for E_7.

The system coexistence equilibrium E_8 proves to be harder for a complete stability analysis, but we give some results. The Jacobian $J^{(8)}$ at this point gets a bit simplified, as the diagonal terms, in view of the equations defining the equilibrium, become

$$J_{11}^{(8)} = -\frac{r}{H}P_8, \quad J_{22}^{(8)} = -\frac{r_1}{K_1}S_1^{(8)}, \quad J_{33}^{(8)} = -\frac{r_2}{K_2}S_2^{(8)}.$$

Thus $-\text{tr}(J^{(8)}) > 0$ is trivially satisfied, but the remaining two Routh-Hurwitz conditions need to be verified, which can be restated as

$$\Gamma^{(8)} > 0, \quad \text{tr}(J^{(8)})M^{(8)} < \Gamma^{(8)}P_8S_1^{(8)}S_2^{(8)}, \tag{12}$$

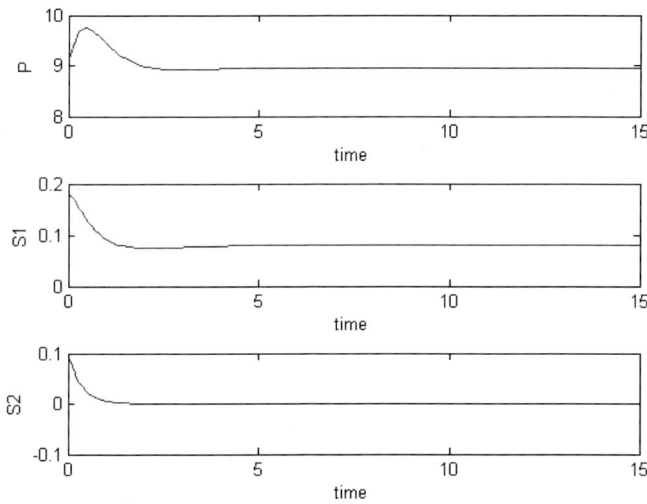

Figure 2. E_7 is stable for the following parameter values $r = 2$, $r_1 = 9$, $r_2 = 1.5$, $H = 8$, $K_1 = 15$, $K_2 = 20$, $b_1 = 1$, $b_2 = 0.45$, $e_1 = 0.1$, $e_2 = 3$, $e_3 = 2.5$, $a = 0.02$.

having set

$$M^{(8)} \equiv \left[\frac{rr_1}{HK_1} + b_1^2 e_2\right] P_8 S_1^{(8)} + \left[\frac{rr_2}{HK_2} + b_2^2 e_3\right] P_8 S_2^{(8)} + \left[\frac{r_1 r_2}{K_1 K_2} - a^2 e_1\right] S_1^{(8)} S_2^{(8)},$$

$$\Gamma^{(8)} \equiv \frac{r}{H}\left(\frac{r_1 r_2}{K_1 K_2} - a^2 e_1\right) + a b_1 b_2 (e_2 + e_1 e_3) + b_2^2 e_3 \frac{r_1}{K_1} + b_1^2 e_2 \frac{r_2}{K_2}.$$

While the first stability condition (12) is easily written in extensive form from the above definition, the second one becomes

$$B + S_1^{(8)} S_2^{(8)} \left[P_8 \left(2\frac{rr_1 r_2}{HK_1 K_2} - a b_1 b_2 (e_2 + e_1 e_3)\right)\right.$$
$$\left. + \left(\frac{r_1 r_2}{K_1 K_2} - a^2 e_1\right)\left(\frac{r_1}{K_1} S_1^{(8)} + \frac{r_2}{K_2} S_2^{(8)}\right)\right] > 0,$$

with

$$B \equiv P_8^2 \frac{r}{H}\left[\left(\frac{rr_1}{HK_1} + b_1^2 e_2\right) S_1^{(8)} + \left(\frac{rr_2}{HK_2} + b_2^2 e_3\right) S_2^{(8)}\right]$$
$$+ P_8 \left[\frac{r_1}{K_1}\left(\frac{rr_1}{HK_1} + b_1^2 e_2\right)\left(S_1^{(8)}\right)^2 + \frac{r_2}{K_2}\left(\frac{rr_2}{HK_2} + b_2^2 e_3\right)\left(S_2^{(8)}\right)^2\right] > 0.$$

In Figure 3 we show a simulation in which this equilibrium is achieved, for the parameter values $r = 2$, $r_1 = 3$, $H = 8$, $K_1 = 15$, $K_2 = 20$, $b = 0.09$, $e = 3$, $r_2 = 3.5$, $e_1 = 1$, $e_2 = 3$, $e_3 = 2.5$, $b_1 = 0.09$, $b_2 = 0.12$.

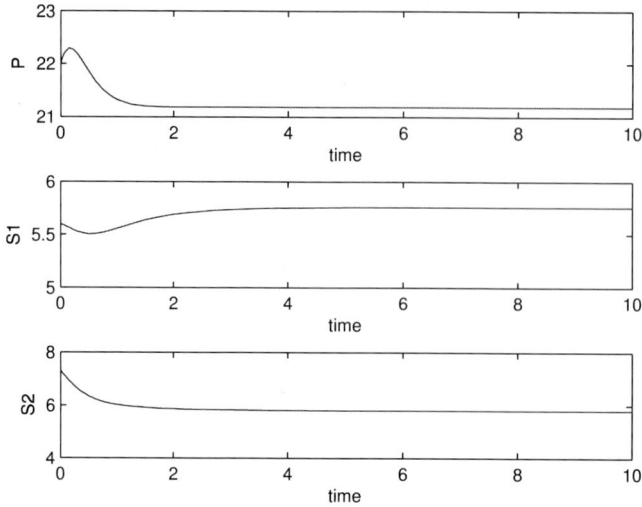

Figure 3. E_8 is stable for the following parameter values $r = 2$, $r_1 = 3$, $H = 8$, $K_1 = 15$, $K_2 = 20$, $b = 0.09$, $e = 3$, $r_2 = 3.5$, $e_1 = 1$, $e_2 = 3$, $e_3 = 2.5$, $b_1 = 0.09$, $b_2 = 0.12$.

The following Table summarizes our findings for the boundary equilibria

Equilibrium	Stability Condition
E_1	Unstable
E_2	Unstable
E_3	Unstable
E_4	Unstable
E_5	(9)
E_6	(10)
E_7	(11)

Note that the stability condition (9) for E_5 is the opposite of the feasibility conditions (4) and (5) for E_6 and E_7, which means that whatever of these two equilibria is feasible, for a larger H, when H increases beyond the critical value ensuring feasibility, then stability of E_5 takes over, i.e. a transcritical bifurcation occurs.

2.3.3.   Discussion

In summary, in view of the above analysis, we can state that the system can evolve only toward four types of situations. In the first one, the predators-only equilibrium, the predators survive due to the other available food sources. Alternatively, the system tends toward situations in which predators thrive together with either one or both the symbiotic species. The disappearance of the symbiotic system in the former case represents a loss from the environmental point of view, but if the mutualistic association is a nuisance, this is a way of eradicating it that can be implemented by making the predators carrying capacity large enough. In other words, if we give the predators large amounts of alternative food, they will increase their population size and thus ultimately wipe out the symbiotic system. This result is kind of unexpected, but it is not uncommon and has been discovered also in other circumstances. See for instance [24] where a phenomenon like this one can be ascribed to the disease affecting the prey. In a sense it is the reverse phenomenon of the paradox of enrichment, [34], with which it should be compared. The latter states that by increasing the food available to the prey the predator's population gets destabilized. The second conclusion is that the predators, under these assumptions, cannot be wiped out. The reason is imbedded in the system's assumptions, in the fact that they can find alternative ways of feeding themselves. Thus they cannot be harmed by a symbiotic system that lies at a lower trophic level. This issue, however, will be further investigated in Section 3 in which we will instead assume a toxic effect by the mutualistic association over the predators.

2.2.   A Particular Case for Symbiosis

We now study a particular case of the model (1), since in this case some information on the most complicated equilibria can be obtained.

Here we assume that the two prey populations have the same reproduction rate i.e., $r_1 = r_2$ and the predator obtains the same benefit from both prey populations i.e., $e \equiv e_2 = e_3$. Moreover P^s predates S_1^s, and S_2^s at the same rate $b \equiv b_1 = b_2$. We assume that $e_1 = 1$, i.e., S_1^s and S_2^s obtain the same benefit by their mutual interactions.

In these conditions, all the previous equilibria of model (1) are found once again. In fact we find the same population values at equilibria and the same feasibility conditions, but for the simplifications. However, as mentioned, the simplifications we made allow us to better discuss the most complicated ones. For the stability, indeed, it is possible to say something more. While equilibria $E_1^s \equiv E_1$, $E_2^s \equiv E_2$, $E_3^s \equiv E_3$ and $E_4^s \equiv E_4$ retain their instability, and $E_5^s \equiv E_5$ its stability conditions, which now however reduce to just

$$H > \frac{r_1}{b}, \tag{13}$$

we can now study E_6^s.

In fact the determinant of $J_6^s = J(E_6^s)$ is given by

$$\det(J_6^s) = \frac{r^2 r_1}{(r_1 r + b^2 H K_2 e)^2}(bH - r_1)^2(r_1 + aK_2)(r + beK_2)$$

which is seen to be always positive. Thus the cubic characteristic polynomial $\Pi_6(\lambda) = \sum_{k=0}^3 a_k \lambda^k$, with $a_3 = 1$ and $a_0 = -\det(J_6^s)$ has a negative value at the origin so that there always exists a positive eigenvalue. Hence E_6^s is always unstable.

Further, an eigenvalue of $J_7^s = J(E_7^s)$ is immediately given by

$$\frac{r(r_1 + aK_1)(r_1 - bH)}{r_1 r + b^2 H K_1 e},$$

which, from the feasibility condition (5), is easily seen to be always positive. Thus also the equilibrium point E_7^s is unconditionally unstable.

For the interior steady state E_8^s, we can now obtain explicit values of the equilibrium populations

$$\widehat{P^s} = \frac{H(r_1^2 r - a^2 K_2 r K_1 + K_1 e b r_1^2 + 2K_1 e b a K_2 r_1 + e b K_2 r_1^2)}{r_1^2 r + r_1 b^2 H K_1 e + b^2 H K_2 e r_1 - a^2 K_2 r K_1 + 2a K_2 K_1 b^2 H e} \equiv H \frac{N_P}{D},$$

$$\widehat{S_1^s} = \frac{K_1 r (r_1^2 - r_1 b H + a K_2 r_1 - b H K_2 a)}{r_1^2 r + r_1 b^2 H K_1 e + b^2 H K_2 e r_1 - a^2 K_2 r K_1 + 2a K_2 K_1 b^2 H e} \equiv K_1 r \frac{N_{S1}}{D},$$

$$\widehat{S_2^s} = \frac{K_2 r (r_1^2 + r_1 a K_1 - r_1 b H - H b a K_1)}{r_1^2 r + r_1 b^2 H K_1 e + b^2 H K_2 e r_1 - a^2 K_2 r K_1 + 2a K_2 K_1 b^2 H e} \equiv K_2 r \frac{N_{S2}}{D},$$

where

$$N_{S1} = (r_1 + a K_2)(r_1 - b H), \quad N_{S2} = (r_1 + a K_1)(r_1 - b H).$$

For feasibility we need $N_{S1}, N_{S2}, D, N_P = r(r_1^2 - a^2 K_1 K_2) + r_1 L$, where $L \equiv K_1(r_1 + a K_2) + K_2(r_1 + a K_1)$, all of the same sign, i.e., with a bit of rearrangement, either

$$r_1 \geq b H, \quad \min\left\{ r r_1^2 + b H L, r_1(r r_1 + L) \right\} \geq a^2 r K_1 K_2,$$

or the opposite inequalities.

Sufficient conditions for the feasibility of E_8^s are given by

$$K_1 K_2 < \frac{r_1^2}{a^2}, \quad H < \frac{r_1}{b}. \tag{14}$$

In this case, equilibria E_6^s and E_7^s are also feasible, but as mentioned always unstable. E_5^s is also feasible but always unstable, (13), as is E_4^s.

In these conditions, namely when (14) hold, we can analyse stability. The characteristic equation of the Jacobian at E_8^s is

$$\lambda^3 + A_1 \lambda^2 + A_2 \lambda + A_3 = 0,$$

where

$$A_1 = \frac{rP}{H} + \frac{r_1 S_1}{K_1} + \frac{r_1 S_2}{K_2} > 0,$$

$$A_2 = \frac{r r_1 P S_1}{H K_1} + b^2 e P S_1 + \frac{r r_1 P S_2}{H K_2} + b^2 e P S_2 + \frac{r_1^2 S_1 S_2}{K_1 K_2} - a^2 e S_1 S_2 > 0,$$

$$A_3 = P S_1 S_2 \left[\frac{r}{H}\left(\frac{r_1^2}{K_1 K_2} - a^2\right) + 2 a b^2 e + b^2 e r_1 \left(\frac{1}{K_1} + \frac{1}{K_2}\right) \right],$$

in which all the population values must be evaluated at E_8^s. We do not emphasize this anymore below. Using condition (14), we easily see that also A_3 is positive. Hence the first two Routh-Hurwitz conditions for stability are satisfied. Stability is thus regulated by the remaining condition, namely

$$A_1 A_2 - A_3 > 0, \tag{15}$$

which leads to the rather complicated explicit expression

$$A_1 A_2 - A_3 \equiv A + 2 P S_1 S_2 \left(\frac{r r_1^2}{H K_1 K_2} - a b^2 e\right) + S_1 S_2 r_1 \left(\frac{S_2}{K_2} + \frac{S_1}{K_1}\right)\left(\frac{r_1^2}{K_1 K_2} - a^2\right)$$

with

$$A = P^2 S_1 \left(\frac{b^2 er}{H} + \frac{r^2 r_1}{H^2 K_1} \right) + P^2 S_2 \left(\frac{b^2 er}{H} + \frac{r^2 r_1}{H^2 K_2} \right)$$

$$+ PS_2^2 \left(\frac{r_1^2 r}{HK_2^2} + \frac{eb^2 r_1}{K_2} \right) + PS_1^2 \left(\frac{r_1^2 r}{HK_1^2} + \frac{eb^2 r_1}{K_1} \right) > 0.$$

But in view of (14) and $0 < e < 1$, (15) holds. Thus if (14) are satisfied, the coexistence equilibrium is unconditionally stable, when feasible.

2.1.1. Discussion

In summary, the simplified case of (1) in consideration here either tends to the predators-only equilibrium E_5^s, i.e. the symbiotic system is wiped out, or toward the coexistence equilibrium. Note that indeed the conditions (13) and (14) are alternative and in fact all the other possible equilibria are unstable. The key parameter is here

$$\rho = \frac{r_1}{bH}.$$

When it is smaller than 1, predators wipe out the symbiotic subsystem, as the system attains the stable equilibrium E_5^s. When instead ρ exceeds 1, the former becomes unstable while instead E_8^s becomes feasible and since it is unconditionally stable, the system settles to coexistence of the three populations.

2.3. The Model with Obligated Mutualism

In this case, predators can still have alternative food sources, but each symbiotic species cannot survive in absence of the other one. Therefore the logistic terms in the symbiotic populations are replaced by Malthusian mortalities

$$\dot{P^O} = r \left(1 - \frac{P^O}{H} \right) P^O + e_2 b_1 S_1^O P^O + e_3 b_2 S_2^O P^O,$$

$$\dot{S_1^O} = -m_1 S_1^O + a S_1^O S_2^O - b_1 S_1^O P^O, \qquad (16)$$

$$\dot{S_2^O} = -m_2 S_2^O + a e_1 S_1^O S_2^O - b_2 S_2^O P^O.$$

The parameters retain their meaning as expressed in model (1), while the new parameters m_1 and m_2 denote here the mortality rate of each of the symbiotic population. For the stability analysis, we also report the Jacobian matrix for this case,

$$J^O = \begin{pmatrix} J_{11}^O & e_2 b_1 P^O & e_3 b_2 P^O \\ -b_1 S_1^O & -m_1 + a S_2^O - b_1 P^O & a S_1^O \\ -b_2 S_2^O & a e_1 S_2^O & -m_2 + a e_1 S_1^O - b_2 P^O \end{pmatrix}.$$

with

$$J_{11}^O = -r \frac{P}{H} + r \left(1 - \frac{P}{H} \right) + e_2 b_1 S_1^O + e_3 b_2 S_2^O.$$

The system (16) has the following feasible equilibria,

Equilibrium Point	P^O	S_1^O	S_2^O
E_1^O	0	0	0
E_2^O	0	$m_2(ae_1)^{-1}$	$m_1 a^{-1}$
E_5^O	H	0	0
E_6^O	$\widehat{P^O}$	$\widehat{S_1^O}$	$\widehat{S_2^O}$

But E_1^O and E_2^O both have one positive eigenvalue, r and $r + b_1 e_2 S_1^O + b_2 e_3 S_2^O$ respectively, so that they are unstable. For E_5^O, the eigenvalues are $-r$, $-m_1 - b_1 H$ and $-m_2 - b_2 H$, so that it is always stable.

The coexistence equilibrium E_6^O, has the following population values

$$\widehat{P_6^O} = \frac{H(m_1 e_1 e_3 b_2 + ae_1 r + m_2 b_1 e_2)}{ae_1 r - e_1 b_1 H e_3 b_2 - b_2 H e_2 b_1},$$

$$\widehat{S_6^{(1)O}} = \frac{ram_2 + Hrab_2 + He_3 b_2^2 m_1 - b_1 H e_3 b_2 m_2}{a(ae_1 r - e_1 b_1 H e_3 b_2 - b_2 H e_2 b_1)},$$

$$\widehat{S_6^{(2)O}} = \frac{ae_1 r m_1 + ae_1 r b_1 H - b_2 H e_2 b_1 m_1 + m_2 b_1^2 H e_2}{a(ae_1 r - e_1 b_1 H e_3 b_2 - b_2 H e_2 b_1)}.$$

It is feasible for

$$r > \frac{(e_1 e_3 + e_2)b_2 b_1 H}{ae_1}. \tag{17}$$

To analyze its stability, we note that the Jacobian $J_6^O \equiv J(E_6^O)$ has the determinant

$$\frac{a}{H} S_1^O S_2^O P^O \left[ae_1 r - (e_1 e_3 + e_2)b_2 b_1 H \right],$$

which in view of (17) is positive, so that E_6^O is always unstable, as $\det(J_6^O) > 0$ ensures the existence of one real positive eigenvalue. The following table summarizes our findings.

Equilibrium Point	Feasibility	Stability
E_1	—	unstable
E_2	—	unstable
E_5	—	always stable
E_6	(17)	unstable

2.1.1. Discussion

Therefore in this case, the system settles always to E_5, i.e. the symbiotic populations are always wiped out. An obligated commensalism is destroyed then by a predator feeding on both, or either one, of the symbiotic species, provided that the latter has other resources for its own foraging.

3. Poisonous Effects of the Symbiosis

3.1. Toxic Facultative Mutualism

We consider here now another special case, namely when the symbiotic system is toxic for its predators. We investigate whether predators can be harmed and possibly wiped out by a symbiotic system that is poisonous.

Replacing e_2 and e_3 by $-e_2$ and $-e_3$ in (1) the first three equilibria do not change their population levels and remain unstable. The fourth one however, E_4^X, retains its population values and feasibility condition (3) as for E_4, but becomes stable if the following condition holds

$$r < b_1 e_2 1 S_1^{X(4)} + b_2 e_3 S_2^{X(4)}. \tag{18}$$

In this case thus, the predators are wiped out while the symbiotic system still thrives.

For E_5^X we find again the same population values as for (1) and the stability condition (9) still holds unchanged.

For E_6^X and E_7^X we observe that some additional constraints appear in their feasibility and stability conditions. In fact for E_6^X we find the following two alternative sets for feasibility:

$$r_2 r > b_2^2 H K_2 e_3, \quad r > e_3 b_2 K_2, \quad r_2 > b_2 H; \tag{19}$$
$$r_2 r < b_2^2 H K_2 e_3, \quad r < e_3 b_2 K_2, \quad r_2 < b_2 H, \tag{20}$$

while similarly for E_7^X we have

$$r_1 r > b_1^2 H K_1 e_2, \quad r > e_2 b_1 K_1, \quad r_1 > b_1 H; \tag{21}$$
$$r_1 r < b_2^2 H K_2 e_3, \quad r < e_2 b_1 K_1, \quad r_1 < b_1 H. \tag{22}$$

For E_8^X, the feasibility condition (6) is retained, while the other ones obviously become

$$H\left[r_2 b_1 r + b_2 K_2(ar + b_2 e_3 r_1)\right] \geq rr_2(aK_2 + r_2) + Hr_2 b_1 b_2 e_3 K_2, \tag{23}$$
$$2b_2 r_1^2 r + b_1^2 H K_1(b_1 e_2 r_2 + ae_1 r) \geq b_1 r_1(b_1 H K_1 b_2 e_2 + rr_2 + ae_1 r K_1 + b_2 r H),$$
$$b_1^3 e_2 r_2 H K_1 + ab_1^2 b_2 H K_1 K_2(e_2 + e_1 e_3) + b_1 b_2^2 e_3 r_1 H K_2 > b_1 rr_1 r_2 + a^2 b_1 e_1 r K_1 K_2.$$

or the converse of all the above inequalities.

3.1.1. Discussion

The fundamental difference of this model from the one in which prey have nutritional value fo the predators, is that toxicity in fact can stabilize equilibrium E_4, thereby the mutualistic prey can eliminate their predators.

3.2. Poisonous Obligated Mutualism

We thus study also the case in which predators do not have other food sources.

In this case, let e_2 and e_3 in (16) again be replaced by $-e_2$ and $-e_3$ respectively. The feasible equilibria remain the same, with only a change in the population values at equilibrium, namely

$$P_6^{OX} = \frac{H(ae_1r - m_1e_1e_3b_2 - m_2b_1e_2)}{ae_1r + e_1b_1He_3b_2 + b_2He_2b_1},$$

$$S_6^{OX(1)} = \frac{ram_2 + Hrab_2 - He_3b_2^2m_1 + b_1He_3b_2m_2}{a(ae_1r + e_1b_1He_3b_2 + b_2He_2b_1)},$$

$$S_6^{OX(2)} = \frac{ae_1rm_1 + ae_1rb_1H + b_2He_2b_1m_1 - m_2b_1^2He_2}{a(ae_1r + e_1b_1He_3b_2 + b_2He_2b_1)},$$

and in this case the feasibility condition becomes

$$r > \frac{m_2b_1e_2 + m_1e_1e_3b_2}{ae_1}. \tag{24}$$

The stability analysis gives surprisingly the very same conclusions of the previous model. Indeed, there is no change in the eigenvalues at the origin, while E_2^{OX} is unstable in view of the eigenvalue $\sqrt{m_1m_2}$, E_6^{OX} has the determinant of the Jacobian positive, namely

$$S_6^{OX(1)}S_6^{OX(2)}P_6^{OX}a\left(b_1b_2e_2 + b_1b_2e_1e_3 + ae_1\frac{r}{H}\right) > 0,$$

making it unstable. E_5^{OX} is once again always stable, in view of the eigenvalues $-r$, $-m_1 - b_1P_6^{OX}$, $-m_2 - b_2P_6^{OX}$.

3.1.1. Discussion

Thus in spite of the fact that the symbiotic system is lethal for the predators, this fact does not prevent them to thrive and wipe out the mutualistic association when the latter is obligated.

Comparing this result with the one in Section 3.1 we recognize the importance of having other predators' food sources available for the survival of toxic obligated symbiotic associations in presence of predators.

4.   Symbiosis Among the Predators

Now we consider different models for symbiotic associations that act as predators on a prey at a lower trophic level. In the first model, the mutualism is facultative, since one of the symbiotic populations can thrive also in absence of the other one. This is expressed by the logistic model, stating that mutualistic predators have other feeding sources. In the second model the hunting activities of the symbiotic predators are expressed via a Leslie-Gower mechanism. Instead in the third model obligated mutualism is considered.

4.1. Facultative Mutualism with Logistic Growth

4.1.1. The Model

Let $S_1(t)$ and $S_2(t)$ be the predators and $Q(t)$ denote the prey population. The equations for the symbiotic populations contain now logistic growth, the term describing their mutual positive interactions, and the reward they have from predation. The prey population grows instead exponentially, and feels the negative impact of the predators' hunting. The model thus reads:

$$
\begin{aligned}
\dot{S}_1(t) &= r_1\left(1 - \frac{S_1(t)}{k_1}\right)S_1(t) + c_1 S_1(t)S_2(t) + a_1 Q(t)S_1(t) \\
\dot{S}_2(t) &= r_2\left(1 - \frac{S_2(t)}{k_2}\right)S_2(t) + c_2 S_1(t)S_2(t) + a_2 Q(t)S_2(t) \qquad (25) \\
\dot{Q}(t) &= Q(t)(r - b_1 S_1(t) - b_2 S_2(t)).
\end{aligned}
$$

Predators grow with net rates r_1 and r_2 respectively and carrying capacities k_1 and k_2, due to the availability of alternative food sources. Mutual benefits from their symbiotic interactions are given by c_1 and c_2; The parameters a_1 and a_2 express the conversion rates of captured prey into new predators. The prey grow following a Malthusian model with growth rate r. The parameters b_1 and b_2 denote the predators' hunting rates.

4.2.2. Equilibrium Points and Stability

The table below contains the possible equilibrium points of system (25),

Equilibrium Point	S_1	S_2	Q
E_1^T	0	0	0
E_2^T	0	k_2	0
E_3^T	0	$+$	$+$
E_4^T	k_1	0	0
E_5^T	$+$	0	$+$
E_6^T	$+$	$+$	0
E_7^T	$+$	$+$	$+$

where, letting $D_7 = r_2 a_1 b_1 k_1 + r_1 a_2 b_2 k_2 + k_1 k_2(a_1 b_2 c_2 + a_2 b_1 c_1)$,

$$
S_2^{T(3)} = \frac{r}{b_2}, \quad Q_3^T = \frac{rr_2 - r_2 b_2 k_2}{a_2 b_2 k_2}; \quad S_1^{T(5)} = \frac{r}{b_1}, \quad Q_5^T = \frac{rr_1 - r_1 b_1 k_1}{a_1 b_1 k_1};
$$

$$
S_1^{T(6)} = \frac{c_1 k_1}{r_1} S_2^{T(6)} + k_1, \quad S_2^{T(6)} = \frac{(c_2 k_1 + r_2)k_2 r_1}{r_1 r_2 - c_1 c_2 k_1 k_2};
$$

$$
S_1^{T(7)} = \frac{r}{b_1} - \frac{a_2}{a_1} S_2^{T(7)}, \quad S_2^{T(7)} = k_2 D_7^{-1}\left[k_1 b_1(a_1 r_2 - a_2 r_1) + k_1 a_1 c_2 r + r_1 a_2 r\right],
$$

$$
Q_7 = \frac{1}{a_1 b_1 k_1}\left[rr_1 - b_1 k_1 r_1 - S_2^{T(7)}(r_1 b_2 + k_1 b_1 c_1)\right].
$$

To study the stability of the points we will use the Jacobian,

$$
J^T = \begin{pmatrix}
r_1 - \frac{2r_1}{k_1}S_1 + c_1 S_2 + a_1 Q & c_1 S_1 & a_1 S_1 \\
c_2 S_2 & r_2 - \frac{2r_2}{k_2}S_2 + c_2 S_1 + a_2 Q & a_2 S_2 \\
-b_1 Q & -b_2 Q & r - b_1 S_1 - b_2 S_2
\end{pmatrix},
$$

(26)

The points E_1^T, E_2^T, E_4^T are unconditionally feasible in view of the assumption on the nonnegativity of the parameters. They however are unstable because linearizing the system, eigenvalues arise with positive real part.

The point E_3^T is feasible for $r > b_2 k_2$ but this condition implies that the following eigenvalue of the linearized system

$$
\lambda = r_1 + \frac{c_1 r}{b_2} + \frac{a_1 r_2 (r - b_2 k_2)}{a_2 b_2 k_2}
$$

is positive, so that the equilibrium is unstable. The same situation arises with E_5^T feasible for $r > b_1 k_1$, which implies instability in view of the positivity of the eigenvalue

$$
\lambda = r_2 + \frac{c_2 r}{b_1} + \frac{a_2 r_1 (r - b_1 k_1)}{a_1 b_1 k_1}.
$$

To study the stability of the points E_6^T and E_7^T we use the the analysis of the purely symbiotic model, the isoclines of which,

$$
\alpha: \quad S_2 = \frac{r_1 S_1}{k_1 c_1} - \frac{r_1}{c_1}, \quad \beta: S_2 = \frac{c_2 k_2 S_1}{r_2} + k_2,
$$

meet in the first quadrant if the slope of β is smaller than that of α i.e. if

$$
k_1 k_2 c_1 c_2 < r_1 r_2.
$$

(27)

In this conditions, E_6^T is feasible.

The characteristic polynomial factors to give one eigenvalue explicitly, which is negative if

$$
r < b_1 \left(\frac{c_1 k_1}{r_1} S_2^{T(6)} + k_1 \right) + b_2 S_2^{T(6)},
$$

(28)

and a quadratic equation of the form $\lambda^2 + B\lambda + C = 0$, whose roots, by Descartes rule or the Routh-Hurwitz conditions, have negative real parts if we impose $B, C > 0$. Explicitly, this becomes

$$
r_1 + c_1 S_2^{T(6)} - r_2 + \frac{2r_2}{k_2} S_2^{T(6)} - c_2 \left(\frac{c_1 k_1}{r_1} S_2^{T(6)} + k_1 \right) > 0,
$$

(29)

$$
(r_1 + c_1 S_2^{T(6)}) \left[-r_2 + \frac{2r_2}{k_2} S_2^{T(6)} - c_2 \left(\frac{c_1 k_1}{r_1} S_2^{T(6)} + k_1 \right) \right]
$$
$$
- c_1 c_2 S_2^{T(6)} \left(\frac{c_1 k_1}{r_1} S_2^{T(6)} + k_1 \right) > 0
$$

These conditions are nonempty, as simulations reveal that equilibrium E_6^T is indeed stable for suitable parameter values, Figure 4.

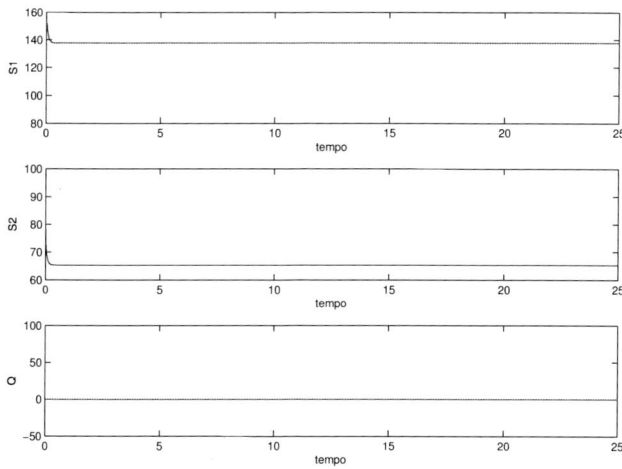

Figure 4. For the parameter values $a_1 = 1.8$, $a_2 = 1.9$, $k_1 = 20$, $k_2 = 15$, $c_1 = 0.9$, $c_2 = 0.5$, $r_1 = 10$, $r_2 = 20.5$, $r = 8$, $b_1 = 2.7$, $b_2 = 2.85$ the equilibrium E_6^T is stable.

The explicit expressions of the other components of E_7^T are

$$S_1^{T(7)} = k_1 D_7^{-1} \left[r_2 r a_1 + k_2 (a_2 r c_1 - a_1 b_2 r_2 + a_2 b_2 r_1) \right],$$
$$Q_7^T = k_2 D_7^{-1} \left[r_1 r_2 (r - b_1 k_1 - b_2 k_2) - k_1 k_2 (b_2 c_2 r_1 + b_1 c_1 r_2 + r c_1 c_2) \right].$$

From them, the feasibility conditions for E_7^T follow:

$$r \geq k_2 b_2 \frac{a_1 r_2 - a_2 r_1}{a_1 r_2 + k_2 a_2 c_1}, \quad r \geq k_1 b_1 \frac{a_2 r_1 - a_1 r_2}{a_2 r_1 + k_1 a_1 c_2}, \tag{30}$$
$$r \geq \frac{r_1 r_2 (b_1 k_1 + b_2 k_2) + k_1 k_2 (b_2 c_2 r_1 + b_1 c_1 r_2 + r c_1 c_2)}{r_1 r_2 - c_1 c_2 k_1 k_2}$$

for $r_1 r_2 > c_1 c_2 k_1 k_2$ or the reverse last inequality otherwise. Note that the two first conditions are alternative of each other; for $a_1 r_2 - a_2 r_1 > 0$ the second condition is always satisfied, while conversely for $a_2 r_1 - a_1 r_2 > 0$ the first condition follows automatically. Stability of this equilibrium is obtained observing that the Jacobian main diagonal simplifies, giving

$$J_{11}^T(E_7) = -\frac{r_1}{K_1} S_1^{T(7)}, \quad J_{22}^T(E_7) = -\frac{r_2}{K_2} S_2^{T(7)}, \quad J_{33}^T(E_7) = 0,$$

from which $-\operatorname{tr}(J^T(E_7)) > 0$. Also,

$$-\det(J^T(E_7)) = S_1^{T(7)} S_2^{T(7)} Q \left[a_1 a_2 e (c_1 + c_2) + e a_1^2 \frac{r_2}{K_2} + e a_2^2 \frac{r_1}{K_1} \right] > 0.$$

Thus two Routh-Hurwitz conditions hold, the remaining one becomes

$$C > S_1^{T(7)} S_2^{T(7)} \left[\left(\frac{r_1}{K_1} S_1^{T(7)} + \frac{r_2}{K_2} S_2^{T(7)} \right) \left(c_1 c_2 - \frac{r_1 r_2}{K_1 K_2} \right) + Q_7 a_1 a_2 e (c_1 + c_2) \right], \tag{31}$$
$$C \equiv e Q_7 \left[a_1^2 \frac{r_1}{K_1} \left(S_1^{T(7)} \right)^2 + a_2^2 \frac{r_2}{K_2} \left(S_2^{T(7)} \right)^2 \right].$$

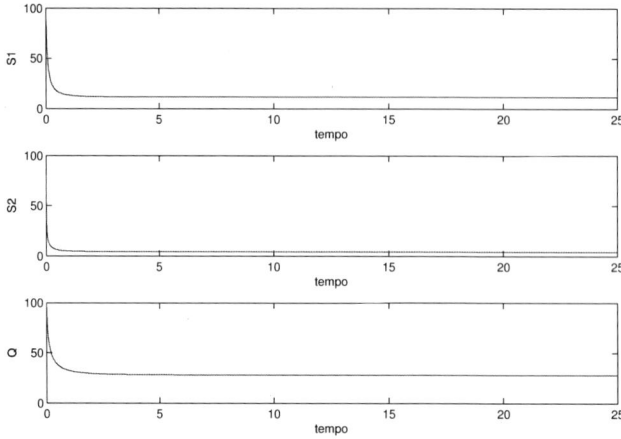

Figure 5. For the parameter values $a_1 = 0.03$, $a_2 = 0.1$, $k_1 = 2$, $k_2 = 1$, $c_1 = 0.9$, $c_2 = 0.5$, $r_1 = 1$, $r_2 = 2.5$, $r = 1.2$, $b_1 = 0.045$, $b_2 = 0.15$ the equilibrium E_7^T is stable.

Stability for E_7^T holds in fact for certain parameter ranges, as it is empirically demonstrated in Figure 5.

4.3.3. Discussion

In this system, only two outcomes are possible. Either the prey are wiped out and the mutualistic populations coexit, at equilibrium E_6^T, or the three populations coexist together. The fact that the mutualistic subsystem thrives also in the absence of the bottom prey is to be ascribed clearly to their mutual beneficial interactions. In fact the presence of alternative food sources in this logistic model is not enough to sustain each one of the symbiotic populations alone, as equilibria E_2^T and are E_4^T unstable.

4.2. A Leslie-Gower Model of Facultative Mutualism

4.1.1. The Model

Denoting by $S_1(t)$, $S_2(t)$, and $Q(t)$ the populations as in (25), the model changes amount to delete the hunting terms in the predators' equations, while preserving them in the prey dynamics, and replace the carrying capacities by the prey populations. In this way, as in the standard Leslie-Gower model, the predators grow toward an asymptotic value proportional to the available food resources. The resulting system is therefore

$$\dot{S}_1(t) = r_1\left(1 - g_1\frac{S_1(t)}{Q(t)}\right)S_1(t) + c_1S_1(t)S_2(t) \qquad (32)$$

$$\dot{S}_2(t) = r_2\left(1 - g_2\frac{S_2(t)}{Q(t)}\right)S_2(t) + c_2S_1(t)S_2(t)$$

$$\dot{Q}(t) = Q(t)(r - e_1S_1(t) - e_2S_2(t)).$$

In this case the two predator populations grow logistically with a carrying capacity proportional to the size of the prey population, respectively in a proportional fashion to g_1 and g_2. The parameters c_1 and c_2 have the same meaning of model (25). The parameters e_1 and e_2 represent the hunting rates on the prey from the two symbiotic predators.

The Jacobian of (32) at a generic point (S_1, S_2, Q) is

$$J = \begin{pmatrix} r_1 - 2r_1g_1\frac{S_1}{Q} + c_1S_2 & c_1S_1 & r_1g_1S_1^2Q^{-2} \\ c_2S_2 & r_2 - 2r_2g_2\frac{S_2}{Q} + c_2S_1 & r_2g_2S_2^2Q^{-2} \\ -e_1Q & -e_2Q & r - e_1S_1 - e_2S_2 \end{pmatrix}. \tag{33}$$

4.2.2.   Equilibrium Points and Stability

The feasible equilibria are listed in the following table

Equilibrium point	S_1	S_2	Q
E_1^{LG}	0	+	+
E_2^{LG}	+	0	+
E_3^{LG}	+	+	+

with

$$S_2^{LG(1)} = \frac{r}{e_2}, \quad Q_1^{LG} = \frac{rb_2}{a_2}, \quad S_1^{LG(2)} = \frac{r}{e_1}, \quad Q_2^{LG} = \frac{rb_1}{a_1},$$

$$S_1^{LG(3)} = \frac{c_1r_2(Q_3^{LG})^2 + r_1r_2g_2Q_3^{LG}}{r_1r_2g_1g_2 - c_1c_2(Q_3^{LG})^2}, \quad S_2^{LG(3)} = \frac{c_2r_1(Q_3^{LG})^2 + r_1r_2g_1Q_3^{LG}}{r_1r_2g_1g_2 - c_1c_2(Q_3^{LG})^2},$$

$$Q_3^{LG} = -\frac{e_1r_1r_2g_2 + e_2g_1r_1r_2}{2(rc_1c_2 + e_1c_1r_2 + e_2c_2r_1)}$$

$$+ \sqrt{\frac{(e_1r_1r_2g_2 + e_2r_1r_2g_1)^2 + 4(rr_1r_2g_1g_2)(rc_1c_2 + e_1c_1r_2 + e_2c_2r_1)}{4(rc_1c_2 + e_1c_1r_2 + e_2c_2r_1)^2}}$$

The point E_1^{LG} is always feasible but unconditionally unstable in view of the positive eigenvalue, $r_1 + c_1re_2^{-1}$.

The point E_2^{LG} is also feasible but again unconditionally unstable due to the presence of the positive eigenvalue $r_2 + c_2re_1^{-1}$.

The feasibility conditions for the interior point E_3^{LG} impose an upper bound on the prey population size,

$$Q_3^{LG} < \sqrt{\frac{g_1g_2r_1r_2}{c_1c_2}}. \tag{34}$$

The value of the prey population is obtained by solving the quadratic

$$(rc_1c_2 + e_1c_1r_2 + e_2c_2r_1)Q^2 + (e_1g_2r_1r_2 + e_2g_1r_1r_2)Q - rg_1g_2r_1r_2 = 0$$

which, by the Descartes rule of signs, has certainly a positive root, which needs to verify (34).

The stability of this equilibrium is established numerically in Fig. 6 for the parameter values $g_1 = 2$, $g_2 = 1.4$, $c_1 = 0.9$, $c_2 = 0.4$, $r_1 = 8$, $r_2 = 10$, $e_1 = 0.8$, $e_2 = 0.9$, $r = 50$.

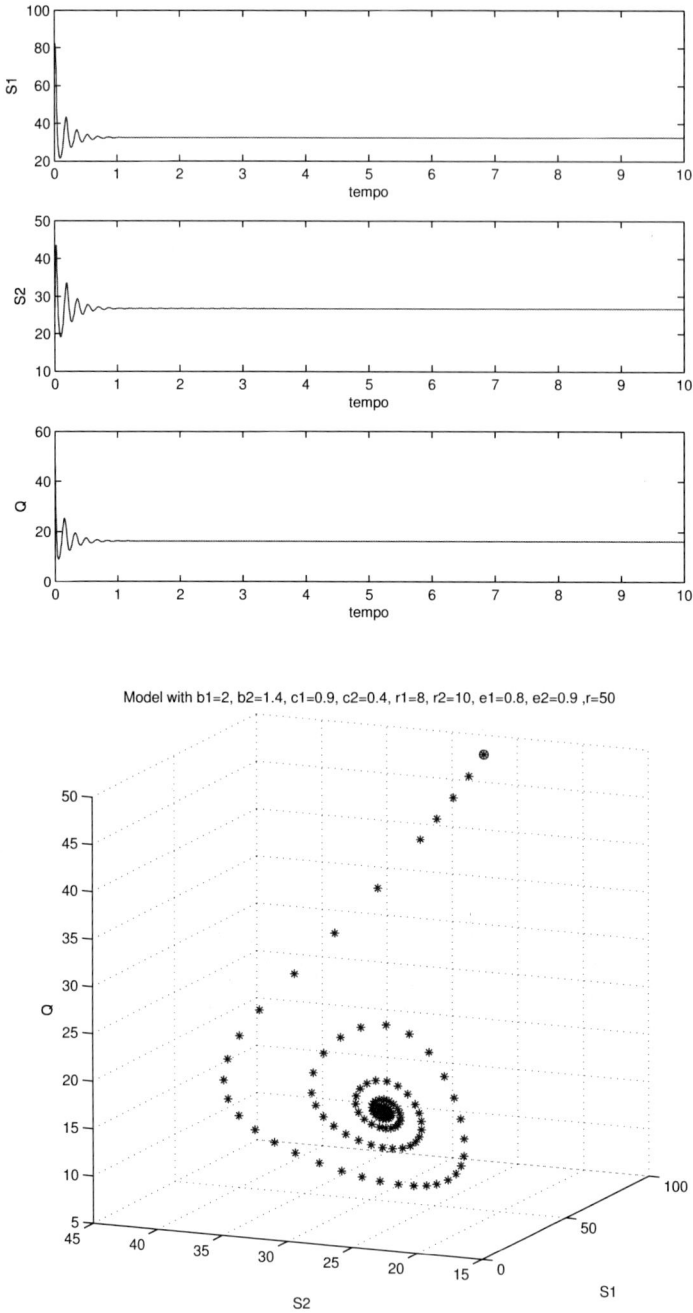

Figure 6. The model (32) settles to the coexistence equilibrium for the parameter values $g_1 = 2$, $g_2 = 1.4$, $c_1 = 0.9$, $c_2 = 0.4$, $r_1 = 8$, $r_2 = 10$, $e_1 = 0.8$, $e_2 = 0.9$, $r = 50$. Top: the populations evolution in time; bottom: the solution trajectory in the phase space.

For the following parameter values, $g_1 = 2$, $g_2 = 1$, $c_1 = 0.3$, $c_2 = 0.4$, $r_1 = 0.005$, $r_2 = 0.005$, $e_1 = 0.1$, $e_2 = 0.1$, $r = 20$, we instead observe that the right hand side of (34) takes the value 0.0204, while Q from solving the above quadratic has the same value. In fact it seems that it is very hard not to verify condition (34), in spite of several tentatives toward this end. In any case the system can exhibit larger and larger oscillations, i.e. the mutualistic association makes all the populations thrive, as shown in Figure 7. This of course cannot happen in practice because of resources limitation, but this constraint is not contained in our model. Thus the simulations show that the association is anyway beneficial for the whole ecosystem.

4.3.3. Discussion

In this situation, the disappearance of the prey is excluded at the onset by the very model set-up. So only few points remain as candidates for being system's equilibria. But when we consider their stability, it turns out that only the coexistence equilibrium is achievable. In fact we have empirically shown that it is attained by suitable parameter values. The three populations can also thrive together by persistent oscillations of wider and wider amplitudes, as shown in our simulations.

4.3. The Model for Obligated Mutualism

Letting once again $S_1(t)$ and $S_2(t)$ denote the two predators and $Q(t)$ the prey population, we resume the system (25) and change its logistic portions with a term denoting pure mortality for each symbiotic population. The resulting model is

$$
\begin{aligned}
\dot{S}_1(t) &= -m_1 S_1(t) + c_1 S_1(t) S_2(t) + a_1 S_1(t) Q(t) \qquad (35)\\
\dot{S}_2(t) &= -m_2 S_2(t) + c_2 S_1(t) S_2(t) + a_2 S_2(t) Q(t) \\
\dot{Q}(t) &= Q(t)(r - b_1 S_1(t) - b_2 S_2(t)).
\end{aligned}
$$

Here the symbiosis is necessary for the survival of the predators populations, in absence of the prey Q, because if also say S_2 is absent, for S_1 there are no resources and therefore it will get extinguished too. The predator's mortality rates m_1 and m_2 are responsible for this situation. The remaining parameters have the same meaning as in system (25).

The Jacobian of (35) is

$$
J^{OM} = \begin{pmatrix}
-m_1 + c_1 S_2 + a_1 Q & c_1 S_1 & a_1 S_1 \\
c_2 S_2 & -m_2 + c_2 S_1 + a_2 Q & a_2 S_2 \\
-b_1 Q & -b_2 Q & r - b_1 S_1 - b_2 S_2
\end{pmatrix}. \qquad (36)
$$

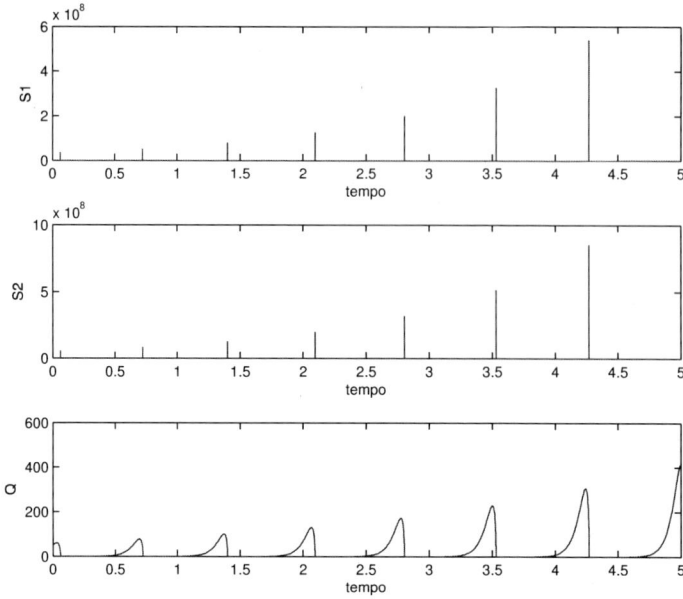

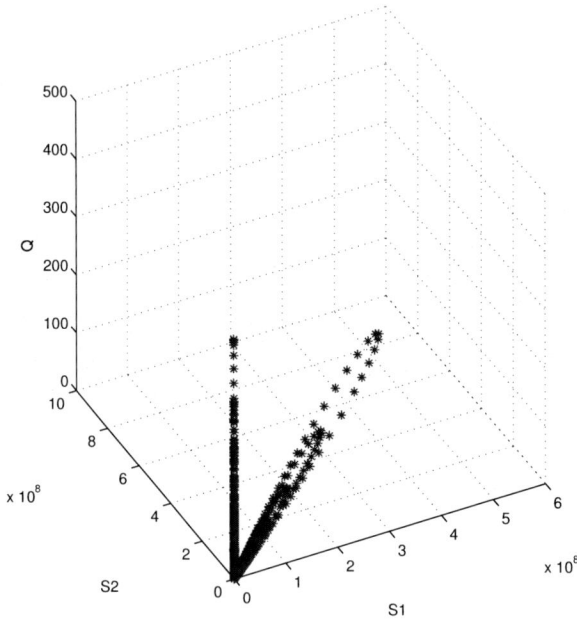

Figure 7. The model (32) shows increasing amplitude oscillations for the parameter values $g_1 = 2$, $g_2 = 1$, $c_1 = 0.3$, $c_2 = 0.4$, $r_1 = 0.005$, $r_2 = 0.005$, $e_1 = 0.1$, $e_2 = 0.1$, $r = 20$. Top: the populations evolution in time; bottom: the solution trajectory in the phase space. Note the large value of the scales on the axes.

4.1.1. Equilibria and Their Stability

The possibly feasible system's equilibria are listed in the following table.

Equilibrium point	S_1^{OM}	S_2^{OM}	Q^{OM}
E_1^{OM}	0	0	0
E_2^{OM}	0	+	+
E_3^{OM}	+	0	+
E_4^{OM}	+	+	0
E_5^{OM}	$\frac{m_2}{c_2} - \frac{a_2}{c_2}Q_5^{OM}$	$\frac{m_1}{c_1} - \frac{a_1}{c_1}Q_5^{OM}$	Q_5^{OM}

with

$$S_2^{OM(2)} = \frac{r}{b_2}, \quad Q_2^{OM} = \frac{m_2}{a_2}, \quad S_1^{OM(3)} = \frac{r}{b_1}, \quad Q_3^{OM} = \frac{m_1}{a_1},$$

$$S_1^{OM(4)} = \frac{m_2}{c_2}, \quad S_2^{OM(4)} = \frac{m_1}{c_1}, \quad Q_5^{OM} = \frac{b_1 m_2 c_1 + b_2 m_1 c_2 - r c_1 c_2}{b_1 a_2 c_1 + b_2 a_1 c_2},$$

$$S_1^{OM(5)} = \frac{r a_2 c_1 + a_1 b_2 m_2 - a_2 b_2 m_1}{b_1 a_2 c_1 + b_2 a_1 c_2}, \quad S_2^{OM(5)} = \frac{r a_1 c_2 + a_2 b_1 m_1 - a_1 b_1 m_2}{b_1 a_2 c_1 + b_2 a_1 c_2}.$$

One eigenvalue of E_1^{OM} is $r > 0$ establishing thus its instability. Similarly E_4^{OM} has the eigenvalue $\sqrt{m_1 m_2}$ and therefore is also unstable. At E_5^{OM} the main diagonal of the Jacobian vanishes, so that the Routh-Hurwitz condition $C_2 C_1 > C_0$ for the cubic polynomial $\sum_{k=0}^{3} C_k \lambda^k$, with $C_3 = 1$, cannot be satisfied since $C_2 = 0$ and $C_0 = -\det(J_5^{OM}) = (a_1 b_2 c_2 + a_2 b_1 c_1) S_1^{OM(5)} S_2^{OM(5)} Q_5^{OM} > 0$. Thus E_5^{OM} is unstable.

Equilibria E_2^{OM} and E_3^{OM} are interesting, since both have a real and a pair of purely imaginary eigenvalues as follows,

$$\lambda_1 = \frac{c_1 r}{b_2} + \frac{m_2 a_1}{a_2} - m_1 \quad \lambda_{2,3} = \pm i \sqrt{m_2 r}$$

for E_2^{OM} and

$$\lambda_1 = \frac{c_2 r}{b_1} + \frac{m_1 a_2}{a_1} - m_2 \quad \lambda_{2,3} = \pm i \sqrt{m_1 r}$$

for E_3^{OM}. By imposing the negativity of the former ones, i.e. respectively for

$$m_1 > \frac{c_1 r}{b_2} + \frac{m_2 a_1}{a_2}, \quad m_2 > \frac{c_2 r}{b_1} + \frac{m_1 a_2}{a_1}, \tag{37}$$

we find limit cycles on the coordinate planes $S_1 = 0$ and $S_2 = 0$ respectively. These correspond to the neutrally stable oscillations of the standard Lotka-Volterra model, see Figures 8 and 9.

4.2.2. Discussion

For the facultative symbiosis model, the predators can always survive, in view of other food sources, even if their prey get extinguished. For the Leslie-Gower symbiotic model

Model with a1=0.85, a2=3, c1=0.5, c2=0.3, m1=69, m2=120 ,b1=1.275, b2=4.5, r=15

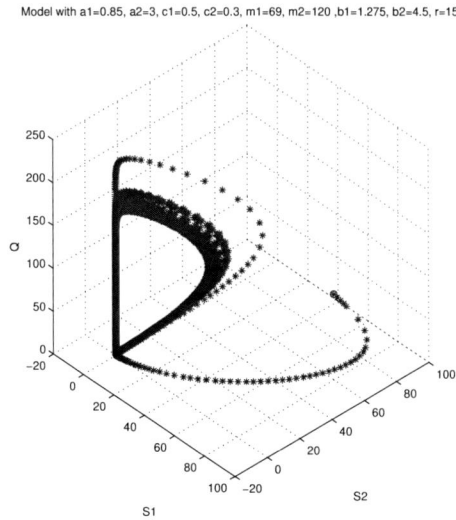

Figure 8. Phase space portrait with limit cycles around E_2^{OM} for the parameter values $a_1 = 0.85$, $a_2 = 3$, $c_1 = 0.5$, $c_2 = 0.3$, $m_1 = 69$, $m_2 = 120$, $b_1 = 1.275$, $b_2 = 4.5$, $r = 15$.

Model with a1=0.85, a2=3, c1=0.5, c2=0.3, m1=120, m2=444.486 ,b1=3.4, b2=12, r=15

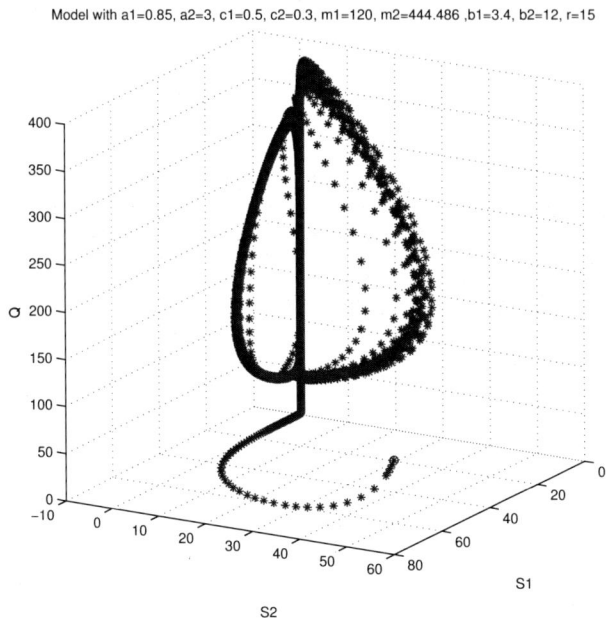

Figure 9. Phase space portrait showing a limit cycle around E_3^{OM}; it is one cycle, compare indeed with the population values of Figure 10. It is obtained for the parameter values $a_1 = 0.85$, $a_2 = 3$, $c_1 = 0.5$, $c_2 = 0.3$, $m_1 = 120$, $m_2 = 444.486$, $b_1 = 3.4$, $b_2 = 12$, $r = 15$.

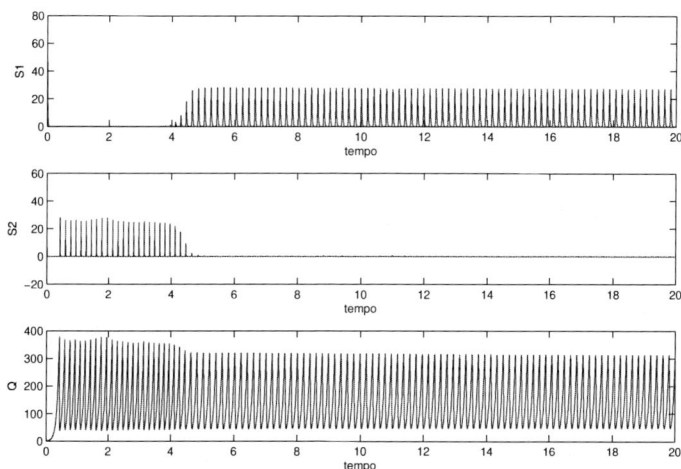

Figure 10. Populations as functions of time, showing a limit cycle around E_3^{OM} for the same parameter values of Figure 9.

with suitable conditions the coexistence equilibrium can be reached. In general in the model of obligated mutualism instead the coexistence of the three populations is not supported. But the most interesting conclusion in this case is that in the conditions (37) the prey wipes out one of its symbiotic predators. These conditions are mutually exclusive, as only one of them can be satisfied. This is clearly seen by rewriting them respectively as follows

$$r < \frac{b_2}{a_2 c_1}(a_2 m_1 - a_1 m_2), \quad r < -\frac{b_1}{a_1 c_2}(a_2 m_1 - a_1 m_2) \tag{38}$$

since only one of the right hand sides can be positive. Then, for a low enough reproductivity of the prey, the first S_1, respectively the second S_2, symbiotic species is wiped out. Again, this is a counterintuitive result, but perhaps it can be explained as follows. Suppose that the ratio of the feeding reward on prey over mortality rate of the symbiotic species is larger for S_2^{OM} than for S_1^{OM}, namely

$$\frac{a_2}{m_2} > \frac{a_1}{m_1},$$

so that we are in the first case (38). Then the prey reproduces slowly, but S_2^{OM} rather than S_1^{OM} gets the major gain from it and therefore no other resources are left for S_1^{OM} to survive.

5.   A Web with Intermediate Symbiotic System

We consider now a more complex food chain, consisting of three trophic levels, in which the two population symbiotic system is present at the intermediate stage. Namely it sits below a top predator, that hunts both the symbiotic populations, and above a bottom prey on which both symbiotic species exert a pressure.

Assume that the primary predators P can feed only on the symbiotic populations S_1 and S_2, so that in the latter's absence they are bound to disappear. Note that this assumption makes the top portion of the systems (40) and (39) below differ from the models of Section 2, so that the analysis of some particular cases here does not completely match the one performed earlier, rather it can be considered an interesting extension to a different situation of the one illustrated in Section 2. Let r_P denote the top predators' mortality rate, while we assume Holling type I interactions with their prey at respective rates ω_1 and ω_2. Let the prey Q at the bottom of the chain reproduce at rate r_q and let their environment have a carrying capacity K. They are hunted only by the symbiotic species, at respective rates γ_i.

Let c and g denote the benefit the symbiotic populations get from their mutual interactions, μ_i the respective benefits from hunting the bottom prey Q at rates γ_i and η_i the damage they suffer by the predation of the P's. For the symbiotic system we can consider again the case of obligated or facultative mutualism. Let a and e denote the net birth rates of S_1 and S_2 respectively, which correspond to negative mortality in the first case and are instead positive in the second one. Further, in the latter case let b and f denote intraspecific competition, again respectively for S_1 and S_2. Since not all the captured prey are turned into new predators, the following restrictions hold $\mu_i < \gamma_i$, $\omega_i < \eta_i$, $i = 1, 2$.

The models then are

$$\dot{P} = -r_p P + P\left[\omega_1 S_1 + \omega_2 S_2\right], \quad (39)$$
$$\dot{S}_1 = S_1\left(-a + cS_2\right) + \mu_1 QS_1 - \eta_1 S_1 P,$$
$$\dot{S}_2 = S_2\left(-e + gS_1\right) + \mu_2 QS_2 - \eta_2 S_2 P,$$
$$\dot{Q} = r_q Q\left[1 - \frac{Q}{K}\right] - Q\left[\gamma_1 S_1 + \gamma_2 S_2\right].$$

and

$$\dot{P} = -r_p P + P\left[\omega_1 S_1 + \omega_2 S_2\right], \quad (40)$$
$$\dot{S}_1 = S_1\left(a - bS_1 + cS_2\right) + \mu_1 QS_1 - \eta_1 S_1 P,$$
$$\dot{S}_2 = S_2\left(e - fS_2 + gS_1\right) + \mu_2 QS_2 - \eta_2 S_2 P,$$
$$\dot{Q} = r_q Q\left[1 - \frac{Q}{K}\right] - Q\left[\gamma_1 S_1 + \gamma_2 S_2\right].$$

5.1. Obligated Mutualism

In this case, the Jacobian of (39) is

$$J^{CO} = \begin{pmatrix} J_{11}^{CO} & P\omega_1 & P\omega_2 & 0 \\ -S_1\eta_1 & J_{22}^{CO} & S_1 c & S_1\mu_1 \\ -S_2\eta_2 & S_2 g & J_{33}^{CO} & S_2\mu_2 \\ 0 & -Q\gamma_1 & -Q\gamma_2 & J_{44}^{CO} \end{pmatrix},$$

where the diagonal elements are given as follows $J_{11}^{CO} = -r_p + \omega_1 S_1 + \omega_2 S_2$, $J_{22}^{CO} = -a + cS_2 + \mu_1 Q - \eta_1 P$, $J_{33}^{CO} = -e + gS_1 + \mu_2 Q - \eta_2 P$, $J_{44}^{CO} = r_q - \frac{2r_q Q}{K} - \gamma_1 S_1 - \gamma_2 S_2$.

We now list the possible feasible equilibria of the system (40), in addition to the coexistence equilibrium E_{11}^{CO}, which appears to be too hard to be analytically investigated.

Equilibrium	P	S_1	S_2	Q	Feasibility condition
E_0^{CO}	0	0	0	0	—
E_2^{CO}	0	+	+	0	—
E_3^{CO}	+	+	+	0	(41)
E_5^{CO}	0	0	0	K	—
E_6^{CO}	0	0	+	+	(44)
E_7^{CO}	+	0	+	+	(48)
E_8^{CO}	0	+	0	+	(45)
E_9^{CO}	+	+	0	+	(49)
E_{10}^{CO}	0	+	+	+	(52) and either (53) or (54)

E_0^{CO} is always unstable, in view of the positive eigenvalue r_p. Also E_2^{CO}, with components $S_1^{CO(2)} = eg^{-1}$ and $S_2^{CO(2)} = ac^{-1}$, is unstable in view of the eigenvalue $\sqrt{ae} > 0$. This is to be compared with the equilibrium E_2^O of Section 2. These two equilibria clearly coincide since in them the predators do not play any influence, and therefore this system and (16) give the same result.

E_3^{CO} with components

$$P_3^{CO} = \frac{1}{\eta_1}(cS_2^{CO} - a) = \frac{1}{\eta_2}(gS_1^{CO} - e),$$

$$S_1^{CO(3)} = \frac{\eta_2 cr_p - a\eta_2\omega_2 + \eta_1 e\omega_2}{\eta_1 g\omega_2 + \eta_2 c\omega_1}, \quad S_2^{CO(3)} = \frac{a\eta_2\omega_1 - \eta_1 e\omega_1 + \eta_1 gr_p}{\eta_1 g\omega_2 + \eta_2 c\omega_1},$$

is feasible for

$$r_p \geq \omega_1\frac{e}{g} + \omega_2\frac{a}{c}. \tag{41}$$

Its stability has been investigated numerically, without success. We conjecture that E_3^{CO} is an unstable equilibrium. It is to be compared with E_6^O of Section 2.

E_5^{CO} is stable for

$$K < \frac{a}{\mu_1}, \quad K < \frac{e}{\mu_2}. \tag{42}$$

These are the opposite feasibility conditions (44) and (45) below, for E_6^{CO} and E_8^{CO}, thus E_5^{CO} is stable if and only if these two equilibria are both infeasible. Note that this equilibrium again is not found in the model (35), since here the prey has a logistic growth, while in (35) Malthusian growth was assumed. Hence, once again, this is a slightly different situation, worth of investigating, and not just a duplicate of the work done in the previous Sections 2 and 4.

The condition (42) can indeed be satisfied, as shown in Figure 11, using the parameter values

$$r_p = 4, \quad r_q = 2, \quad a = 2, \quad \mu_1 = 1, \quad \mu_2 = 2, \quad \eta_1 = 3, \quad \gamma_2 = 3, \quad \gamma_1 = 2, \tag{43}$$
$$\omega_1 = 1.2, \quad g = 0.37, \quad \omega_2 = 0.7, \quad c = 0.5, \quad e = 1.3, \quad \eta_2 = 2, \quad K = 0.53.$$

E_6^{CO} with components

$$Q_6^{CO} = \frac{e}{\mu_2}, \quad S_2^{CO(6)} = \frac{r_q}{\gamma_2}\left(1 - \frac{e}{\mu_2 K}\right),$$

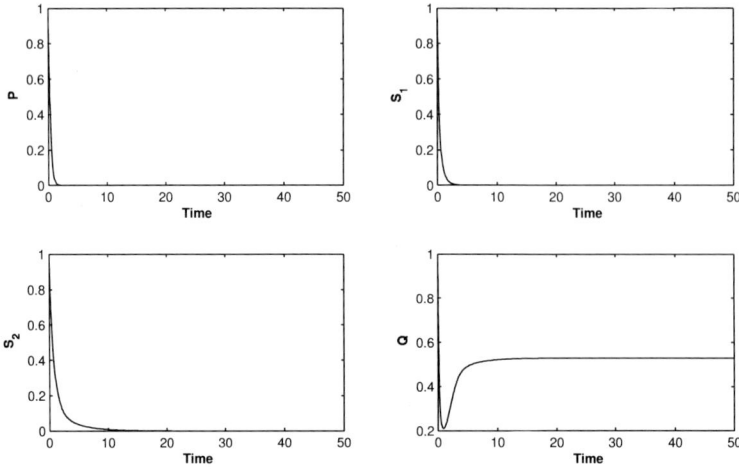

Figure 11. Stability of E_5^{CO} is reached for the parameter values $r_p = 4$, $\omega_1 = 1.2$, $\omega_2 = 0.7$, $a = 2$, $c = 0.5$, $\mu_1 = 1$, $\eta_1 = 3$, $e = 1.3$, $g = 0.37$, $\mu_2 = 2$, $\eta_2 = 2$, $r_q = 2.0$, $K = 0.53$, $\gamma_1 = 2$, $\gamma_2 = 3$.

and E_8, with components

$$Q_8^{CO} = \frac{a}{\mu_1}, \quad S_1^{CO(8)} = \frac{r_q}{\gamma_1}\left(1 - \frac{a}{\mu_1 K}\right),$$

are equilibria at which respectively thrive the second symbiotic population and the prey, and the first mutualistic population and the prey. They are feasible respectively for

$$K \geq \frac{e}{\mu_2} \tag{44}$$

and

$$K \geq \frac{a}{\mu_1}. \tag{45}$$

Both will collide into E_5^{CO} when K decreases. For their stability, two eigenvalues in each case are immediate, while the Routh-Hurwitz conditions on the remaining quadratic are always satisfied. Specifically for E_6^{CO} we find the eigenvalues $\omega_2 S_2^{CO(6)} - r_p$ and $c S_2^{CO(6)} + \mu_1 Q_6^{CO} - a$ from which the stability conditions

$$\omega_2 r_q (\mu_2 K - e) < r_p \gamma_2 \mu_2 K, \quad c r_q (\mu_2 K - e) + e\gamma_2 \mu_2 K < a\gamma_2 \mu_2 K, \tag{46}$$

and for E_8^{CO} we have the eigenvalues $\omega_1 S_1^{CO(8)} - r_p$ and $g S_1^{CO(8)} + \mu_2 Q_8^{CO} - e$ from which stability is obtained if

$$\omega_1 r_q (\mu_1 K - e) < r_p \gamma_1 \mu_1 K, \quad g r_q (\mu_1 K - a) + a\gamma_1 \mu_2 K > e\gamma_1 \mu_1 K. \tag{47}$$

These results should be compared with those of Section 4, namely equilibria E_2^{OM} and E_3^{OM}. See also the discussion about the limit cycles and (37).

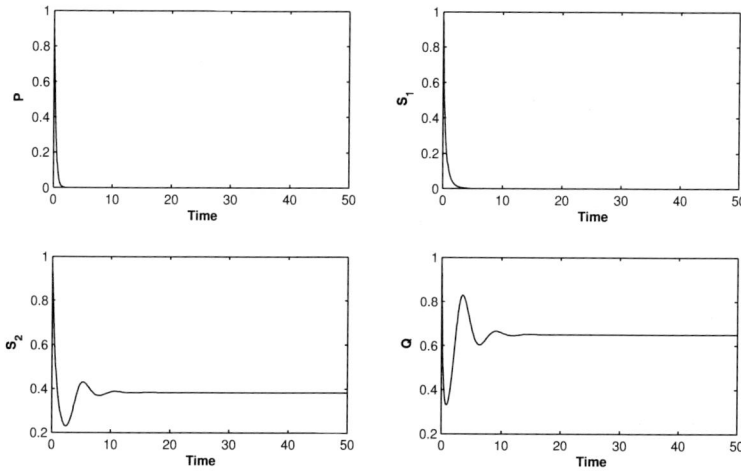

Figure 12. Stability of E_6^{CO} is reached for the parameter values $r_p = 4, \omega_1 = 1.2, \omega_2 = 0.7,$ $a = 2, c = 0.5, \mu_1 = 1, \eta_1 = 3, e = 1.3, g = 0.37, \mu_2 = 2, \eta_2 = 2, r_q = 2.0, K = 1.53,$ $\gamma_1 = 2, \gamma_2 = 3$.

For E_6^{CO} we run simulations starting from the set of parameter values (43). We increase the value of K to 1.53 so that condition (44) is violated and obtain its stability, see Figure 12. Note that instead a simulation for E_3^{CO} can be obtained by raising the value of c and g to 2.5 and 2.37 respectively, not reported here.

Taking the same parameter values as in Figure 11 but reducing a to 0.2 so that the condition (45) holds, we have stability for E_8^{CO}, as shown in Figure 13.

The most interesting points are however E_7^{CO} and E_9^{CO} because these are really equilibria intrinsic of this model, they do not appear in Sections 2 and 4 in view of the fact that they involve all the three trophic levels.

The point E_7^{CO}, with components

$$S_2^{CO(7)} = \frac{r_p}{\omega_2}, \quad Q_7^{CO} = K\left(1 - \frac{\gamma_2 r_p}{r_q \omega_2}\right), \quad P_7^{CO} = \frac{1}{\eta_2}\left(\mu_2 Q_7^{CO} - e\right),$$

is feasible for

$$K \geq \frac{\omega_2 r_q e}{(\omega_2 r_q - r_p \gamma_2)\mu_2} > 0. \tag{48}$$

For E_9^{CO}, with components

$$S_1^{CO(9)} = \frac{r_p}{\omega_1}, \quad Q_9^{CO} = K\left(1 - \frac{\gamma_1 r_p}{r_q \omega_1}\right), \quad P_9^{CO} = \frac{1}{\eta_1}\left(\mu_1 Q_9^{CO} - a\right),$$

the feasibility condition is

$$K \geq \frac{\omega_1 r_q a}{(\omega_1 r_q - r_p \gamma_1)\mu_1} > 0. \tag{49}$$

As for stability, E_7^{CO} has an immediate eigenvalue, giving the stability condition

$$cS_2^{CO(7)} + \mu_1 Q_7 < \eta_1 P_7 + a. \tag{50}$$

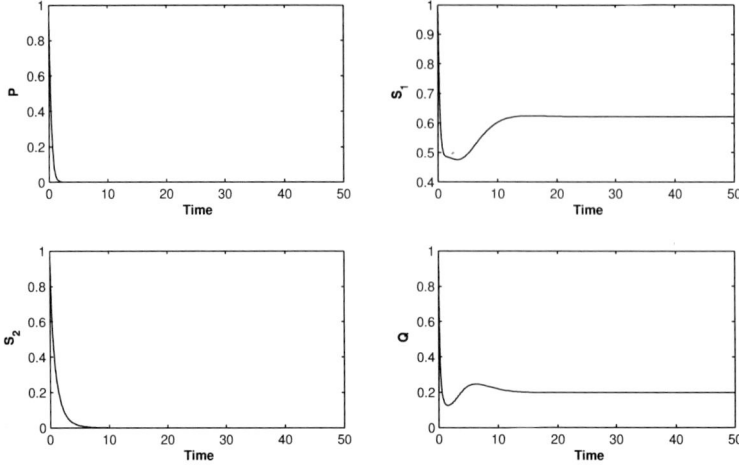

Figure 13. Stability of E_8^{CO} is reached for the parameter values $r_p = 4$, $\omega_1 = 1.2$, $\omega_2 = 0.7$, $a = 0.2$, $c = 0.5$, $\mu_1 = 1$, $\eta_1 = 3$, $e = 1.3$, $g = 0.37$, $\mu_2 = 2$, $\eta_2 = 2$, $r_q = 2.0$, $K = 0.53$, $\gamma_1 = 2$, $\gamma_2 = 3$.

For the remaining cubic characteristic equation $\sum_{k=0}^{3} C_k \lambda^k = 0$, $C_3 = 1$, two of the Routh-Hurwitz conditions hold immediately, as

$$C_2 = \frac{r_q}{K} Q_7^{CO} > 0, \quad C_0 = \eta_2 \omega_2 \frac{r_q}{K} P_7^{CO} Q_7^{CO} S_2^{CO(7)} > 0,$$

while the remaining one is also always satisfied, as it reduces to

$$\left(\eta_2 \omega_2 P_7^{CO} + \gamma_2 \mu_2 Q_7^{CO} \right) - \eta_2 \omega_2 P_7^{CO} = \gamma_2 \mu_2 Q_7^{CO} > 0.$$

E_7^{CO} is then stable if just (50) holds.

The stability calculations for E_9^{CO} are very similar. Stability hinges on an explicit eigenvalue, from which the condition

$$gS_1^{CO(9)} + \mu_2 Q_9 < \eta_2 P_9 + e, \tag{51}$$

and of the eigenvalues of a submatrix $\widehat{J}_9$ of order 3, for which

$$\operatorname{tr}(\widehat{J}_9) = \frac{r_q}{K} Q_9^{CO} < 0, \quad \det(\widehat{J}_9) = \eta_1 \omega_1 \frac{r_q}{K} P_9^{CO} Q_9^{CO} S_1^{CO(9)} < 0.$$

The sum M_9^{CO} of the minors of order 2 of $\widehat{J}_9$ is

$$M_9^{CO} = \left(\eta_1 \omega_1 P_9^{CO} + \gamma_1 \mu_1 Q_9^{CO} \right) S_1^{CO(9)},$$

so that the third Routh-Hurwitz condition, upon simplification and collecting the term $r_q K^{-1} Q_9^{CO} S_1^{CO(9)}$, gives

$$\left(\eta_1 \omega_1 P_9^{CO} + \gamma_1 \mu_1 Q_9^{CO} \right) - \eta_1 \omega_1 Q_9^{CO} = \gamma_1 \mu_1 P_9^{CO} > 0$$

which is trivially satisfied, so that the stability condition reduces only to (51).

For E_{10}^{CO}, from the last system's equation we obtain

$$Q_{10}^{CO} = \frac{K}{r_q}(r_q - \gamma_1 S_1^{CO(10)} - \gamma_2 S_2^{CO(10)}),$$

from which the first feasibility condition

$$r_q \geq \gamma_1 S_1^{CO(10)} + \gamma_2 S_2^{CO(10)}. \tag{52}$$

Again, this situation differs from the one of Section 4 in that here the prey grow logistically, while previously we had an exponential growth. Substitution into the second and third equations leads to a linear system whose solution gives the remaining components. Letting $\Delta_* = c\mu_2\gamma_1 K + g\mu_1\gamma_2 K - cgr_q$, we find

$$S_1^{CO(10)} = \frac{1}{\Delta_*}[\gamma_2 K(e\mu_1 - a\mu_2) + cr_q(\mu_2 K - e)],$$

$$S_2^{CO(10)} = \frac{1}{\Delta_*}[\gamma_1 K(a\mu_2 - e\gamma_1) + gr_q(K\mu_1 - a)].$$

The other feasibility conditions are then either

$$c\mu_2\gamma_1 K + g\mu_1\gamma_2 K > cgr_q, \quad \mu_1 K\gamma_2 e + c\mu_2 r_q K \geq cr_q e + a\mu_2 K\gamma_2, \tag{53}$$
$$\mu_2 K\gamma_1 a + g\mu_1 r_q K \geq gr_q a + e\mu_1 K\gamma_1,$$

or the reverse inequalities,

$$c\mu_2\gamma_1 K + g\mu_1\gamma_2 K < cgr_q, \quad \mu_1 K\gamma_2 e + c\mu_2 r_q K \leq cr_q e + a\mu_2 K\gamma_2, \tag{54}$$
$$\mu_2 K\gamma_1 a + g\mu_1 r_q K \leq gr_q a + e\mu_1 K\gamma_1.$$

Stability is regulated by one explicit eigenvalue, which gives the condition

$$r_p \geq \omega_1 S_1^{CO(10)} + \omega_2 S_2^{CO(10)}, \tag{55}$$

and by a cubic, for which one of the Routh-Hurwitz conditions holds trivially, as it reduces to $r_q K^{-1} Q_{10}^{CO} > 0$, while the remaining ones give

$$gcr_q < K(\mu_2\gamma_1 c + g\mu_1\gamma_2), \tag{56}$$
$$K S_1^{CO(10)} S_2^{CO(10)} (c\gamma_1\mu_2 + g\gamma_2\mu_1)$$
$$< r_q Q_{10}^{CO} (\gamma_2\mu_2 S_2^{CO(10)} + \gamma_1\mu_1 S_1^{CO(10)}).$$

A simulation showing E_{10}^{CO} is contained in Figure 14 for the parameter values $r_p = 1.25, \omega_1 = 1.2, \omega_2 = 0.7, a = 0.12, c = 0.5, \mu_1 = 2, \eta_1 = 3, e = 0.5, g = 0.4, \mu_2 = 3.2, \eta_2 = 1.2, r_q = 2.0, K = 0.53, \gamma_1 = 2, \gamma_2 = 3$.

The feasibility of the remaining coexistence equilibrium is hard to assess. But we run simulations to show that it can be achieved. For the following parameter values, we have been able to empirically show coexistence of the ecosystem, see Figure 15: $r_p = 0.4, \omega_1 = 1.2, \omega_2 = 0.7, a = 0.12, c = 0.5, \mu_1 = 1, \eta_1 = 3, e = 0.5, g = 0.37, \mu_2 = 2, \eta_2 = 1.2, r_q = 3.0, K = 0.53, \gamma_1 = 2, \gamma_2 = 3$.

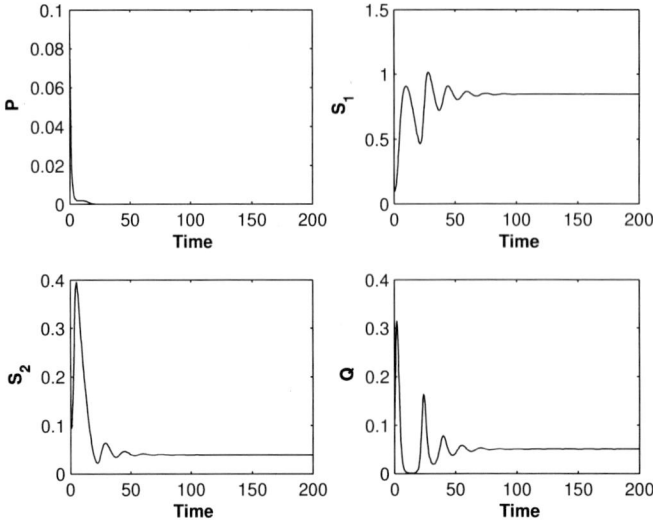

Figure 14. Equilibrium E_{10}^{CO} of the model (39) obtained with the parameter values $r_p = 1.25, \omega_1 = 1.2, \omega_2 = 0.7, a = 0.12, c = 0.5, \mu_1 = 2, \eta_1 = 3, e = 0.5, g = 0.4, \mu_2 = 3.2, \eta_2 = 1.2, r_q = 2.0, K = 0.53, \gamma_1 = 2, \gamma_2 = 3.$

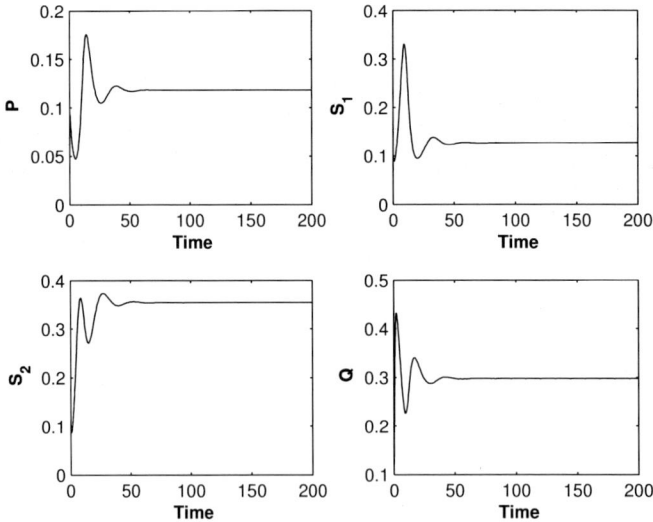

Figure 15. Coexistence equilibrium for the model (39) obtained with the parameter values $r_p = 0.4, \omega_1 = 1.2, \omega_2 = 0.7, a = 0.12, c = 0.5, \mu_1 = 1, \eta_1 = 3, e = 0.5, g = 0.37, \mu_2 = 2, \eta_2 = 1.2, r_q = 3.0, K = 0.53, \gamma_1 = 2, \gamma_2 = 3.$

5.1.1. Discussion

In this situation, there are several possible system's outcomes. Only its disappearance is forbidden, together with the thriving of just its mutualistic part. This seems the most striking result, which matches the result for the same equilibrium E_2^O in Section 2. But its real motivation is not to be sought in its interactions with the top predator (or the bottom prey), but rather on the symbiotic association itself. In fact, the positive eigenvalue does not contain parameters that are related with these interactions, but only those pertaining to the mutualism. And ultimately, the reason is that also in absence of the other populations in the food chain, the symbiotic system does have a coexistence equilibrium, which however is always a saddle, in case of an obligated association. In fact, it has the very same eigenvalues of opposite signs, $\pm\sqrt{ae}$.

5.2. Facultative Mutualism

Again, here let us recall that the model differs from both (1) and (25) of the previous Sections 2 and 4. Therefore we carry out the whole analysis, as not every equilibrium reduces to the ones of the previous cases.

In this case, the Jacobian of (40) becomes

$$J^{CF} = \begin{pmatrix} J_{11}^{CF} & P\omega_1 & P\omega_2 & 0 \\ -S_1\eta_1 & J_{22}^{CF} & S_1c & S_1\mu_1 \\ -S_2\eta_2 & S_2g & J_{33}^{CF} & S_2\mu_2 \\ 0 & -Q\gamma_1 & -Q\gamma_2 & J_{44}^{CF} \end{pmatrix}, \tag{57}$$

with $J_{11}^{CF} = -r_p + \omega_1 S_1 + \omega_2 S_2$, $J_{22}^{CF} = a - 2bS_1 + cS_2 + \mu_1 Q - \eta_1 P$, $J_{33}^{CF} = e - 2fS_2 + gS_1 + \mu_2 Q - \eta_2 P$, $J_{44}^{CF} = r_q - \frac{2r_q Q}{K} - \gamma_1 S_1 - \gamma_2 S_2$.

There are a number of equilibria which are possibly feasible, listed below, in addition to the coexistence equilibrium E_{13}^{CF}.

Equilibrium	P	S_1	S_2	Q	Feasibility condition
E_0^{CF}	0	0	0	0	—
E_1^{CF}	0	0	+	0	—
E_2^{CF}	+	0	+	0	(58)
E_3^{CF}	0	+	0	0	—
E_4^{CF}	+	+	0	0	(60)
E_5^{CF}	0	+	+	0	(62)
E_6^{CF}	+	+	+	0	(64)
E_7^{CF}	0	0	0	K	—
E_8^{CF}	0	0	+	+	(66)
E_9^{CF}	+	0	+	+	(71)
E_{10}^{CF}	0	+	0	+	(67)
E_{11}^{CF}	+	+	0	+	(73)
E_{12}^{CF}	0	+	+	+	(68)

We now study each one of these equilibria individually.

E_0^{CF} is always unstable in view of the eigenvalues $-r_p$, a, e, r_q. Also E_1^{CF} with nonvanishing component $S_2^{CO(1)} = ef^{-1}$ is always unstable, since it has the eigenvalues

$$\frac{1}{f}(e\omega_2 - fr_p), \quad \frac{1}{f}(af + ce) > 0, \quad -e, \quad \frac{1}{f}(fr_q - e\gamma_2).$$

Similarly, E_3^{CF} with $S_1^{CO(3)} = ab^{-1}$ is unstable, with eigenvalues

$$\frac{1}{b}(a\omega_1 - br_p), \quad -a, \quad \frac{1}{b}(ag + be) > 0, \quad \frac{1}{b}(br_q - a\gamma_1).$$

These results agree with those of Section 2, compare respectively the stability of E_2 and E_3. The same instability result holds for E_7^{CF}; the eigenvalues are indeed $-r_p$, $a - K\mu_1$, $e + K\mu_2 > 0$, $-r_q$. Note that for this equilibrium there is no counterpart in Section 4, in view of the two models' different assumptions on the prey Q dynamics.

The point E_2^{CF} with

$$P_2^{CO} = \frac{\omega_2 e - r_p f}{\omega_2 \eta_2}, \quad S_2^{CO(2)} = \frac{r_p}{\omega_2},$$

is feasible for

$$\frac{e}{f} \geq \frac{r_p}{\omega_2}. \tag{58}$$

Stability is achieved, since the characteristic polynomial factors to give two explicit eigenvalues and the remaining quadratic satisfies the Routh-Hurwitz conditions. The stability conditions are

$$r_q\omega_2 < r_p\gamma_2, \quad a\omega_2\eta_2 + fr_p\eta_1 + cr_p\eta_2 < e\eta_1\omega_2. \tag{59}$$

There is a transcritical bifurcation with E_2^{CF} when P_2^{CF} vanishes, i.e. when (58) becomes an equality. E_2^{CF} then coalesces with E_1^{CF}. Compare these results with those for E_7 in Section 2.

The equilibrium E_4^{CF} with

$$P_4^{CO} = \frac{\omega_1 a - r_p b}{\omega_1 \eta_1}, \quad S_1^{CO(4)} = \frac{r_p}{\omega_1},$$

is feasible for

$$\frac{a}{b} \geq \frac{r_p}{\omega_1}. \tag{60}$$

Again the characteristic equation factors to give a quadratic for which the Routh-Hurwitz conditions hold, and two explicit eigenvalues from which the stability conditions are derived:

$$r_q\omega_1 < r_p\gamma_1, \quad e\omega_1\eta_1 + gr_p\eta_1 + br_p\eta_2 < a\eta_2\omega_1. \tag{61}$$

Here again a transcritical bifurcation arises with the equilibrium E_3^{CO}, when say $\omega_1 = a^{-1}br_p$, i.e. when (60) becomes an equality. The results for this equilibrium should be compared with those of E_7 and E_7^T in Section 2, namely (11) and (30).

At E_5^{CF} with

$$S_1^{CO(5)} = \frac{af + ce}{fb - gc}, \quad S_2^{CO(5)} = \frac{eb + ga}{fb - gc}$$

the feasibility condition is

$$fb > gc, \qquad (62)$$

and its use makes the Routh-Hurwitz conditions for the quadratic arising from the factorization of the Jacobian hold. Stability is governed by the remaining explicit eigenvalues, giving

$$r_q < \gamma_1 S_1^{CF(5)} + \gamma_2 S_2^{CF(5)}, \quad r_p > \omega_1 S_1^{CF(5)} + \omega_2 S_2^{CF(5)}. \qquad (63)$$

This equilibrium should be compared with E_4 of Section 2, which is always unstable, and E_6^T of Section 4, which is stable if (28) and (29) hold. Compare also the feasibility condition (62) with (3) for E_4 and (27) for E_6^T. These differences again should not surprise, in view of the different assumptions in each one of these models.

Letting $\Delta_6 = (g\eta_1 + b\eta_2)\omega_2 + (f\eta_1 + c\eta_2)\omega_1$, the equilibrium E_6^{CF} with

$$P_6^{CF} = \frac{1}{\eta_1}(a - bS_1^{CF(6)} - cS_2^{CF(6)}),$$

$$S_1^{CO(6)} = \frac{1}{\Delta_6}[(a\eta_2 - e\eta_1)\omega_2 + r_p(f\eta_1 + c\eta_2)],$$

$$S_2^{CO(6)} = \frac{1}{\Delta_6}[(e\eta_1 - a\eta_2)\omega_1 + r_p(g\eta_1 + b\eta_2)],$$

is feasible if

$$a\eta_2\omega_2 + r_p(f\eta_1 + c\eta_2) > e\eta_1\omega_2, \quad e\eta_1\omega_1 + r_p(g\eta_1 + b\eta_2) > a\eta_2\omega_1, \qquad (64)$$
$$a + cS_2^{CF(6)} > bS_1^{CF(6)}.$$

Stability is ensured using the Routh-Hurwitz conditions for a cubic, two of which are always true. Since the Jacobian explicitly gives one eigenvalue we are left with the corresponding nonnegativity condition and the third Routh-Hurwitz condition. The final outcome are the inequalities

$$r_q < \gamma_1 S_1^{CF(6)} + \gamma_2 S_2^{CF(6)}, \qquad (65)$$
$$\left[b\eta_1\omega_1(S_1^{CF(6)})^2 + f\eta_2\omega_2(S_2^{CF(6)})^2\right] P_6^{CF}$$
$$+ \left[bS_1^{CF(6)} + fS_2^{CF(6)}\right](bf - cg)S_1^{CF(6)}S_2^{CF(6)} >$$
$$P_6^{CF} S_1^{CF(6)} S_2^{CF(6)}(c\eta_2\omega_1 + g\eta_1\omega_2).$$

Note that the first condition is the opposite of the one for E_5^{CF}, so that the latter and this equilibrium E_6^{CF} can never stably coexist. This equilibrium corresponds to the coexistence equilibrium E_8 of Section 2. Here, in view of the assumptions underlying the model (40), it has been possible to work out a bit more the stability conditions.

The equilibrium E_8^{CF} with

$$S_1^{CF(8)} = \frac{r_q}{\Delta_8}(e + \mu_2 K), \quad Q^{CF(8)} = \frac{K}{\Delta_8}(fr_q - e\gamma_2), \quad \Delta_8 = fr_q + K\gamma_2\mu_2$$

is feasible for

$$fr_q \geq e\gamma_2 \qquad (66)$$

but in view of the eigenvalue $a + cS_2^{CF(8)} + \mu_1 Q_8 > 0$ it is always unstable. This result is to be compared with the one for E_3^T of Section 4, which is also unstable.

Similarly E_{10}^{CF}, with

$$S_1^{CF(10)} = \frac{r_q}{\Delta}(a + \mu_1 K), \quad Q^{CF(10)} = \frac{K}{\Delta}(br_q - a\gamma_1), \quad \Delta = br_q + K\gamma_1\mu_1,$$

feasible for

$$br_q \geq a\gamma_1, \tag{67}$$

is unstable since it has the positive eigenvalue $e + gS_1^{CF(10)} + \mu_2 Q_{10}^{CF}$. Compare it with E_5^T of Section 4, also unstable.

For E_{12}^{CF} the population values are

$$Q_{12}^{CF} = \frac{1}{\mu_1}(bS_1^{CF(12)} - a - cS_2^{CF(12)}) = \frac{1}{\mu_2}(fS_2^{CF(12)} - e - gS_1^{CF(12)}),$$

$$\Delta = \left(K\frac{\gamma_1}{r_q} + \frac{b}{\mu_1}\right)\left(K\frac{\gamma_2}{r_q} + \frac{f}{\mu_2}\right) - \left(K\frac{\gamma_1}{r_q} - \frac{g}{\mu_2}\right)\left(K\frac{\gamma_2}{r_q} + \frac{c}{\mu_1}\right),$$

$$S_1^{CF(12)} = \frac{1}{\Delta}\left[\left(K + \frac{a}{\mu_1}\right)\left(K\frac{\gamma_2}{r_q} + \frac{f}{\mu_2}\right) - \left(K + \frac{e}{\mu_2}\right)\left(K\frac{\gamma_2}{r_q} - \frac{c}{\mu_1}\right)\right],$$

$$S_2^{CF(12)} = \frac{1}{\Delta}\left[\left(K + \frac{e}{\mu_2}\right)\left(K\frac{\gamma_1}{r_q} + \frac{b}{\mu_1}\right) - \left(K + \frac{a}{\mu_1}\right)\left(K\frac{\gamma_1}{r_q} - \frac{g}{\mu_2}\right)\right].$$

Note that combining the requests for nonnegativity from both the above equations for Q_{12}^{CF}, the condition $bf > cg$ follows from

$$bfS_1^{CF(12)} > af + cfS_2^{CF(12)} > af + c(e + gS_1^{CF(12)}),$$

which gives $(bf - cg)S_1^{CF(12)} > af + ce > 0$. It is needed below to show that $\det(J^{CF(12)}) > 0$. Feasibility conditions are the set

$$\Delta > 0, \quad S_1^{CF(12)} \geq 0, \quad S_2^{CF(12)} \geq 0, \tag{68}$$

or their opposite inequalities. Stability imposes the negativity of one explicit eigenvalue, giving

$$r_p > \omega_1 S_1^{CF(12)} + \omega_2 S_2^{CF(12)} \tag{69}$$

and from the remaining 3 by 3 submatrix $J^{CF(12)}$, two Routh-Hurwitz conditions are easily satified, $-\mathrm{tr}J^{CF(12)} > 0$ and $-\det J^{CF(12)} > 0$, having used $bf > cg$, while the last one gives

$$S_1^{CF(12)}S_2^{CF(12)}(bf - cg)\left(bS_1^{CF(12)} + fS_2^{CF(12)}\right) \tag{70}$$
$$+Q^{CF(12)}S_2^{CF(12)}\left(f\frac{r_q}{K} + \gamma_2\mu_2\right)\left(fS_2^{CF(12)} + \frac{r_q}{K}Q^{CF(12)}\right)$$
$$+S_1^{CF(12)}Q^{CF(12)}\left(b\frac{r_q}{K} + \gamma_1\mu_1\right)\left(bS_1^{CF(12)} + \frac{r_q}{K}Q^{CF(12)}\right)$$
$$+S_1^{CF(12)}S_2^{CF(12)}Q^{CF(12)}\left(2bf\frac{r_q}{K} - c\gamma_1\mu_2 - g\gamma_2\mu_1\right) > 0.$$

This would be the counterpart of equilibrium E_7^T in Section 4. Compare also the feasibility conditions (68) and (30).

Again the equilibria involving all the three trophic levels, that therefore cannot arise in the previous models of Sections 2 and 4, are E_9^{CF} and E_{11}^{CF}.

For E_9^{CF} the components are

$$S_2^{CF(9)} = \frac{r_p}{\omega_2}, \quad Q_9^{CF} = \frac{K}{r_q}\left(r_q - \gamma_2 S_2^{CF(9)}\right),$$

$$P_9^{CF} = \frac{1}{\eta_2}(e - f S_2^{CF(9)} + \mu_2 Q_9^{CF}),$$

from which the feasibility conditions follow

$$r_q \omega_2 \geq r_p \gamma_2, \quad r_p(f r_q + K \gamma_2 \mu_2) \leq r_q \omega_2(e + K \mu_2). \tag{71}$$

Stability is ensured by factorization of the characteristic equation, giving one explicit eigenvalue if the next condition holds

$$a + c S_2^{CF(9)} + \mu_1 Q^{CF(9)} < \eta_1 P^{CF(9)} \tag{72}$$

and from a 3 by 3 submatrix $J^{CF(9)}$, giving the remaining cubic polynomial, for which the Routh-Hurwitz conditions are satisfied, namely

$$-\mathrm{tr}(J^{CF(9)}) = f S_2^{CF(9)} + \frac{1}{K} r_q Q_9 > 0,$$

$$-\det(J^{CF(9)}) = \frac{1}{K} r_q \eta_2 \omega_2 P_9 Q_9 S_2^{CF(9)} > 0,$$

$$-\mathrm{tr}(J^{CF(9)}) M_2(J^{CF(9)}) + \det(J^{CF(9)})$$

$$= S_2^{CF(9)} (Q_9^{CF})^2 \frac{r_q}{K^2}(f r_q + K \gamma_2 \mu_2) + S_2^{CF(9)} P_9^{CF} \eta_2 \omega_2$$

$$+ S_2^{CF(9)} Q_9^{CF} \frac{1}{K}(f r_q + K \gamma_2 \mu_2) > 0.$$

At E_{11}^{CF} the population values are

$$S_1^{CF(11)} = \frac{r_p}{\omega_1}, \quad Q_{11}^{CF(11)} = \frac{K}{r_q \omega_1}(r_q \omega_1 - \gamma_1 r_p),$$

$$P_{11}^{CF} = \frac{1}{\eta_1}(a - \frac{b r_p}{\omega_1} + \mu_1 Q_{11}^{CF(11)}),$$

from which the feasibility conditions

$$r_q \omega_1 \geq r_p \gamma_1, \quad r_q \omega_1(a + K \mu_1) \geq r_p(\mu_1 K \gamma_1 + b r_q). \tag{73}$$

Again stability is obtained from an explicit eigenvalue, since the three Routh-Hurwitz conditions for a cubic hold unconditionally with expressions similar to the ones as above, giving

$$e + g S_1^{CF(11)} + \mu_2 Q_{11}^{CF} < \eta_2 P_{11}^{CF}. \tag{74}$$

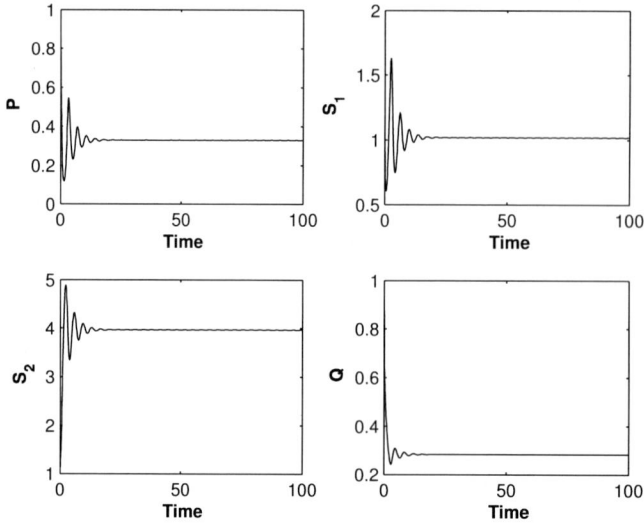

Figure 16. Coexistence equilibrium for the model (40) obtained for the parameter values $r_p = 4$, $w_1 = 1.2$, $w_2 = 0.7$, $a = 0.12$, $b = 0.4$, $c = 0.25$, $f = 0.4$, $\mu_1 = 1$, $\eta_1 = 3$, $e = 1.3$, $g = 0.37$, $\mu_2 = 2$, $\eta_2 = 2$, $r_q = 3.0$, $K = 0.53$, $\gamma_1 = 0.2$, $\gamma_2 = 0.3$.

For these two last equilibria, transcritical bifurcations can occur. In particular, E_9^{CF} could coalesce with E_2^{CF}, while E_{11}^{CF} would collapse onto E_4^{CF}.

Finally, the feasibility and stability of the coexistence equilibrium are shown empirically in Figure 16, for the following parameter values $r_p = 4$, $w_1 = 1.2$, $w_2 = 0.7$, $a = 0.12$, $b = 0.4$, $c = 0.25$, $f = 0.4$, $\mu_1 = 1$, $\eta_1 = 3$, $e = 1.3$, $g = 0.37$, $\mu_2 = 2$, $\eta_2 = 2$, $r_q = 3.0$, $K = 0.53$, $\gamma_1 = 0.2$, $\gamma_2 = 0.3$.

5.1.1. Discussion

In this system, a number of equilibria are not sustainable. The system does not disappear, nor can anyone of the two symbiotic populations thrive alone, in spite of the other available food sources. This result is unexpected, but it stems again from the underlying symbiotic subsystem. In fact, each of these equilibria would be a boundary equilibrium in the two populations symbiotic subsystem, and each of them would have the positive eigenvalues $f^{-1}(af+ce)$ for $(0, ef^{-1})$ and $b^{-1}(eb+ag)$ for $(ab^{-1}, 0)$. For this subsystem, all the trajectories would in fact tend to the coexistence equilibrium $(\Lambda_*^{-1}(af+ec), \Lambda_*^{-1}(be+ag))$, where $\Lambda_* = bf - cg$, with the same feasibility condition as for E_5^{CF}, (62). Letting J^* denote its Jacobian, the Routh-Hurwitz conditions are always satisfied, since they are given by $-\text{tr}(J^*) = bfS_1^*S_2^* > 0$, $\det(J^*) = \Lambda_* S_1^* S_2^* > 0$. This very same coexistence equilibrium can also be achieved in the present situation, see equilibrium E_5^{CF}, but in the food chain it is subject to an additional stability condition, namely (63). Thus the influence of the either one of the top predator or the bottom prey can destabilize this equilibrium. Indeed it is enough that the reproduction rates of the bottom prey increases past a threshold, see the first condition of (63), or the mortality rate of the top predator falls below another threshold, see the second (63). These come as no surprise, if the prey can reproduce enough, they will establish themselves in the ecosystem in spite of the symbiotic predators exerting a pressure on them, and similarly if the top predators have a low mortality rate, they will thrive.

The other unstable configurations are the bottom prey alone, in spite of the fact that they grow logistically and the environment could support them at carrying capacity K, and these very bottom prey with just either one of their symbiotic predators.

The top predators in suitable conditions instead can thrive with one or both of their prey living in mutualistic association. Note that for latter case, see (65), a low bottom prey reproduction rate is needed. Both three trophic level equilibria can be achieved, containing either one of the intermediate symbiotic populations. This cannot occur simultaneously with equilibria E_2^{CF} and E_4^{CF} in view of the opposite conditions, compare the first feasibility condition (71) for E_9^{CF} with the first stability condition (59) of E_2^{CF}, and the similar situation of E_{11}^{CF} and E_4^{CF}, namely (73) with (61). Also the subsystem formed by the symbiotic association and the bottom prey can thrive, but the mortality rate of the top predators should be high enough, see (69).

6. A Disturbance on a Synergetic Association

We want to consider finally a model in which a symbiotic community is subject to the action of an external pressure, to which in part it reacts and in part it succumbs. Such a situation is very uncommon in biological terms, but it may have an interpretation in the economic or financial world. As mentioned in the Introduction, the activities of a company for instance may outcompete one of two synergetic partners, but may be outcompeted by the other one. Again we distinguish a situation in which the synergy is obligated and one in which instead it is facultative.

6.1. Obligated Synergy

More specifically, we assume here in fact that two entities X and Y are synergetic, and each one of them cannot survive in absence of the other one and, at least for X, also if the third entity Z is missing. With rispect to a third party Z, which may be a competitor, one of them, Y, might be weaker and therefore will feel negatively the presence of Z, while the other one, X, as mentioned instead may be stronger and therefore hinder the growth of the third party Z and gain from the interaction with it. The minimal model has thus the following form,

$$\frac{dX}{dt} = X\left(-\delta + aZ + bY\right), \qquad (75)$$
$$\frac{dY}{dt} = Y\left(-cZ - \gamma + bX\right),$$
$$\frac{dZ}{dt} = Z\left(-\theta_1 aX + \theta_2 cY - \mu\right).$$

The first two equations express the obligated mutualism of X and Y, in this case for simplicity taking their mutual benefits, expressed by the common parameter b, to be equal. The last equation states that Z thrives only because of its plundering on Y, otherwise it would disappear due to the action of the "mortality" μ. This is reflected by a corresponding term in the Y equation, negatively influencing it. The converse holds for the interaction of Z with X since it is the latter to benefit from it, and the former that suffers.

6.1.1. Analysis

The system (75) has only three equilibria: the origin E_0^{DO}, the boundary equilibrium point $E_1^{DO} = (\gamma b^{-1}, \delta b^{-1}, 0)$ and the coexistence equilibrium E_2^{DO}. The equilibrium points E_0^{DO} and E_1^{DO} are always feasible. The coordinates of E_2^{DO} are

$$E_2^{DO} = \left(\frac{\theta_2 a\gamma + \theta_2 c\delta - b\mu}{(\theta_1 + \theta_2)ab}, \; \frac{b\mu + \theta_1 a\gamma + \theta_1 c\delta}{(\theta_1 + \theta_2)bc}, \; \frac{\theta_2 c\delta - \theta_1 a\gamma - b\mu}{(\theta_1 + \theta_2)ac} \right),$$

and therefore its feasibility condition is

$$b\mu + \theta_1 \gamma a < \theta_2 c\delta. \tag{76}$$

The Jacobian matrix of the system (75) has the form

$$J^{DO} \equiv \begin{pmatrix} -\delta + aZ_i + bY_i & bX_i & aX_i \\ bY_i & -cZ_i - \gamma + bX_i & -cY_i \\ -\theta_1 aZ_i & \theta_2 cZ_i & -\theta_1 aX_i + \theta_2 cY_i - \mu \end{pmatrix}. \tag{77}$$

At the origin, the eigenvalues $-\delta$, $-\gamma$, $-\mu$ are found, showing its stability. Again, this is not a nice result, as it means that the system can in fact disappear. At E_1^{DO}, we have the eigenvalues $\lambda_1^{DO} = \sqrt{\delta\gamma}$, $\lambda_2^{DO} = -\sqrt{\delta\gamma}$, $\lambda_3^{DO} = -\mu + b^{-1}(\theta_2 c\delta - \theta_1 a\gamma)$. Thus, E_1^{DO} is always unstable. Hence the presence of the disturbance Z does not help in stabilizing the equilibrium that the synergetic system shows in its absence. Note in fact that E_1^{DO} cannot be achieved even in absence of Z since it is a saddle, its eigenvalues are those of the 2 by 2 top principal minor of J^{DO}, i.e. λ_1^{DO} and λ_2^{DO}. Thus in the $X - Y$ phase plane the point E_1^{DO} is a saddle, and trajectories either tend to the origin or grow up to infinity. The latter is a phenomenon that cannot occur in practice in view of lack of resources, it is mainly due to our assumption of working with a minimal model, but it illustrates the fact that the synergy can ultimately benefit the two entities X and Y beyond any optimistic expectations. In the phase plane there must then also exist a separatrix, for which a basin of attraction of the origin is identified, that for suitable initial conditions attracts trajectories. But note that with higher returns for the parties, i.e. increasing b, the equilibrium E_1^{DO} moves toward the origin, thereby most likely decreasing the latter's basin of attaction, thus benefiting the mutualistic interations.

At the interior equilibrium point the characteristic equation is

$$\Psi \equiv \lambda^3 + \left(\theta_1 a^2 X_2 Z_2 + \theta_2 c^2 Y_2 Z_2 - b^2 X_2 Y_2 \right) \lambda - (\theta_1 + \theta_2)abc X_2 Y_2 Z_2 = 0. \tag{78}$$

$\Psi(\lambda)$ is a cubic tending to infinity for large values of λ. But at the origin, it has a negative value, so that a real positive root must exist. Hence one eigenvalue is positive, and the equilibrium E_2^{DO} is unstable.

6.2.2. Discussion

Thus also in the three dimensional situation the coexistence equilibrium acts like a saddle. We conjecture the existence of a separating surface, that divides the basin of attraction of the origin from another region away from it, on which trajectories eventually expand without bounds. Therefore these complex interactions of synergetic and predatory types appear still to be beneficial for the system as a whole, or at least for some of its parts, if the initial status of the system lies ouside the basin of attraction of the origin.

Further, the fact that all the boundary equilibria are either unfeasible or unstable is a positive result, in the sense that no agent can disappear alone. Either all survive, or all are wiped out. The synergetic nature of part of the association favors the long term presence of competition. Instead, for two competing populations the principle of competitive exclusion is well known, only one of the two competitors will ultimately survive, and the final system's outcome is already contained in the initial conditions of the dynamics. The winner is the nonvanishing population at the boundary equilibrium in the domain of attaction of which the initial state of the system lies.

6.2. Facultative Synergy

As already done in all previous Sections, we consider also the alternative formulation. Assuming that all variables have other ways of supporting themselves, modeled by the logistic terms, the system in this case reads

$$\dot{X} = X\left[r\left(1 - \frac{X}{K}\right) + aZ + bY\right], \tag{79}$$

$$\dot{Y} = Y\left[s\left(1 - \frac{Y}{H}\right) - cZ + gX\right],$$

$$\dot{Z} = Z\left[u\left(1 - \frac{Z}{L}\right) - eX + fY\right],$$

with the parameter assumptions $a < e$ and $f < c$. Its Jacobian J^{DF} is

$$\begin{pmatrix} J_{11}^{DF} & bX & aX \\ gY & J_{22}^{DF} & -cY \\ -eZ & fZ & J_{33}^{DF} \end{pmatrix}$$

with

$$J_{11}^{DF} = r\left(1 - \frac{X}{K}\right) + aZ + bY - \frac{r}{K}X,$$

$$J_{22}^{DF} = s\left(1 - \frac{Y}{H}\right) - cZ + gX - \frac{s}{H}Y,$$

$$J_{33}^{DF} = u\left(1 - \frac{Z}{L}\right) - eX + fY - \frac{u}{L}Z.$$

6.1.1. Analysis

We have gained enough experience by now to state from the onset of the analysis that here all the 8 equilibria of the system are possible. The boundary ones are summarized in the following table

Equilibrium	X	Y	Z	Feasibility condition
E_0^{DF}	0	0	0	—
E_1^{DF}	0	0	L	—
E_2^{DF}	0	H	0	—
E_3^{DF}	0	+	+	(80)
E_4^{DF}	K	0	0	—
E_5^{DF}	+	0	+	(81)
E_6^{DF}	+	+	0	(82)

and then we have the coexistence equilibrium E_7^{DF}. The components of E_3^{DF} are

$$Y_3^{DF} = uH\frac{s - cL}{us + cfHL}, \quad Z_3^{DF} = \frac{s}{c}\left(1 - \frac{Y_3^{DF}}{H}\right),$$

those of E_5^{DF} are instead

$$X_5^{DF} = \frac{u}{e}\left(1 - \frac{Z_5^{DF}}{L}\right), \quad Z_5^{DF} = rL\frac{u - eK}{ru + aeKL},$$

and finally the ones of E_6^{DF} are

$$X_6^{DF} = sK\frac{r + bH}{rs - bgHK}, \quad Y_6^{DF} = \frac{r}{b}\left(\frac{X_6^{DF}}{K} - 1\right).$$

Feasibility for E_3^{DF} requires

$$s > cL, \tag{80}$$

for E_5^{DF} we have instead

$$u > eK, \tag{81}$$

and finally for E_6^{DF} we find

$$sr > bgHK. \tag{82}$$

For the coexistence equilibrium we have instead

$$X_7^{DF} = K\frac{\Gamma}{\Omega}, \quad Y_7^{DF} = H\frac{\Lambda}{\Omega}, \quad Z_7^{DF} = L\frac{\Pi}{\Omega},$$

with

$$\Gamma = -rsu - rfLcH - aLsu - aLfsH + bHLcu - bHsu,$$
$$\Lambda = -gkru - gkLau - kreLc - seLak + Lrcu - rsu,$$
$$\Pi = -rsu + rsek - rfgHk - rfsH + gHbku + ebksH,$$
$$\Omega = -rsu - rfLcH + gHbku - seLak + fLgHak + eLbkcH,$$

with feasibility conditions given by one of the two alternative sets of inequalities

$$\Gamma > 0, \quad \Lambda > 0, \quad \Pi > 0, \quad \Omega > 0, \tag{83}$$
$$\Gamma < 0, \quad \Lambda < 0, \quad \Pi < 0, \quad \Omega < 0.$$

As far as their stability is concerned it is easy to see that equilibria E_0^{DF} to E_4^{DF} are all unstable, since they have all at least one positive eigenvalue, in fact r, s and u for E_0^{DF} and then respectively $r + aL$ for E_1^{DF}, $r + bH$ for E_2^{DF}, $s + gH$ for E_3^{DF}, $r + aZ_4^{DF} + bY_4^{DF}$ for E_4^{DF}.

For E_5^{DF}, one eigenvalue is explicit from the factorization of the characteristic equation and gives the stability condition $s + gX_5^{DF} < cZ_5^{DF}$ which explicitly becomes

$$rsu + aesKL + aguKL + gruK + cerKL < cruL, \qquad (84)$$

while the other two stem from a quadratic with positive coefficients that thus leads to eigenvalues with negative real parts, thus not altering the stability.

For E_6^{DF} we have a similar situation, $u - eX_6^{DF} + fY_6^{DF}$ being the explicit eigenvalue and the remaining quadratic with positive coefficients if we make use of (82). Explicitly the stability condition is

$$frH + fgrHK + rsu < bguHK + ersK + besHK. \qquad (85)$$

For E_7^{DF} the entries on the main diagonal of the Jacobian J^{DF} become

$$J_{11}^{DF} = -\frac{r}{K}X, \quad J_{22}^{DF} = -\frac{s}{H}Y, \quad J_{33}^{DF} = -\frac{u}{L}Z.$$

Now clearly $-\mathrm{tr}(J^{DF}(E_7^{DF})) > 0$. Also, we have

$$\det(J^{DF}(E_7^{DF})) = X_7^{DF}Y_7^{DF}Z_7^{DF}\left[bce + afg - \frac{rsu}{HKL} - ae\frac{s}{H} - cf\frac{r}{K} + bg\frac{u}{L}\right],$$

thus the first stability condition follows

$$bce + afg + bg\frac{u}{L} < \frac{rsu}{HKL} + ae\frac{s}{H} + cf\frac{r}{K}. \qquad (86)$$

The third Routh-Hurwitz condition, setting

$$D \equiv X_7^{DF}Y_7^{DF}Z_7^{DF}\left(bce + afg + 2\frac{rsu}{HKL}\right)$$
$$+ X_7^{DF}Z_7^{DF}\left(ae + \frac{ru}{KL}\right)\left(X_7^{DF}\frac{r}{K} + Z_7^{DF}\frac{u}{L}\right)$$
$$+ Y_7^{DF}Z_7^{DF}\left(cf + \frac{su}{HL}\right)\left(Y_7^{DF}\frac{s}{H} + Z_7^{DF}\frac{u}{L}\right)$$

instead leads to the following inequality

$$D + X_7^{DF}Y_7^{DF}\left(\frac{rs}{HK} - bg\right)\left(X_7^{DF}\frac{r}{K} + Y_7^{DF}\frac{s}{H}\right) > 0. \qquad (87)$$

Thus stability at E_7^{DF} holds whenever (86) and (87) are satisfied.

6.2.2. Discussion

In this situation only two system's outcomes are possible in addition to the coexistence of the three populations, for suitable parameter values satisfying (83), (86) and (87). The populations that can survive together in absence of the third one are the pairs X with Z and X with Y, thereby excluding that Y with Z can coexist. In this case, in principle Y and Z could coexist in an isolated environment, as they constitute a predator-prey system, but with X acting as a top predator for both, their stable coexistence is destroyed. The same does not occur in the other two cases, since the vanishing population never acts as a top predator.

References

[1] K.V.L.N. Acharyulu, N.Ch. Pattabhi Ramacharyulu, 2011, An Ammensal-Prey with Three Species Ecosystem, *International Journal of Computational Cognition* 9, 54-63.

[2] J.F. Addicott, Competition in mutualistic systems, 1985, in D.H. Boucher (Editor), *The Biology of Mutualism: Ecology and Evolution*, Croom Helm, London, 217-247.

[3] V. Ajraldi, M. Pittavino, E. Venturino, 2011, Modelling herd behavior in population systems, *Nonlinear Analysis: Real World Applications*, 12, 2319-2338.

[4] P. Auger, J.-C. Poggiale, 1998, Aggregation and emergence in systems of ordinary differential equations, *Math. Comput. Modelling*, 27, 1-21.

[5] I. Barbalat, 1959, Systèmes d'équations différentielles d'oscillation non lineares, *Rev. Math. Pure et Appl.*, 4, 267.

[6] D.H. Boucher (Editor), 1985, *The Biology of Mutualism: Ecology and Evolution*, Croom Helm, London.

[7] P.A. Braza, 2003, The bifurcations structure for the Holling Tanner model for predator-prey interactions using two-timing, *SIAM. J. Appl. Math.*, 63, 889-904.

[8] J. Chaston, H. Goodrich-Blair, 2010, Common Trends in Mutualism Revealed by Model Associations Between Invertebrates and Bacteria, *FEMS Microbiol Rev.*, 34, 41-58.

[9] J.R. Christie, K. Gopalsamy, J. Li, 1995, Chaos in sociobiology, *Bull. Austral. Math. Soc.*, 51, 439-451.

[10] C. Cioffi-Revilla, 1989, Mathematical contributions to the scientific understanding of war, *Math. Computer Modelling*, 12, 561-575.

[11] R.K. Colwell, B.J. Betts, P. Bunnell, F.L. Carpenter, P. Feisinger, 1974, Competition for the nectar of Centropogon valerii by the hummingbird Colibri thalassinus and the flower-piercer Diglossa plumbea, and its evolutionary implications, *Condor*, 76 447-452.

[12] C. Combes, 2001, Les associations du vivant, Flammarion, Paris.

[13] C.R. Currie, A.N.M. Bot, J.J. Boomsma, 2003, Experimental evidence of a tripartite mutualism: bacteria protect ant fungus gardens from specialized parasites, *OIKOS*, 101, 91-102.

[14] A.P. Dobson, 1985, The population dynamics of competition between parasites, *Parasitology*, 91, 317-347.

[15] M.M.A. El-Sheikh, S.A.A. Mahrouf, 2005, Stability and bifurcation of a simple food chain in chemostat with removal rates, *Chaos Solitons Fractals*, 23, 1475-1489.

[16] M. Genkai-Kato, N. Yamamura, 1999, Evolution of Mutualistic Symbiosis without Vertical Transmission, *Theoretical Population Biology*, 55, 309-323.

[17] B.S. Goh, 1979, Stability in models of mutualism, *The American Naturalist*, 113, 261-275.

[18] M. Haque, E. Venturino, 2008, Effect of parasitic infection in the Leslie-Gower predator-prey model, *Journal of Biological Systems*, 16, 425-444.

[19] M. Haque, E. Venturino, 2009, Mathematical models of diseases spreading in symbiotic communities, in J.D. Harris, P.L. Brown (Editors), *Wildlife: Destruction, Conservation and Biodiversity*, NOVA Science Publishers, New York, 135-179.

[20] S.E. Hartley, A.C. Gange, 2009, Impacts of Plant Symbiotic Fungi on Insect Herbivores: Mutualism in a Multitrophic Context, *Annu. Rev. Entomol.*, 54, 323-342.

[21] M.Hirsch, S. Smale, 1974, *Differential Equations, Dynamical Systems and Linear Algebra*, Academic Press, New York.

[22] R.D. Holt, 1977, Predation, apparent competition and the structure of prey communities, *Theoretical Population Biology*, 12, 197-229.

[23] A. Horovitz, T. Ben-Hur, 1983, A model for the kinectis of crime, *J. Theor. Biol.*, 103, 609-617.

[24] I. S. Hotopp, H. Malchow, E. Venturino, 2010, Switching feeding among sound and infected prey in ecoepidemic systems, *Journal of Biological Systems*, 18, 727-747.

[25] S.B. Hsu, T.W., Hwang, Y., Kuang, 2003, A ratio-dependent food chain model and its applications to biological control, *Math. Biosci.* 181, 55-83.

[26] P.K.R. Kambam, M.A. Henson, L. Sun, 2008, Design and mathematical modelling of a synthetic symbiotic ecosystem, *Systems Biology, IET*, 2, 33-38.

[27] Y. Kang, M. Makiyama, R. Clark, J. Fewell, 2011, Mathematical Modeling on Obligate Mutualism: Interactions between leaf-cutter ants and their fungus garden, *Journal of Theoretical Biology*, 289, 116-127.

[28] B.W. Kooi, M.P. Boer, S.A.L.M. Kooijman, 1998, On the use of the logistic equation in food chains, *Bulletin of Mathematical Biology*, 60, 231-246.

[29] M. Kwiatkowski, C. Vorburger, 2012, Modeling the Ecology of Symbiont-Mediated Protection against Parasites, *The American Naturalist*, 179, 595-605.

[30] S.A. Levin, 1983, Coevolution, in *Population Biology*, H.I. Freedman and C. Strobeck (Editors), Lecture Notes in Biomathematics 52, Springer Verlag, Heidelberg.

[31] H. Malchow, S. Petrovskii, E. Venturino, 2008, *Spatiotemporal patterns in Ecology and Epidemiology*, CRC, Boca Raton.

[32] L. Muscatine, J. W. Porter, 1971, Reef corals: mutualistic symbioses adapted to nutrient-poor environments, *Bioscience* 27, 454-460.

[33] N. Nagumo, 1942, Über die Lage der Integralkurven gewönlicher Differentialgleichungen, *Proc. Phys. Math. Soc. Japan*, 24, 551.

[34] M. L. Rosenzweig, 1971, Paradox of Enrichment: Destabilization of Exploitation Ecosystems in Ecological Time, *Science*, New Series, 171, No. 3969, 385-387.

[35] E. Venturino, 2007, How diseases affect symbiotic communities, *Math. Biosc.*, 206, 11-30.

[36] C. Zhang, G. Yan, 2011, Simulation Research in Mutualism Behavior of Automobile Industry Cluster, *Journal of Management and Strategy,* 2, 41-47.

In: Symbiosis: Evolution, Biology and Ecological Effects   ISBN: 978-1-62257-211-3
Editors: A. F. Camiso and C. C. Pedroso   © 2013 Nova Science Publishers, Inc.

Chapter 10

THE BISTABLE BRAIN: A NEURONAL MODEL WITH SYMBIOTIC INTERACTIONS

Ricardo López-Ruiz[1,*] *and Danièle Fournier-Prunaret*[2,†]
[1]Department of Computer Science and BIFI,
Universidad de Zaragoza, Spain
[2]LAAS-CNRS, 7 Avenue du Colonel Roche, and INSA,
Université de Toulouse, Cedex, France

Abstract

In general, the behavior of large and complex aggregates of elementary components can not be understood nor extrapolated from the properties of a few components. The brain is a good example of this type of networked systems where some patterns of behavior are observed independently of the topology and of the number of coupled units. Following this insight, we have studied the dynamics of different aggregates of logistic maps according to a particular *symbiotic* coupling scheme that imitates the neuronal excitation coupling. All these aggregates show some common dynamical properties, concretely a bistable behavior that is reported here with a certain detail. Thus, the qualitative relationship with neural systems is suggested through a naive model of many of such networked logistic maps whose behavior mimics the waking-sleeping bistability displayed by brain systems. Due to its relevance, some regions of multistability are determined and sketched for all these logistic models.

PACS 07.05.Mh, 05.45.Ra, 05.45.Xt

Keywords: Bistalibity, coupled logistic oscillators, neural networks

1. Introduction

The brain is a natural networked system [1, 2]. The understanding of this complex system is one of the most fascinating scientific tasks today, concretely how this set of millions of neurons can *symbiotically* interact among them to give rise to the collective phenomenon

---

*E-mail address: rilopez@unizar.es
†E-mail address: Daniele.Fournier@insa-toulouse.fr

of human thinking [3], or, in a simpler and more realistic approach, what neural features can make possible, for example, the birdsongs [4]. Different aspects of neurocomputation take contact on this problem: how brain stores information and how brain processes it to take decisions or to create new information. Other universal properties of this system are more evident. One of them is the existence of a regular daily behavior: the sleep-wake cycle [5, 6].

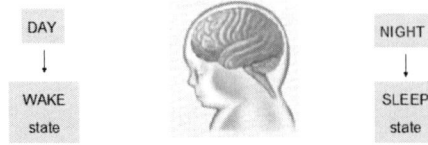

Figure 1. Brain bistability provoked by the solar light cycle.

The internal circadian rhythm is closely synchronized with the cycle of sun light (Fig. 1). Roughly speaking and depending on the particular species, the brain is awake during the day and it is slept during the night, or vice versa. All mammals and birds sleep. There is not a well established law relating the size of the animal with the daily time it spends sleeping, but, in general, large animals tend to sleep less than small animals (Fig. 2). Hence, at first sight, the emergent bistable sleep-wake behavior seems not depend on the precise architecture of the brain nor on its size. This structural property would mean that, if we represent the brain as a complex assembly of units [7, 8, 9], the possible bistability, where large groups of neurons can show some kind of synchronization, should not depend on the topology (structure) nor on the number of nodes (size) of the network (Fig. 3).

ANIMAL	BRAIN WEIGHT (gr.)	SLEEP TIME (hours)
Cat	30	13
Dog	70	11
Human (-)	350	16
Human (+)	1300	8
Horse	530	2
Elephant	4500	4

Figure 2. Brain size and mean sleeping time for different animals. (Humans(-) represent new born humans and Humans(+) represent middle age humans).

Then, it is essential the type of local dynamics and the excitation/inhibition coupling among the nodes that must be implemented in order to get bistability as a possible dynamical state in a complex network. So, on one side, it has been recently argued [10] that the distribution of functional connections $p(k)$ in the human brain, where $p(k)$ represents the probability of finding an element with k connections to other elements of the network, follows the same distribution of a scale-free network [11], that is a power law behavior, $p(k) \sim k^{-\gamma}$, with γ around 2. This finding means that there are regions in the brain that participate in a large number of tasks while most of the regions are only involved in a tiny fraction of the brain's activities. On the other side, it has been shown by Kuhn et al. [12] the nonlinear processing of synaptic inputs in cortical neurons. They studied the response of

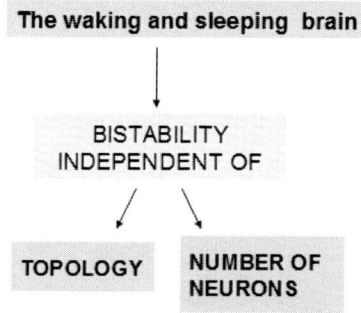

Figure 3. The waking-sleeping brain.

a model neuron with a simultaneous increase of excitation and inhibition. They found that the firing rate of the model neuron first increases, reaches a maximum, and then decreases at higher input rates. Functionally, this means that the firing rate, commonly assumed to be the carrier of information in the brain, is a non-monotonic function of balanced input. These findings do not depend on details of the model and, hence, they are relevant to cells of other cortical areas as well.

Figure 4. QUESTION: Is it possible to implement some kind of coupling and nonlinear dynamics in each node of a complex network in order to get bistability?. ANSWER: Yes.

Putting together all these facts, we arrive to the central question that we want to bring to the reader: Is it possible to reproduce the bistability in a complex network independently of the topology and of the number of nodes?. The answer is 'Yes' (Fig. 4). What kind of local dynamics and coupling among nodes must be implemented in order to get this behavior? In Section 2, we give different possible strategies for the coupling and the local dynamics which should be implemented in a few coupled functional units in order to get a bistable behavior. Then, in Section 3, the same model is exported on a many units network where the desired bistability is also retained. In view of the results, the possibility of constructivism in the world of complex systems in general, and in the neural networks in particular, is suggested in the Conclusion.

2.   Models of a Few Coupled Functional Units

2.1.   General Model

Our approach considers the so called *functional unit*, i.e. a neuron or group of neurons (voxels), as a discrete nonlinear oscillator with two possible states: active (meaning one type of activity) or not (meaning other type of activity). Hence, in this naive vision of the brain as a networked system, if x_n^i, with $0 < x_n^i < 1$, represents a measurement of the ith functional unit activity at time n, it can be reasonable to take the most elemental local nonlinearity, for instance, a logistic evolution [13], which presents a quadratic term, as a first toy-model for the local neuronal activity:

$$x_{n+1}^i = \bar{p}_i \, x_n^i (1 - x_n^i). \tag{1}$$

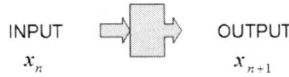

Figure 5. Discrete nonlinear model for the local evolution of a functional unit.

It presents only one stable state for each $\bar{p}_i$. Then, there is no bistability in the basic component of our models. For $0 < \bar{p}_i < 1$, the dynamics dissipates to zero, $x_n^i = 0$, then it can represent the functional unit with no activity. For $1 < \bar{p}_i < 4$, the dynamics is non null and it would represent an active functional unit.

We can suppose that this local parameter $\bar{p}_i$ is controlled by the signals of neighbor units, simulating in some way the effect of the synapses among neurons. Excitatory and inhibitory synaptic couplings have been shown to be determinant on the synchronization of neuronal firing. For instance, facilitatory connections are important to explain the neural mechanisms that make possible the object representation by synchronization in the visual cortex [14]. While it seems clear that excitatory (*symbiotic*) coupling can lead to synchronization, frequently inhibition rather than excitation synchronizes firing [15]. The importance of these two kinds of coupling mechanisms has also been studied for other types of neurons, v.g., motor neurons [16].

If a neuron unit simultaneously processes a plurality of binary input signals, we can think that this local information processing is reflected by the parameter $\bar{p}_i$. The functional dependence of this local coupling on the neighbor states is essential in order to get a good brain-like behavior (i.e., as far as the bistability of the sleep-wake cycle is concerned) of the network. As a first approach, we can take $\bar{p}_i$ as a linear function depending on the actual mean value, X_n^i, of the neighboring signal activity and expanding the interval $(0, 4)$ in the form:

$$\bar{p}_i \;=\; p_i \, (3X_n^i + 1), \qquad (excitation \;\; coupling) \tag{2}$$
$$or$$
$$\bar{p}_i \;=\; p_i \, (-3X_n^i + 4), \qquad (inhibition \;\; coupling) \tag{3}$$

with

$$X_n^i = \frac{1}{N_i} \sum_{j=1}^{N_i} x_n^j. \tag{4}$$

N_i is the number of neighbors of the ith functional unit, and p_i, which gives us an idea of the interaction of the functional unit with its first-neighbor functional units, is the control parameter. This parameter runs in the range $0 < p_i < p_{max}$, where $p_{max} \geq 1$. When $p_i = p$

for all i, the dynamical behavior of these networks with the excitation type coupling [7, 8] presents an attractive global null configuration that has been identified as the *turned off* state of the network. Also they show a completely synchronized non-null stable configuration that represents the *turned on* state of the network. Moreover, a robust bistability between these two perfect synchronized states is found in that particular model. For different models with a few coupled functional units we sketch in the next subsections the regions where they present a bistable behavior. The details of the complete unfolding [17] of these dynamical systems can be found in the references [18, 19, 20].

2.2. Models of Two Functional Units

Let us start with the simplest case of two interconnected (x_n, y_n) functional units. Three different combinations of couplings are possible: $(excitation, excitation)$, $(excitation, inhibition)$ and $(inhibition, inhibition)$.

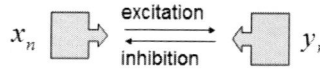

$$x_n \quad \overset{\text{excitation}}{\underset{\text{inhibition}}{\longleftrightarrow}} \quad y_n$$

Figure 6. Two functional coupled units.

As we are concerned with interactions with a certain degree of symbiosis between the coupled units, the regions of parameter space where bistability is found for the first two cases are presented here. Experimental systems where similar couplings have been found or implemented for two cells systems have been reported in the literature [21, 22].

2.2.1. Model with Mutual Excitation

The dynamics of the $(excitation, excitation)$ case [18] is given by the coupled equations:

$$x_{n+1} = p\,(3y_n + 1)x_n(1 - x_n), \tag{5}$$
$$y_{n+1} = p\,(3x_n + 1)y_n(1 - y_n). \tag{6}$$

The regions of the parameter space (Fig. 7) where we can find bistability are:

- For $0.75 < p < 0.86$, the synchronized state, $x_+ = (\bar{x}, \bar{x}) = P_4$, with $\bar{x} = \frac{1}{3}\{1 + (4 - \frac{3}{p})^{\frac{1}{2}}\}$, which arises from a saddle-node bifurcation for the critical value $p = 0.75$, is a stable *turned on* state. This state coexists with the *turned off* state $x_\theta = 0$. The system presents now bistability and depending on the initial conditions, the final state can be x_θ or x_+. Switching on the system from x_θ requires a level of noise in both functional units sufficient to render the activity on the basin of attraction of x_+. On the contrary, switching off the two functional units network can be done, for instance, by making zero the activity of one functional unit, or by doing the coupling p lower than 0.75.

- For $0.86 < p < 0.95$, the active state of the network is now a period-2 oscillation, namely the period-2 cycle (P_5, P_6) in Fig. 7. This new dynamical state bifurcates from x_+ for $p = 0.86$. A smaller noise is necessary to activate the system from x_θ. Making zero the activity of one functional unit continues to be a good strategy to turn off the network.

- For $0.95 < p < 1$, the active state acquires a new frequency and presents quasiperi-odicity (the invariant closed curves of Fig. 7). It is still possible to switch off the network by putting to zero one of the functional units.

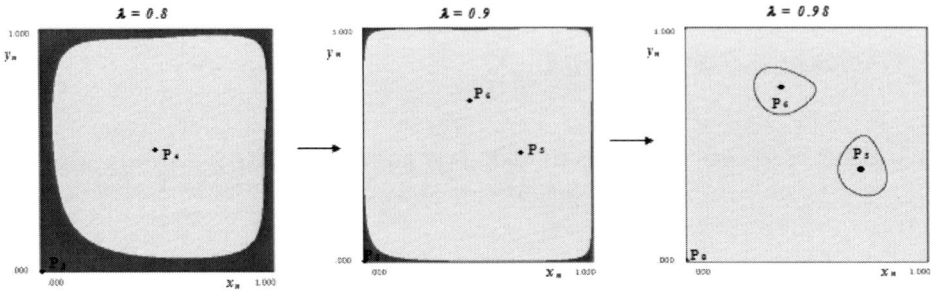

Figure 7. Bistability in 2 functional units with excitation type coupling ($p = \lambda$).

2.2.2. Model with Excitation + Inhibition

The dynamics of the $(excitation, inhibition)$ case [19] is given by the coupled equations:

$$x_{n+1} = p\,(3y_n + 1)x_n(1 - x_n), \tag{7}$$
$$y_{n+1} = p\,(-3x_n + 4)y_n(1 - y_n). \tag{8}$$

The regions of the parameter space (Fig. 8) where we can find bistability are:

- For $1.051 < p < 1.0851$, a stable period three cycle (Q_1, Q_2, Q_3) appears in the system. It coexists with the fixed point P_4. When p is increased, a period-doubling cascade takes place and generates successive cycles of higher periods $3 \cdot 2^n$. The system presents bistability. Depending on the initial conditions, both populations (x_n, y_n) oscillate in a periodic orbit or, alternatively, settle down in the fixed point. The borders between the two basins are complex.

- For $1.0851 < p < 1.0997$, an aperiodic dynamics is possible. The period-doubling cascade has finally given birth to an order three cyclic chaotic band(s) (A_{31}, A_{32}, A_{33}). The system can now present an irregular oscillation besides the stable equilibrium with final fixed populations. The two basins are now fractal.

2.3. Models of Three Functional Units

Following the strategy given by relation (2-3) several models with three functional units can be established. We have studied in some detail three of them [9, 20] and their bistable behavior is reported here.

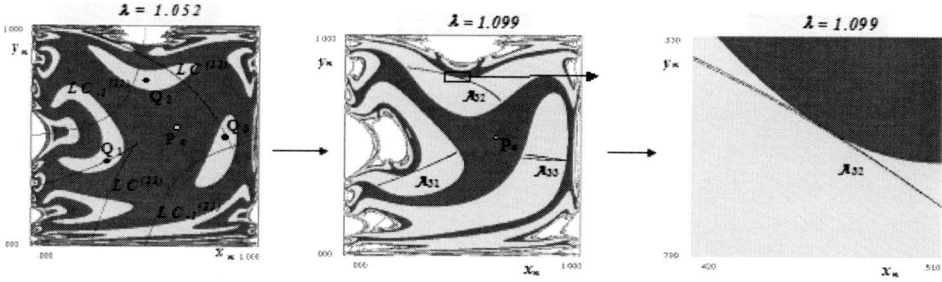

Figure 8. Bistability in 2 functional units with excitation+inhibition type coupling ($p = \lambda$).

2.3.1.   Model with Local Mutual Excitation

Let us start with the case of three alternatively interconnected (x_n, y_n, z_n) functional units under a mutual excitation scheme.

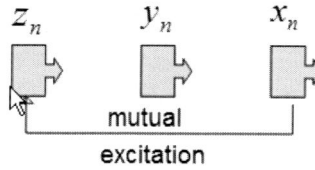

Figure 9. Three alternatively coupled functional units under the excitation scheme.

Then the dynamics of the system is given by the coupled equations:

$$x_{n+1} = p\,(3y_n + 1)x_n(1 - x_n), \tag{9}$$
$$y_{n+1} = p\,(3z_n + 1)y_n(1 - y_n), \tag{10}$$
$$z_{n+1} = p\,(3x_n + 1)z_n(1 - z_n). \tag{11}$$

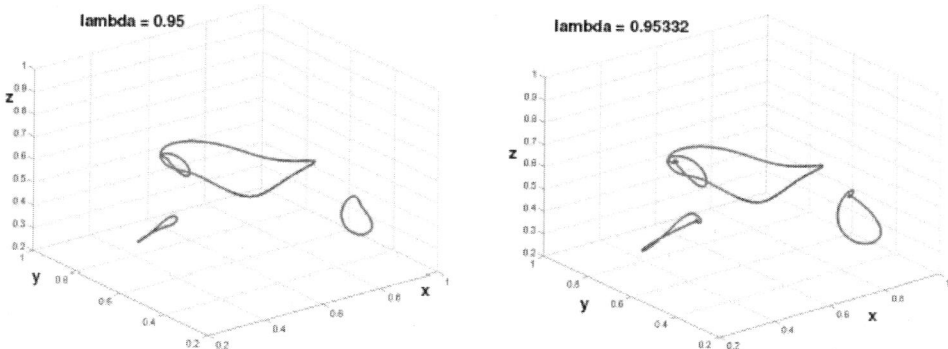

Figure 10. Bistability in 3 functional units with local excitation type coupling ($p = \lambda$).

The regions of the parameter space where we have found bistability are:

- For $0.93310 < p < 0.95334$, a big invariant closed curve (ICC) $C1$ coexists with a period-3 orbit that bifurcates, first to an order 3-cyclic ICC (Fig. 10), and finally to an order-3 weakly chaotic ring (WCR) before disappearing.

- For $0.98418 < p < 0.98763$, the ICC $C1$ coexists with another ICC $C2$ (see Ref. [20]) that becomes chaotic, by following a period doubling cascade of tori, before disappearing.

- For $1.00360 < p < 1.00402$, the ICC $C1$ coexists with a high period orbit that gives rise to an ICC $C3$. This ICC also becomes a chaotic band (see Ref. [20]) by following a period doubling cascade of tori before disappearing.

2.3.2. Model with Global Mutual Excitation

We expose now the case of three globally interconnected (x_n, y_n, z_n) functional units under a mutual excitation scheme.

Then the dynamics of the system is given by the coupled equations:

$$x_{n+1} = p\,(x_n + y_n + z_n + 1)x_n(1 - x_n), \qquad (12)$$
$$y_{n+1} = p\,(x_n + y_n + z_n + 1)y_n(1 - y_n), \qquad (13)$$
$$z_{n+1} = p\,(x_n + y_n + z_n + 1)z_n(1 - z_n). \qquad (14)$$

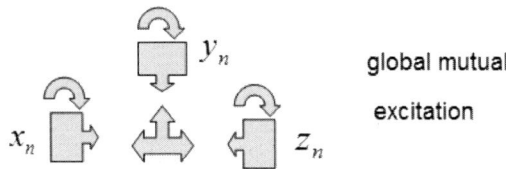

Figure 11. Three globally coupled functional units under the excitation scheme.

For the whole range of the parameter, $0 < p < 1.17$, bistability is present in this system:

- Firstly, two order-2 cyclic ICC coexist before becoming two order-2 cyclic chaotic attractors (Fig. 12) by contact bifurcations of heteroclinic type. Finally the two chaotic attractors become a single one before disappearing.

2.3.3. Model with Partial Mutual Excitation

The new case [9] of three partially interconnected (x_n, y_n, z_n) functional units under a mutual excitation scheme is represented in Fig. 13.

The dynamics of the system is given by the coupled equations:

$$x_{n+1} = p\,(3(y_n + z_n)/2 + 1)x_n(1 - x_n), \qquad (15)$$
$$y_{n+1} = p\,(3(x_n + z_n)/2 + 1)y_n(1 - y_n), \qquad (16)$$
$$z_{n+1} = p\,(3(x_n + y_n)/2 + 1)z_n(1 - z_n). \qquad (17)$$

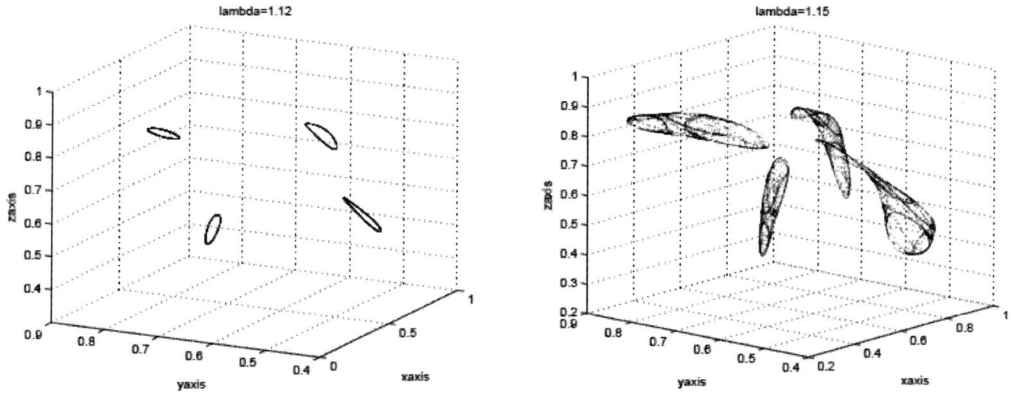

Figure 12. Bistability in 3 functional units with global excitation type coupling ($p = \lambda$).

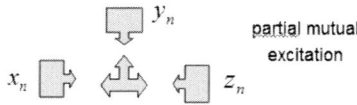

Figure 13. Three partially coupled functional units under the excitation scheme.

The rough inspection of this system puts in evidence the existence of different regions of multistability in the parameter space. These are:

- For $0.93 < p < 1.04$, there is coexistence among three cycles of period-2.

- For $1.04 < p < 1.06$, the three cycles bifurcate giving rise to three order-2 ICC (Fig. 14).

- For $1.06 < p < 1.08$, the system can present three mode-locked periodic orbits, each one with period multiple of 6, displaying period doubling cascades giving rise to three chaotic cyclic attractors. Three chaotic cyclic attractors of order 2 are also possible in this region (Fig. 15).

- For $p > 1.08$, the chaotic cyclic attractors collapse in an unique chaotic attractor (Fig. 16).

Also, other multistable situations can be found for some particular values of the parameter p in the former intervals, such as it can be seen in Fig. 17, where the x-projection of a generic bifurcation diagram is plotted. Similar diagrams are found for the y- and z-projections.

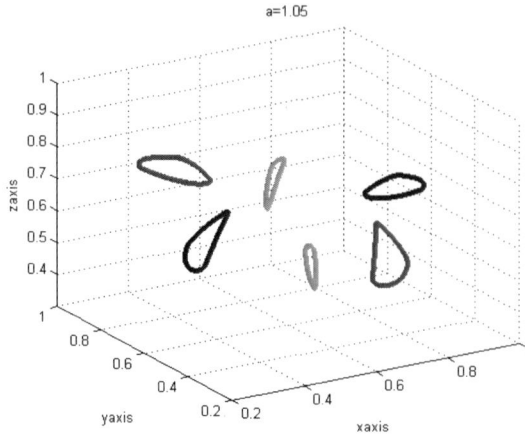

Figure 14. Multistability in 3 functional units with partial excitation type coupling. The system presents three order-2 ICC for $p = a = 1.05$.

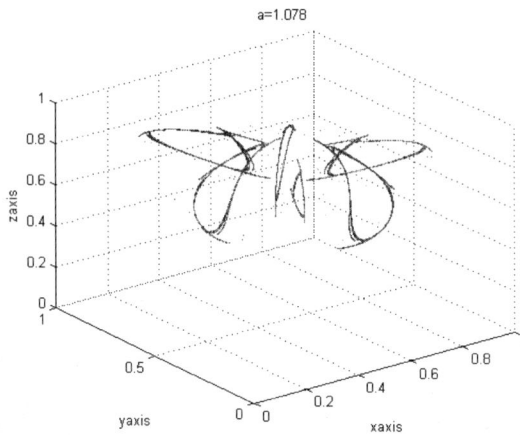

Figure 15. Multistability of three chaotic cyclic attractors of order 2 for $p = a = 1.078$.

3.   Model of Many Coupled Functional Units

3.1.   The Model

The complete synchronization [23, 24] of the network means that $x_n^i = x_n$ for all i , with $i = 1, 2, \ldots, N$ and $N \gg 1$ [25, 26]. In this regime, we also have $X_n^i = x_n$. The time evolution of the network [8] on the synchronization manifold is then given by the cubic mapping:

$$x_{n+1} = p\,(3x_n + 1)\,x_n(1 - x_n). \qquad (18)$$

The fixed points of this system are found by solving $x_{n+1} = x_n$. The solutions are $x_\theta = 0$ and $x_\pm = \frac{1}{3}\{1 \pm (4 - \frac{3}{p})^{\frac{1}{2}}\}$. The first state x_θ is stable for $0 < p < 1$ and $x_\pm$ take birth after a saddle-node bifurcation for $p = p_0 = 0.75$. The node x_+ is stable for $0.75 < p < 1.157$

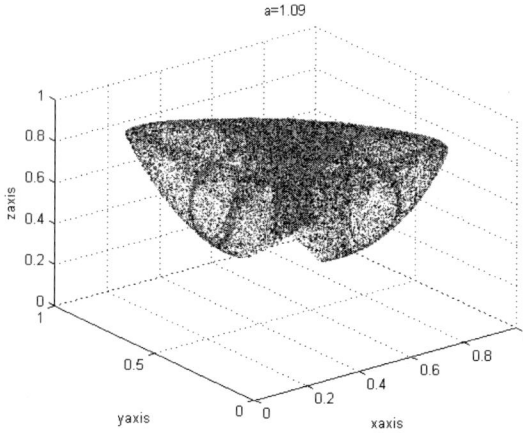

a=1.09

Figure 16. An unique chaotic attractor for $p = a = 1.09$.

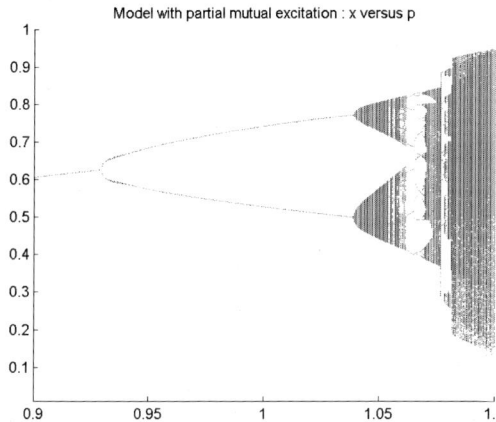

Model with partial mutual excitation : x versus p

Figure 17. Bifurcation diagram projected in the x coordinate for the initial conditions ($x_0 = 0.37, y_0 = 0.36, z_0 = 0.33$) as a function of the parameter p.

and the saddle x_- is unstable. Therefore bistability between the states

$$x_n^i = x_\theta, \quad \forall i \longrightarrow \quad TURNED \quad OFF \quad STATE, \tag{19}$$
$$x_n^i = x_+, \quad \forall i \longrightarrow \quad TURNED \quad ON \quad STATE, \tag{20}$$

seems to be also possible for $p > p_0 = 0.75$ in the case of many interacting units. But stability on the synchronization manifold does not imply the global stability of it. Small transverse perturbations to this manifold can make unstable the synchronized states. Let us suppose then a general local perturbation δx_n^i of the element activity,

$$x_n^i = x_* + \delta x_n^i, \tag{21}$$

with x_* representing a synchronized state, x_θ or x_+. We define the perturbation of the local mean-field as

$$\delta X_n^i = \frac{3}{N_i} \sum_{j=1}^{N_i} \delta x_n^j. \tag{22}$$

If these expressions are introduced into equation (1), the time evolution of the local perturbations are found:

$$\delta x_{n+1}^i = p\,(3x_* + 1)(1 - 2x_*)\delta x_n^i + p\,x_*(1 - x_*)\delta X_n^i. \tag{23}$$

The dynamics for the local mean-field perturbation is derived by substituting this last expression in relation (22). We obtain:

$$\delta X_{n+1}^i = p\,(3x_* + 1)(1 - 2x_*)\delta X_n^i + 3p\,x_*(1 - x_*)\frac{1}{N_i} \sum_{j=1}^{N_i} \delta X_n^j. \tag{24}$$

We express now the local mean-field perturbations of the first-neighbors as function of the local mean-field perturbation δX_n^i by defining the local operational quantity σ_i^n,

$$\frac{1}{N_i} \sum_{j=1}^{N_i} \delta X_n^j = \sigma_n^i\,\delta X_n^i, \tag{25}$$

which is determined by the dynamics itself. If we put together the equations (23-24), the linear stability of the synchronized states holds as follows:

$$\begin{pmatrix} \delta x_{n+1}^i \\ \delta X_{n+1}^i \end{pmatrix} = \begin{pmatrix} p\,(3x_* + 1)(1 - 2x_*) & p\,x_*(1 - x_*) \\ 0 & p\,(3x_* + 1)(1 - 2x_*) + 3p\,\sigma_n^i\,x_*(1 - x_*) \end{pmatrix} \\ \times \begin{pmatrix} \delta x_n^i \\ \delta X_n^i \end{pmatrix}. \tag{26}$$

Let us observe that the only dependency on the network topology is included in the quantity σ_n^i. The rest of the stability matrix is the same for all the nodes and therefore it is independent of the local and global network organization.

The turned off state is $x_* = x_\theta = 0$. The eigenvalues of the stability matrix are in this case $\lambda_1 = \lambda_2 = p$. Then, this state is an attractive state in the interval $0 < p < 1$. It loses stability for $p = 1$, then the highest value p_f of the parameter p where bistability is still possible satisfies $p_f \leq 1$.

The turned on state x_+ verifies $x_* = x_+ = \frac{1}{3}\{1 + (4 - \frac{3}{p})^{\frac{1}{2}}\}$. If we suppose $\sigma_n^i = \sigma$, the eigenvalues of the stability matrix are $\lambda_1 = 2 - 2p - p(4 - \frac{3}{p})^{\frac{1}{2}}$ and $\lambda_2 = \lambda_1 + \frac{\sigma}{3}(3 - 2p + p(4 - \frac{3}{p})^{\frac{1}{2}})$. Let us observe that $\lambda_1 = -1$ for $p = 1$. This implies that the parameter p_c for which the synchronized state x_+ looses stability verifies $p_c \leq 1$. Depending on the sign of σ, we can distinguish two cases in the behavior of p_c:

- If $0 < \sigma < 1$, we find that $|\lambda_2| < 1$. Then x_+ bifurcates through a global flip bifurcation for $p = p_c = 1$. In this case, the bifurcation of the synchronized state x_+ for $p_c = 1$ coincides with the loss of the network bistability for $p_f = 1$. Hence

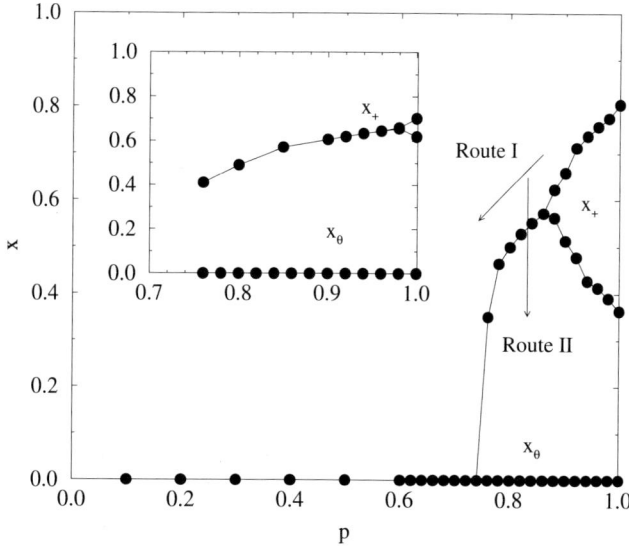

Figure 18. Stable states (x_θ, x_+) of the network for $0 < p < 1$. Let us observe the two zones of bistability: $p_0 < p < p_c$ and $p_c < p < p_f$. The main figure corresponds to a scale free network made up of $N = 10^4$ elements: $p_0 = 0.75$, $p_c = 0.87 \pm 0.01$ and $p_f = 1$. The inset shows the same graph but in an all-to-all network of the same size: $p_0 = 0.75$, $p_c = p_f = 1$. Initial conditions for the x_i's were drawn from a uniform probability distribution in the interval $(0, 1)$.

$p_c = p_f = 1$ for this kind of networks, and the bistability holds between x_θ and x_+ in the parameter interval $p_0 = 0.75 < p < p_c = p_f = 1$. As an example, an all-to-all network shows this behavior because $\sigma = 1$. This is represented in the inset of Fig. 18.

- If $-1 < \sigma < 0$, then $\lambda_2 = -1$ is obtained for a $p = p_c$ smaller than 1. Therefore it is now possible to obtain an active state different from x_+ in the interval $p_c < p < p_f$. For instance, simulations show that the global flip bifurcation of the synchronized state for a scale free network occurs for $p_c = 0.87 \pm 0.01$. A value of $p = 0.866$ is obtained from the stability matrix by taking $\sigma = -1$. For this particular network it is also found that $p_f = 1$. Then, bistability is possible in the range $p_0 = 0.75 < p < p_f = 1$ for this kind of configuration. But now an active state with different dynamical regimes is observed in the interval $p_c = 0.87 < p < p_f = 1$. If we identify the capacity of information storing with the possibility of the system to access to complex dynamical states, then, we could assert, in this sense, that a scale free network has the possibility of storing more elaborated information in the bistable region that an all-to-all network.

Let us note that σ also indicates a different behavior of local dissipation, as expression (25) suggests. A positive σ means a local in-phase oscillation of the node signal and mean-field perturbations. A negative σ is meaning a local out of phase oscillation between those signal perturbations. Hence, σ also brings some kind of structural network information. In all the cases the stability loss of the completely synchronized state is mediated by a global flip bifurcation. The new dynamical state arising from that active state for $p = p_c$ is a periodic pattern with a local period-2 oscillation. The increasing of the coupling parameter monitors other global bifurcations that can lead the system towards a pattern of local chaotic oscillations.

3.2.   Transition between On-Off States

We proceed now to show the different strategies for switching on and off a random scale free network. The choice of this network is suggested by the recent work [10] on the connections distribution among functional units in the brain. They find it to be a power-law distribution. Following this insight [8], a scale-free network following the Barabási-Albert (BA) recipe [11] is generated. In this model, starting from a set of m_0 nodes, one preferentially attaches each time step a newly introduced node to m older nodes. The procedure is repeated $N - m_0$ times and a network of size N with a power law degree distribution $P(k) \sim k^{-\gamma}$ with $\gamma = 3$ and average connectivity $\langle k \rangle = 2m$ builds up. This network is a clear example of a highly heterogenous network, in that the degree distribution has unbounded fluctuations when $N \to \infty$. The exponent reported for the brain functional network has $\gamma < 3$. However, studies of percolation and epidemic spreading [27, 28] on top of scale-free networks have shown that the results obtained for $\gamma = 3$ are consistent with those corresponding to lower values of γ, with $\gamma > 2$. As explained before, network bistability between the active and non active states is here possible in the interval $p_0 = 0.75 < p < p_f = 1$ (Fig. 18).

Switching off the Network

Two different strategies [8] can be followed to carry the network from the active state to that with no activity (Fig. 18).

- *Route I*: By doing the coupling p lower than p_0. This is the easiest and more natural way of performing such an operation. In our naive picture of a brain-like system, it could represent the decrease (or increase, it depends on the specific function) of the synaptic substances that provokes the transition from the awake to the sleep state. The flux of these chemical activators is controlled by the internal circadian clock, which is present in all animals, and which seems to be the result of living during millions of years under the day/night cycle.

- *Route II*: By switching off a critical fraction of functional units for a fixed p. Evidently, at first sight, this strategy has no relation with the behavior of a real brain-like system. Thus, this is done by looking over all the elements of the network, and considering that the element activity is set to zero with probability λ (which implies that on average λN elements are reset to zero). The result of this operation is shown in Fig. 19. Here, it is plotted for different p's the relative size of the biggest (giant) cluster of connected active nodes in the network versus λ. Note that this procedure does not take into account the existence of connectivity classes, but all nodes are equally treated. The procedure is thus equivalent to simulations of random failure in percolation studies [27]. The strategy in which highly connected functional units are first put to zero is more aggressive and leads to quite different results.

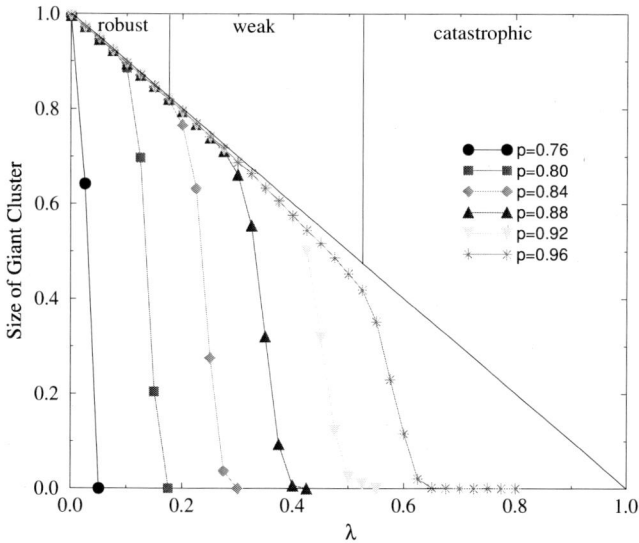

Figure 19. Turning off a scale free network. Three different phases in the behavior of the giant cluster size versus λ (fraction of switched off nodes) are observed. These three phases are illustrated for $p = 0.96$: the robust phase, the weak phase and the catastrophic phase (see the text). Other network parameters are as those of Fig. 18.

Each curve presents three different zones depending on λ:

- the *robust phase*: For small λ, the network is stable and only those states put to zero have no activity. There is a linear dependence on the giant cluster size with λ. In this stage, the switched off nodes do not have the capacity to transmit its actual state to its active neighbors.

- the *weak phase*: For an intermediate λ, the nodes with null activity can influence its neighborhood and switch off some of them. The linearity between the size of the giant cluster and λ shows a higher absolute value of the slope than in the robust zone.

- the *catastrophic phase*: When a critical λ_c is reached, the system undergoes a crisis. The sudden drop in this zone means that a small increase of the non active nodes leads the system to a catastrophe; that is, the null activity is propagated through all the network and it becomes completely down.

It is worth noticing that when the system is outside the bistability region for $p > 1$, the catastrophic phase does not take place. Instead, the turned off nodes do not spread its dynamical state and the neighboring nodes do not die out. This is because the dynamics of an isolated node is self-sustained when $p > 1$. Consequently, it is observed that the network breaks down in many small clusters and the transition resembles that of percolation in scale free nets [27, 29].

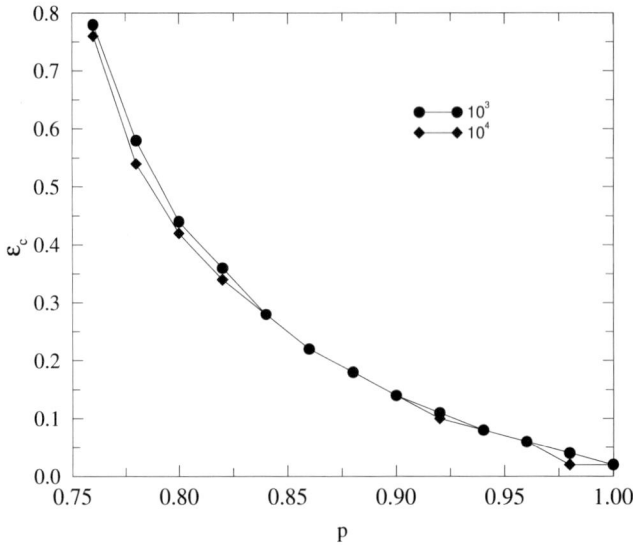

Figure 20. Turning on a scale free network. For a fixed p, a noisy signal randomly distributed in the interval $(0, \epsilon)$ is assigned to every node. When ϵ reaches the critical level ϵ_c the network becomes switched on. Other network parameters are as those of Fig. 18.

Switching on the Network

Two equivalent strategies [8] can be followed for the case of turning on the network (Fig. 20):

- (I) For a fixed p, we can increase the maximum value ϵ of the noisy signal, which is randomly distributed in the interval $(0, \epsilon)$ over the whole system. When ϵ attains a critical value ϵ_c, the noisy configuration can leave the basin of attraction of x_θ, whose boundary seems to have the form in phase space of a "hollow cane" around it, and then the network rapidly evolves toward the turned on state;

- (II) If this operation is executed by letting ϵ to be fixed and by increasing the coupling parameter p, the final result of switching on the network is reached when p takes the value for which $\epsilon = \epsilon_c$. The final result is identical in both cases.

Let us remark that the strategy equivalent to the former Route II, that is, the switching on of a critical fraction of functional units, is not possible in this case. It is a consequence of the fact that a switched off functional unit can not be excited by its neighbors and it will maintain indefinitely the same dynamical state ($x_i = 0$).

Finally, let us note that, from Fig. 20, a bigger p requires a smaller ϵ_c to switch on the net. Observe that this behavior could be interpreted in our approach as the smaller level of noise that is needed for awaking a brain-like system that it is departing from the sleeping state.

Conclusion

One of the more challenging problems in nonlinear science is the goal of understanding the properties of neuronal circuits [30]. Synchrony and multistability are two important dynamical behaviors found in those circuits [31, 32, 33]. In this work, different coupling schemes for networks with local logistic dynamics are proposed [7]. It is observed that these types of couplings generate a global bistability between two different dynamical states [8, 9]. This property seems to be topology and size independent. This is a direct consequence of the local mean-field multiplicative coupling among the first-neighbors. If a formal and naive relationship is established between these two states and the sleep-wake states of a brain, respectively, one would be tempted to assert that these types of couplings in a network, regardless of its simplicity, give us a good qualitative model for explaining that specific bistability. Following this insight, several low-dimensional systems with logistic components coupled under these schemes have been presented and the regions where the dynamics shows bistability have been identified. The extension of this type of coupling to a general network with local logistic dynamics has also been achieved. This system presents global bistability between an active synchronized state and another synchronized state with no activity. This property is topology and size independent. This is a direct consequence of the local mean-field multiplicative coupling among the first-neighbors. If a formal and naive relationship is established between the switched off and switched on states of that network, and the sleep-wake states of a brain, respectively, one would be tempted to assert that this model, regardless of its simplicity, is a good qualitative representation for explaining that specific bistability. Furthermore, on more theoretical grounds, the results obtained here point out the entangled interplay between topology and function in networked systems [34] where complex structures coexist with nonlinear dynamics.

References

[1] Cajal, S.R. *The structure and connections of neurons*; Nobel Lecture; December 12, Stockholm, 1906.

[2] Llinás, R.R. *Nat. Rev. Neurosci.* 2003, 4, 77-80.

[3] Carey J.(Editor) *Brain Facts*; Fifth revised edition, Publications of the Society for Neuroscience; Washington, 2006.

[4] Abarbanel H.D.I.; Gibb L.; Mindlin G.B.; Talathi S. *J. Neurophysiol.* 2004, 92, pp. 96-110.

[5] Winfree A.T. *The timing of biological clocks*; Scientific American Library; New York, 1986.

[6] Bar-Yam Y. *Dynamics of Complex Systems*; Westview Press; New York, 1997.

[7] Lopez-Ruiz R.; Fournier-Prunaret D. MEDYFINOL'06 Conference, *AIP Proceedings* 2007, 913, 89-95.

[8] Lopez-Ruiz R.; Moreno Y.; Pacheco, A.F.; Boccaletti, S.; Hwang, D.-U. *Neural Networks* 2007, 20, 102-108.

[9] Lopez-Ruiz R.; Fournier-Prunaret, D. *NOMA'09 Conference, Proceedings* 2009, 6-9.

[10] Eguiluz V.M.; Chialvo, D.R.; Cecchi, G.; Baliki, M.; Apkarian, A.V. *Phys. Rev. Lett.* 2005, 94, 018102(4).

[11] Barabási A.L.; Albert R. *Science* 1999, 286, 509-512.

[12] Kuhn A.; Aertsen, Ad.; Rotter, S. *J. Neurosci.* 2004, 24, 2345-2356.

[13] May R.M. *Nature* 1976, 261, 459-467.

[14] Eckhorn, R.; Gail, A.M.; Bruns, A.; Gabriel, A.; Al-Shaikhli, B.; Saam, M. *IEEE Trans. Neural Netw.* 2004, 15, 1039-1052.

[15] Vreeswijk1 C.V.; Abbott, L.F.; Ermentrout, G.B. *J. Comput. Neurosci.* 1994, 1, 313-321.

[16] Koenig, J.H.; Ikeda, K. *J. Comp. Physiol. A.* 1983, 150, 305-317.

[17] Mira C.; Gardini, L.; Barugola, A.; Cathala, J.-C. *Chaotic Dynamics in Two-Dimensional Noninvertible Maps*; series A, vol. 20; World Scientific Publishing; Singapore, 1996.

[18] Lopez-Ruiz, R.; Fournier-Prunaret, D. *Math. Biosci. Eng.* 2004, 1, 307-324.

[19] Lopez-Ruiz, R.; Fournier-Prunaret, D. *Chaos, Solitons and Fractals* 2005, 24, 85-101.

[20] Fournier-Prunaret, D.; Lopez-Ruiz, R.; Taha, A.K. *Grazer Mathematische Berichte* 2006; 350, 82-95.

[21] Graham, D.W.; Knapp, C.W.; Van Vleck, E.S.; Bloor, K.; Lane, T.B.; Graham, C.E. *ISME Journal* 2007, 1, 385-393.

[22] Mihailovic, D.T.; Balaz, I. *Mod. Phys. Lett. B* 2012, 26, 1150031(9).

[23] Lopez-Ruiz, R.; Perez-García, C. *Chaos, Solitons and Fractals* 1991, 1, 511-528.

[24] Boccaletti, S.; Kurths, J.; Osipov, G.; Valladares, D.L.; Zhou, C.S. *Phys. Rep.* 2002, 366, 1-101.

[25] Jalan, S.; Amritkar, R.E. *Phys. Rev. Lett.* 2003, 90, 014101(4).

[26] Oprisan S.A. *Int. J. Neurosci.* 2009, 119, 482-491.

[27] Callaway, D.S.; Newman, M.E.J.; Strogatz, S.H.; Watts, D.J. *Phys. Rev. Lett.* 2000, 85, 5468-5471.

[28] Dorogovtsev, S.N.; Mendes, J.F.F. *Evolution of Networks. From Biological Nets to the Internet and the WWW*; Oxford University Press; Oxford, 2003

[29] Vázquez, A.; Moreno, Y. *Phys. Rev. E* 2003, 67, 015101(R).

[30] Rabinovich, M.I.; Varona, P.; Selverston, A.I.; Abarbanel, H.D.I. *Rev. Mod. Phys.* 2006, 78, 1213-1265.

[31] Borgers, C.; Kopell, N. *Neural Comput.* 2003, 15, 509-538.

[32] Hansel, D.; Mato, G. *Neural Comput.* 2003, 15, 1-56.

[33] Buzsàki, G.; Geisler, C.; Henze, D.A.; Wang, X.-J. *Trends in Neurosciences* 2004, 27, 186-193.

[34] Strogatz, S.H. *Nature (London)* 2001, 410, 268-276.

INDEX

D

E

F

flavonol, 163
flexibility, 148
flight, 146
flora, 4, 19, 22
flowers, 81, 84
fluctuations, 148, 248
fluid, 136, 140
food, x, 96, 99, 100, 120, 121, 147, 149, 150, 178,
    179, 186, 189, 190, 193, 196, 197, 199, 201, 202,
    203, 206, 211, 213, 221, 226, 233, 234
food chain, x, 189, 190, 213, 221, 233, 234
food production, 178
food security, 179, 186
forage crops, 68, 75
force, ix, 143, 182
formation, vii, 1, 3, 6, 7, 10, 11, 12, 13, 14, 18, 21,
    22, 23, 36, 42, 43, 48, 51, 54, 55, 56, 58, 59, 65,
    66, 84, 86, 87, 88, 89, 111, 120, 123, 124, 130,
    133, 134, 159, 161, 162, 168, 170, 172
formula, 121
fossils, 3, 4, 5, 6
fouling, 148
fragments, 10, 18
France, 159, 164, 235
free radicals, 17
freshwater, 147, 148
fruits, 69, 188
functional analysis, 160
fungi, vii, 1, 2, 3, 4, 5, 6, 7, 8, 9, 10, 11, 12, 13, 14,
    15, 16, 17, 18, 19, 20, 21, 22, 23, 24, 25, 26, 27,
    28, 29, 30, 31, 32, 33, 36, 79, 121, 146, 151, 168,
    169, 171
fungus, 2, 5, 7, 8, 9, 10, 11, 12, 13, 14, 15, 17, 24,
    29, 30, 31, 32, 33, 80, 156, 187, 233
fusion, 14

G

gel, 33, 133
gene expression, 13, 36, 50, 58, 60, 64, 65, 162, 167,
    170, 172
gene promoter, 174
gene silencing, 171
gene transfer, x, 159
genes, vii, viii, 1, 4, 8, 10, 12, 13, 23, 26, 33, 35, 36,
    37, 39, 41, 42, 48, 49, 52, 53, 55, 56, 57, 58, 59,
    60, 61, 63, 64, 70, 71, 124, 127, 129, 138, 160,
    163, 164, 165, 166, 167, 168, 169, 171, 174
genetic diversity, 78
genetic information, 11
genetic predisposition, 166
genetics, 6, 8, 24, 30, 32, 33, 135, 172, 174
genome, 7, 8, 11, 29, 30, 160, 166, 167
genomics, x, 8, 26, 27, 159, 173
genotype, 44, 45, 46, 47, 62, 76
genus, 39, 99, 135, 168, 183
geotropism, 162
Germany, 24, 26, 30, 32, 33, 139

germination, 10, 11, 21, 26, 27, 75, 77, 84, 88
gibberellin, 53, 59, 63
global scale, 7, 105
global warming, 22
Glomus intraradices, 7, 8, 30, 165
glucose, 15, 136
glutamate, 38, 80
glutamic acid, 86
glutathione, 17
glycine, 86, 136
graft technique, 43
graph, 247
grass, 79, 182, 186
grasses, 83, 91, 179, 181, 182, 187, 188
grasslands, 186
gravity, 44, 107, 108
grazers, 146
grazing, 79, 83, 146, 180, 181
Greece, 95, 117
greening, 73
groundwater, 178
growth, 3, 10, 11, 12, 13, 14, 16, 21, 22, 27, 29, 33,
    36, 38, 43, 48, 50, 53, 56, 59, 61, 63, 69, 70, 75,
    76, 77, 78, 79, 80, 81, 82, 83, 84, 85, 86, 88, 89,
    90, 91, 108, 112, 117, 120, 121, 122, 123, 124,
    125, 126, 127, 132, 133, 134, 135, 136, 137, 138,
    152, 160, 162, 163, 167, 170, 174, 179, 181, 182,
    183, 185, 203, 215, 219, 227
growth arrest, 11
growth hormone, 70
growth rate, 75, 108, 112, 117, 124, 160, 203
growth temperature, 122, 123, 125, 126, 127, 132,
    133, 134, 138

H

habitat, 10, 17, 18, 26, 76, 96, 107, 110, 111, 144,
    150, 156, 158
hair, 70, 145, 147, 151, 164, 170
hair follicle, 145
hairless, 145
harmful effects, 144
Hawaii, 116
health, 3, 17, 18, 28, 120, 121, 145
heavy metals, vii, 1, 17, 20, 87, 88
height, 179
Helicobacter pylori, 153, 157
heme, 73
hemoglobin, 174
hermit crabs, viii, 95, 96, 97, 98, 99, 100, 101, 105,
    106, 107, 108, 109, 110, 111, 112, 113, 114, 115,
    116, 117
heterogeneity, 121, 167
histone, 54
hives, 146
homeostasis, 17, 141, 153
honey bees, 146, 154, 157
hormone, 16, 21

host, vii, viii, ix, 1, 2, 7, 9, 10, 11, 12, 13, 14, 16, 19,
    20, 21, 25, 26, 29, 35, 36, 38, 39, 40, 43, 48, 51,
    52, 62, 66, 70, 71, 72, 73, 76, 77, 80, 84, 86, 97,
    99, 100, 101, 106, 107, 114, 119, 120, 121, 124,
    127, 134, 135, 136, 139, 143, 144, 146, 147, 149,
    150, 151, 152, 153, 154, 155, 156, 157, 159, 160,
    161, 162, 164, 166, 167, 172, 173, 175
human, 22, 80, 84, 123, 145, 153, 186, 236
human brain, 236
hunting, 190, 191, 193, 202, 203, 206, 207, 214
hybridization, 55, 127
hydrogen, 131
hydrogenase, 167
hypothesis, ix, 4, 5, 11, 41, 75, 110, 111, 119, 143,
    144, 165
hypoxia, 113

I

ICC, 242, 243
image analysis, 169
immobilization, 185
immune defense, 144
immune response, 16, 150
immunity, 13, 154, 157
improvements, 89
in vitro, 24, 88, 127, 128
in vivo, 124
India, 88, 92, 121, 189
individuals, 8, 31, 106, 107, 108, 109, 144, 149, 152
Indonesia, 121
induction, viii, 16, 35, 38, 40, 55, 63, 70, 87, 166
industries, 121
industry, ix, 27, 119, 120
ineffectiveness, 52
inequality, 205, 231
infection, 27, 36, 38, 48, 55, 61, 62, 63, 65, 66, 70,
    77, 79, 87, 88, 89, 120, 144, 160, 161, 162, 163,
    164, 165, 166, 168, 170, 171, 172, 173, 175
infestations, 146
information processing, 238
ingest, 153
inhibition, 16, 50, 56, 59, 75, 78, 88, 91, 162, 236,
    237, 238, 239, 240, 241
inhibitor, 40, 41, 54, 58
initial state, 229
initiation, 22, 23, 36, 38, 42, 43, 51, 53, 54, 56, 66,
    79, 80, 81, 84, 152, 164, 170, 172, 173, 174
inoculation, 18, 19, 20, 22, 34, 52, 74, 76, 77, 82, 84,
    85, 87, 89, 120, 161, 163, 165
inoculum, 3, 19, 74, 81, 131
insects, 16, 145, 146, 154, 156
insulation, 148
integration, 56
interdependence, 55
interface, 13, 15
interference, 61, 143, 165, 166
intermediate trophic level, x, 189

internode, 53
intracellular accommodation, vii, 1, 9, 10, 23
invasions, 13
invertebrates, 115, 147, 149, 150
ions, 19, 122
iron, 72, 80, 91, 92
irrigation, 178
isoflavone, 163
isolation, 59, 164
isopods, 155
isotope, 90, 186, 187
issues, ix, 12, 21, 95, 169
Italy, 91, 189

J

Japan, 234
joint ventures, 26

K

kaempferol, 163
Kenya, 179
kidney, 88
kill, 150, 151

L

labeling, 127, 134
lactobacillus, 133
larva, 151
larvae, 146, 150, 151
lateral roots, 77
Latin America, 85
leaching, 120
lead, 54, 131, 144, 151, 161, 238, 247
legislation, 87
legs, 146
legume, vii, viii, ix, x, 19, 22, 24, 35, 36, 39, 42, 43,
    48, 49, 52, 53, 54, 56, 57, 58, 59, 60, 61, 64, 65,
    66, 67, 69, 72, 73, 75, 76, 78, 79, 81, 83, 84, 85,
    86, 89, 90, 91, 92, 93, 119, 120, 138, 139, 165,
    168, 172, 173, 174, 177, 180, 181, 182, 183, 184,
    185, 187
lesions, 147
leucine, 44, 49, 62
lice, 145
life cycle, 5, 7, 10, 97, 150
light, 23, 39, 44, 49, 75, 81, 83, 84, 88, 115, 130,
    150, 171, 179, 180, 182, 236
light cycle, 236
lignin, 184, 185
limestone, 85
linear dependence, 249
linear function, 238
lipid metabolism, 123, 126

M

Q

R

T

DATE DUE

GAYLORD			PRINTED IN U.S.A.